PROCEEDINGS OF THE 12th
ASIAN LOGIC
CONFERENCE

PROCEEDINGS OF THE 12th

ASIAN LOGIC
CONFERENCE

Wellington, New Zealand   15 – 20 December 2011

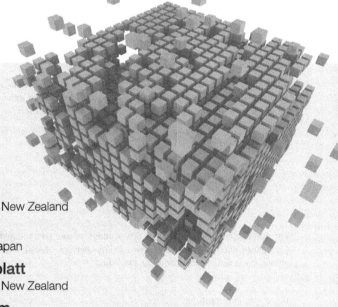

edited by

Rod Downey
Victoria University, New Zealand

Jörg Brendle
Kobe University, Japan

Robert Goldblatt
Victoria University, New Zealand

Byunghan Kim
Yonsei University, Korea

World Scientific

NEW JERSEY · LONDON · SINGAPORE · BEIJING · SHANGHAI · HONG KONG · TAIPEI · CHENNAI

Published by

World Scientific Publishing Co. Pte. Ltd.

5 Toh Tuck Link, Singapore 596224

USA office: 27 Warren Street, Suite 401-402, Hackensack, NJ 07601

UK office: 57 Shelton Street, Covent Garden, London WC2H 9HE

British Library Cataloguing-in-Publication Data
A catalogue record for this book is available from the British Library.

PROCEEDINGS OF THE 12TH ASIAN LOGIC CONFERENCE

ISBN 978-981-4449-26-7

Printed in Singapore

Preface

The 12th Asian Logic Conference was held in Wellington, New Zealand, from 15–20 December 2011. It was held jointly with a meeting of the Australasian Logic Association (AAL) organized by Edwin Mares and Robert Goldblatt. The meetings had a student tutorial day on the 14th where the two tutorial speakers were invited to give preliminary lectures to the graduate students, and there was an associated satellite conference *Analysis and Randomness* in Auckland, New Zealand, 12–13 December 2011 organized by Andre Nies.

From 1981 to 2008, the Asian Logic Conference has been held triennially and rotated among countries in the Asia-Pacific region. The previous meetings took place in Singapore (1981), Bangkok (1984), Beijing (1987), Tokyo (1990), Singapore (1993), Beijing (1996), Hsi-Tou (1999), Chongqing (2002), Novosibirsk (2005), Kobe (2008), and Singapore (2009). In 2008, the *East Asian and Australasian Committees of the Association of Symbolic Logic* decided to shorten the three-year cycle to two, following an initiative of the Association for Symbolic Logic.

The Asian Logic Conference is the most significant logic meeting outside of North America and Europe. Being held in Wellington was a sign of the strong ties of the New Zealand groups to those in Asia proper, and the acknowledgment of New Zealand's place in the Asia-Pacific region.

In 2011 the Asian Logic Conference was an exciting event with 110 attendees, and two stellar tutorials of seven hours each (including student days) by Zlil Sela (Hebrew University) on *Logical Aspects of Groups*, and Martin Grohe (Humboldt Universität) on *Logical Aspects of Graphs*. Plenary lectures were given by Hiroakira Ono (JAIST) (jointly with AAL), Mic Detlefsen (Notre Dame) (jointly with AAL), Akito Tsuboi (University of Tsukuba), Noam Greenberg (Victoria University of Wellington), Simon Thomas (Rutgers University), Isaac Goldbring (UCLA), Grigor Sargsyan (Rutgers University), and Guohua Wu (NTU, Singapore). Special sessions were organized by Feng Qi (NUS, Singapore, and Academica Sinica, China) (set theory), Geoff Whittle (Victoria University) (logical aspects of graphs and matroids), Andre Nies (Auckland) and Rod Downey (Victoria University) (algorithmic randomness), Rod Downey (computability and algebraic structures), and Edwin Mares and Rob Goldblatt (Victoria University) (modal logic). Details can be found in

http://msor.victoria.ac.nz/Events/ALC2011/WebHome.

Abstracts of the talks can be found in the *Bulletin of Symbolic Logic*.

The conference committee consisted of Toshiyasu Arai (Chiba University), Byunghan Kim (Yonsei), Feng Qi (NUS and Academica Sinica), Sergei S. Goncharov (Novosibirsk), Greg Restall (Melbourne), Rod Downey (VUW), and Yang Yue (NUS). The local organizing committee consisted of Ginny Whatarau, Noam Greenberg and Rod Downey.

2011 was not an easy year to be hosting conferences as many historical sources of support were subject to significant budget restrictions. This meeting was supported via a number of grants being assembled. The conference was generously supported by the US National Science Foundation, the Association for Symbolic Logic, the Royal Society of New Zealand, the New Zealand Mathematical Society, the Marsden Fund of New Zealand, the New Zealand Institute for Mathematics and its Applications, Victoria University School of Mathematics, Statistics and Operations Research, and Victoria's Research Office. A number of the speakers generously covered their own costs, also helping to make such an excellent conference economically viable.

As is the tradition of this meeting, World Scientific agreed to publish a post conference fully refereed volume reflecting the business of the Asian Logic Conference. The organizing committee voted the four editors you see here to be responsible for this volume. The result is the attractive volume you have now in your hands.

The Editors: *Rod Downey* (chair)
Robert Goldblatt
Byunghan Kim
Joerg Brendle

Contents

Preface   v

Resolute Sequences in Initial Segment Complexity   1
    G. Barmpalias and R. G. Downey

Approximating Functions and Measuring Distance on a Graph   24
    W. Calvert, R. Miller and J. Chubb Reimann

Carnap and McKinsey: Topics in the Pre-History of Possible-Worlds
Semantics   53
    M. J. Cresswell

Limits to Joining with Generics and Randoms   76
    A. R. Day and D. D. Dzhafarov

Freedom & Consistency   89
    M. Detlefsen

A van Lambalgen Theorem for Demuth Randomness   115
    D. Diamondstone, N. Greenberg and D. Turetsky

Faithful Representations of Polishable Ideals   125
    S. Gao

Further Thoughts on Definability in the Urysohn Sphere   144
    I. Goldbring

Simple Completeness Proofs for Some Spatial Logics of the
Real Line   155
    I. Hodkinson

On a Question of Csima on Computation-Time Domination   178
  X. Hua, J. Liu and G. Wu

A Generalization of Beth Model to Functionals of High Types   185
  F. Kachapova

A Computational Framework for the Study of Partition
Functions and Graph Polynomials   210
  T. Kotek, J. A. Makowsky and E. V. Ravve

Relation Algebras and **R**   231
  T. Kowalski

Van Lambalgen's Theorem for Uniformly Relative Schnorr and
Computable Randomness   251
  K. Miyabe and J. Rute

Computational Aspects of the Hyperimmune-Free Degrees   271
  K. M. Ng, F. Stephan, Y. Yang and L. Yu

Calibrating the Complexity of Δ_2^0 Sets via Their Changes   285
  A. Nies

Topological Full Groups of Minimal Subshifts and
Just-Infinite Groups   298
  S. Thomas

TW-Models for Logic of Knowledge-cum-Belief   314
  S. C.-M. Yang

Resolute sequences in initial segment complexity

George Barmpalias

State Key Lab of Computer Science
Institute of Software, Chinese Academy of Sciences
Beijing 100190, China.
E-mail: barmpalias@gmail.com
www.barmpalias.net

Rod G. Downey

School of Mathematics, Statistics and Operations Research
Victoria University, P.O. Box 600
Wellington, New Zealand
Email: rod.downey@vuw.ac.nz

We study infinite sequences whose initial segment complexity is invariant under effective insertions of blocks of zeros in-between their digits. Surprisingly, such *resolute* sequences may have nontrivial initial segment complexity. In fact, we show that they occur in many well known classes from computability theory, e.g. in every jump class and every high degree. Moreover there are degrees which consist entirely of resolute sequences, while there are degrees which do not contain any. Finally we establish connections with the contiguous c.e. degrees, the ultracompressible sequences, the anti-complex sequences thus demonstrating that this class is an interesting superclass of the sequences with trivial initial segment complexity.

Keywords: Kolmogorov complexity, Computably enumerable sets, trivial reals.

1. Introduction

Given an infinite random binary sequence X we may reduce its initial segment complexity by inserting blocks of zeros between its original digits. Even a single zero in-between every other digit of X will reduce its complexity dramatically. But what if X is not random? Can we always alter the complexity of its initial segments by 'spreading out' its digits in an effective manner? Clearly if X has trivial initial segment complexity, the simplification of its initial segments will not result in a 'measurable' reduction of their complexity. Surprisingly, there are nontrivial sequences X whose initial segment complexity is invariant under such effective 'block inserting' operations. Intuitively, these sequences have the property that

it is very hard to locate bits of significant information in their initial segments.

In this article we exhibit such examples in a variety of classes from computability theory and study this proper superclass of the family of sequences with trivial initial segment complexity. In particular, we establish connections with a number of notions from computability and Kolmogorov complexity like the jump hierarchy, the contiguous degrees, the ultracompressible sets of [1], the facile sets of [2, Section 8.2] and the anti-complex sets of [3]. Another notion of computational weakness was studied in [4] but we have not considered any possible connections with the latter class.

1.1. *Formal expressions of resoluteness*

We measure the complexity of binary strings σ via the plain Kolmogorov complexity $C(\sigma)$ prefix-free Kolmogorov complexity $K(\sigma)$; this is the length of the shortest program that produces σ in an underlying plain or prefix-free machine respectively. For background on Kolmogorov complexity we refer to [5]. Let $X \leq_K Y$ denote $\exists c \forall n \, K(X \restriction_n) \leq K(Y \restriction_n) + c$ and similarly $X \leq_C Y$ for the plain complexity. These preorders induce equivalence relations $\equiv_K$, $\equiv_C$ and corresponding degree structures that are known as the K-degrees and the C-degrees respectively. Intuitively, two sequences in the same degree have the same initial segment complexity.

The operation of inserting 0s between various digits of a given sequence is equivalent to shifting the bits of the sequence at various places and filling in the gaps with 0s. Let us refer to increasing functions $f : \mathbb{N} \to \mathbb{N}$ as *shifts*. If we view an infinite binary sequence X as a set of natural numbers, then the result of such a shift operation may be expressed as the image of X under f.

Definition 1.1 (Shifts). *An increasing function $f : \mathbb{N} \to \mathbb{N}$ is called shift. For each set Z we let $Z_f = \{f(n) \mid n \in Z\}$. A shift is called trivial if $\forall n \, (f(n) < n + c)$ for some constant c.*

Invariance under shift operations with respect to the plain and the prefix-free complexity can be defined as follows.

Definition 1.2 (Invariance). *A set Z is K-invariant under f if $Z \equiv_K Z_f$ and is C-invariant under f if $Z \equiv_C Z_f$.*

For every sequence Z and every computable shift f we have $Z_f \leq_K Z$, so the application of a computable shift on a sequence may only reduce its initial segment complexity. Note that if a shift f is trivial then for every sequence Z the sequence Z_f is (modulo finitely many bits) merely the result of shifting the bits of Z by a fixed number of places. Trivial shifts are not very interesting from our point of view as they preserve most notions of complexity on all sequences.

Definition 1.3 (K-resolute sequences). *An infinite sequence Z is called K-resolute if $Z \equiv_K Z_f$ for all computable shifts f. The C-resolute sequences are defined analogously.*

This definition is arguably a faithful formalization of the property that we discussed earlier, i.e. the ability of a sequence to preserve its initial segment complexity despite any computable insertion of blocks of 0s in-between its digits. This is an expression of 'resoluteness' of a sequence, i.e. the *inability to locate significant amounts of information in its initial segments*. There are other, similar ways to express this informal concept. For example, consider property (1).

$$\text{For all computable shifts } f, \ \exists c \forall n \ K(Z \restriction_{f(n)}) \leq K(Z \restriction_n) + c. \tag{1}$$

This also expresses a form of 'resoluteness' of a sequence. Moreover it is not hard to see that K-resolute sequences meet (1). Indeed, since there exists a constant c such that $\forall n \ |K(Z_f \restriction_{f(n)}) - K(Z \restriction_n)| < c$, for any computable shift f,

$$\text{if } Z \equiv_K Z_f, \text{ then } \exists c \forall n \ |K(Z \restriction_{f(n)}) - K(Z \restriction_n)| < c. \tag{2}$$

We say that a set is *weakly K-resolute* if it meets condition (1).

Yet another form of 'resoluteness' may be expressed in terms of conditional complexity, as in (3). Here an *order* is a nondecreasing and unbounded function.

$$\text{For all computable orders } g, \ \exists c \forall n \ K(Z \restriction_n \mid n) \leq K(Z \restriction_{g(n)} \mid n) + c. \tag{3}$$

Moreover each of the above notions has a version with respect to plain complexity.

As interesting as it may be, we will not be concerned with the technical question about the relationship between the above resoluteness notions. Instead, we focus on the notion of Definition 1.3 and note that our main results also hold for the two other notions (also with respect to plain complexity).

In the following, *we use the term 'resolute' to refer collectively to any of the above three formal variations on this concept and the versions with respect to plain complexity*, while 'K-resolute' is reserved for the notion of Definition 1.3. A degree is K-resolute if it contains a K-resolute set.

1.2. Resoluteness and complexity

Intuitively, sequences with 'consistently high complexity' cannot be resolute. On the other hand, sequences with trivial complexity are resolute. We give an overview of the relationship between complexity and resoluteness in more precise terms. In our context, trivial sequences are the K-trivial sequences, i.e. the sequences X such that $\exists c \forall n K(X \upharpoonright_n) \le K(n) + c$. It was shown in [6] that this class is downward closed under Turing reducibility. Hence, if X is K-trivial and f is a computable shift then $X_f \equiv_K X$. In other words, K-trivial sequences are K-resolute.

On the other end of the spectrum, a sequence X is called *random* if there exists a constant c such that $\forall n \ K(X \upharpoonright_n) \ge n - c$. It is clear that random sequences are not K-resolute. In fact, much more is true. A set is called *complex* if $\forall n \ K(X \upharpoonright_{f(n)}) \ge n$ for some computable function f. This definition is from [7,8] where it was shown to be equivalent to the condition that a diagonally noncomputable function is weak truth table reducible to X. Clearly complex sets are not weakly K-resolute (i.e. they do not meet (1)). It follows that complex sets are not K-resolute. Similar considerations apply to the C-resolute sets.

In turns out that K-resolute sets have very low initial segment complexity, but not necessarily trivial complexity. A class of sequences of 'ultralow' initial segment complexity was introduced in [1]. We say that X is *ultracompressible* if for all computable orders h, there exists c such that $K(X \upharpoonright_n) \le K(n) + h(n)$ for all sufficiently large n. A related class of sequences of low complexity was introduced in [3]. A set X is *anti-complex* if for all computable orders f we have $C(X \upharpoonright_{f(n)}) \le n$ for all but finitely many n. It is not hard to see that in this definition it does not matter if we use prefix-free complexity instead of plain complexity. Also, it is not hard to see that *every ultracompressible set is anti-complex*.

The proof of the following observation uses two notions from computability theory. A set X is called *superlow* if the jump X' of X is truth-table reducible to the halting problem $\emptyset'$. Also, a degree **a** is called *array computable* if there exists a function that can be computed from the halting problem with computable use of this oracle, which dominates all **a**-computable functions.

Proposition 1.1. *Every K-resolute set is ultracompressible (hence, anticomplex). The converse is not true, even for c.e. sets.*

Proof. Let X be a K-resolute set. In order to show that it is ultracompressible, let g be a computable order. Without loss of generality we may assume that g is onto. Let f be a computable increasing function such that $g(f(n)) = n^2$ for all n. Since g is onto, $f(n)$ can be defined by searching for the least number m such that $g(m) = n^2$. Then there exists some constant c such that $K(X_f \restriction_n) \leq K(n) + 2\sqrt{g(n)} + c$ for all n, since $X_f \restriction_n$ has at most $\sqrt{g(n)}$ nonzero bits. Since X is K-resolute, $\exists d \forall n$, $K(X \restriction_n) \leq K(n) + 2\sqrt{g(n)} + d$. Since $\lim_n (g(n) - 2\sqrt{g(n)}) = \infty$ this implies that $K(X \restriction_n) \leq K(n) + g(n)$ for almost all n. Hence X is ultracompressible.

For the second clause we note that by [2, Theorem 8.2.29], every set with array computable c.e. degree is ultracompressible. Also, by [3, Theorem 1.3] (and the fact that the array computable c.e. degrees are exactly the c.e. traceable degrees) every set with array computable c.e. degree is anticomplex. On the other hand, by [2, Exercise 8.2.10], every superlow set is array computable. Hence it suffices to construct a superlow c.e. set which is not K-resolute. This is entirely similar to the typical construction of a superlow c.e. set which is not K-trivial (e.g. see [2, Exercise 5.2.10]) where K-triviality is replaced by (1). We leave this argument as an exercise for the motivated reader, as it does not present any novel features. □

A variation of ultracompressible sets was introduced in [2, Section 8.2] in terms of conditional complexity. A sequence X is called *facile* if for each order h and all sufficiently large n we have $K(X \restriction_n \mid n) \leq h(n)$. It is not hard to see that every facile set is ultracompressible.

Proposition 1.2. *All sequences that meet resoluteness condition (3) are facile, but the converse does not hold (even for c.e. sets).*

The first clause of this proposition is straightforward while the proof of the second clause is entirely analogous to the argument in the proof of Proposition 1.1.

The analogues of Propositions 1.1 and 1.2 with respect to plain complexity also hold (with similar proofs). We illustrate some of the above observations In Figure 1, where one may interpret 'resolute' with any of the three notions of resoluteness that we considered (i.e. K-resoluteness or one of (1), (3) and the plain complexity versions of these notions). Note that in the case of (3), one may also replace 'ultracompressible' with 'facile' since the latter property is guaranteed by (3). Sparse sets will be defined in Section 2.

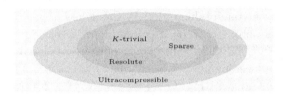

Fig. 1.   Classes of sequences of low initial segment complexity.

2. Resoluteness and sparseness

Intuitively, any information in a resolute set is coded in a very sparse way. In other words, a block of high complexity in a sequence may be used in order to reduce its initial segment complexity significantly, by 'spreading out' the bits of this block. In this section we formulate a notion of sparseness that is sufficient to guarantee resoluteness, and flexible enough to provide examples in many classes from computability theory. A concrete motivation for this notion as a tool for the study of resoluteness is the following observation. By direct coding on the values of the iterations of a given computable shift f we show that there are many sequences whose initial segment complexity is invariant under the application of f.

Proposition 2.1. *Let f be a computable shift. Every many-one degree contains a set X such that $X \equiv_K X_f$.*

Proof. Let Y be a set and $g(n) = f^n(0)$. Define $X = \{g(n) \mid n \in Y\}$ so that $Y \equiv_m X$. It remains to show that $K_M(X \upharpoonright n) \le K(X_f \upharpoonright n) + 1$ for a prefix-free machine M and all n. Let $F = \{f^i(0) \mid i \in \mathbb{N}\}$ and given $n > 0$ let $t_0, \ldots t_k$ be the first $k + 1$ members of F that are less than n. Then all bits of $X \upharpoonright n$ are 0 except perhaps $t_i, i \le k$. Moreover $X(t_i) = X_f(t_{i+1})$ for all $i < k$. Hence in order to describe $X \upharpoonright n$ we just need a description of $X_f \upharpoonright n$ and the value of $X(t_k)$. This shows that there is a machine M such that $\forall n \ (K_M(X \upharpoonright n) \le K(X_f \upharpoonright n) + 1)$. $\qquad\square$

The bits of X of Proposition 2.1 that carry some information ('significant bits') are far apart with respect to f. In particular, the image of the position of each such bit under f is the position of the next significant bit. In fact, for the equivalence $X \equiv_K X_f$ it suffices that the image under f of each significant bit is at most as large as the position of the next significant bit. This property is illustrated in Figure 2 (where 'diamonds' indicate the significant bits and 'dots' indicate the rest of the bits) and is the motivation for Definition 2.1. Given an increasing function g and two sets (viewed as infinite binary sequences) X, Y we denote by $X \otimes_g Y$ the set that is obtained by replacing the $g(i)$th bit of X with the ith bit of Y, for each i.

Definition 2.1 (Sparse sets). *Given an increasing function g we say that a set A is g-sparse if $A = E \otimes_f X$ for a computable set E, a computable function f with $g(f(i)) < f(i + 1)$ for all i, and some set X. A set B is called sparse if it is g-sparse for all computable increasing functions g.*

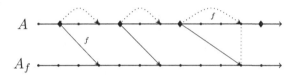

Fig. 2.   Construction of sparse and resolute sets.

Traditional notions of sparseness are based on the feature that the 1s have (in a certain sense) 'low density' in the initial segment of a sequence. An example here is the various immunity notions from classical computability theory (immune, hyperimmune, hyperhyperimmune etc.). These notions are closer to the special case of Definition 2.1 with $E = \emptyset$. Definition 2.1 also involves a notion of domination. For example, if we require $E = \emptyset$ in the definition then we only get sets A with $A' \geq_T \emptyset''$. Such sets compute a function that dominates all computable functions. By considering sparseness modulo computable sets (i.e. allowing E to be a computable set that depends on the choice of g) we obtain a much richer class, as we demonstrate in the following sections. For example, in Section 3 we show that sparse sets occur in all jump classes.

We show that sparseness indeed guarantees resoluteness. In order to do this, we need two technical observations.

Lemma 2.1. *Let $f, g : \mathbb{N} \to \mathbb{N}$ be computable increasing functions with the property that $g(n + 1) > f(g(n))$ for all but finitely many n. If E is a* *computable set then $E \otimes_g X \equiv_K (E \otimes_g X)_f$ for all sets X.*

Proof. It suffices to find a prefix-free machine M such that

$$\forall n \ K_M\big((E \otimes_g X) \restriction_n \big) \leq K\big((E \otimes_g X)_f \restriction_n \big) + 1$$

By the choice of f, g, for each sufficiently large number n we have $|g(\mathbb{N}) \restriction_n$ $| \leq |f(g(\mathbb{N})) \restriction_n | + 1$. Hence for each sufficiently large n, in order to describe the first n bits of $E \otimes_g X$ we only need a description of the first n bits of $(E \otimes_g X)_f$ and at most one extra bit. It follows that the machine M with the desired property exists. $\square$

The following lemma can be proved similarly. ˙

Lemma 2.2. *Let g be an order and let f be a computable increasing function such that $f(n) \leq g(f(n+1))$ for all but finitely many n. If $Z = E \otimes_f X$ where E is a computable set and X is any set, then $\exists c \forall n \ K(Z \restriction_n \mid n) \leq K(Z \restriction_{g(n)} \mid n) + c$.*

If a set A is sparse, then given any any computable order h we have $A = E \otimes_f X$ for some computable set E, some set X and some computable increasing function f such that $f(i) < h(f(i+1))$. Indeed, consider the increasing function $h_* : n \mapsto \min\{i : \max\{n, h(n-1)\} < h(i)\}$. Since A is sparse, it can be written as $E \otimes_f X$ for some computable set E, some set X and some computable increasing function f such that $h_*(f(i)) < f(i+1)$. In particular, $f(i) < h(f(i+1))$ for all i.

The following is a direct consequence of the above discussion, Definitions 1.3, 2.1, and Lemmas 2.1, 2.2. Moreover it holds in terms of plain complexity by the same arguments.

Corollary 2.1. *Every sparse set Z is K-resolute and meets (3).*

We can show that there are Turing complete sparse (hence resolute) sets. Since (by [9]) complete sets are not K-trivial, these are the first (and most easily produced) nontrivial examples of resolute sets. Without additional effort, we take a step further and show the following stronger statement. A degree $\mathbf{a}$ is called high if $\mathbf{a}' \geq \mathbf{0}''$.

Theorem 2.1. *Every high c.e. degree contains a sparse (hence resolute) c.e. set.*

Proof. Let $\mathbf{a}$ be a high c.e. degree and let A be a c.e. set in $\mathbf{a}$. By a simple variation of Martin's characterization of high c.e. degrees in terms of dominating functions in [10], **a** computes a sequence (x_i) with computable approximations $\lim_s x_i[s] = x_i$ satisfying the following properties for each i, s:

(a) $f(x_i) < x_{i+1}$ for all computable shifts f and all but finitely many i;

(b) $x_i[s] < x_{i+1}[s]$ and $x_i[s] \leq x_i[s+1]$;

(c) if $x_n[s] \neq x_n[s+1]$ then $x_n[s+1] > s+1$;

(d) $x_i[s] \in \mathbb{N}^{[i]}$.

Consider the set D consisting of the numbers n such that $n = x_i[s]$ for some i, s such that $x_i \neq x_i[s]$. Clearly D is c.e. and is computable from A. Moreover it is easy to see that it computes A, so that it has degree **a**. It remains to show that D is sparse. We define a computable set E and a computable function g such that $D = E \otimes_g X$ for some set X and $f(g(n)) < g(n+1)$ for all n. Let i_0 be a number such that $f(x_i) < x_{i+1}$ for all $i \geq i_0$ and let s_0 be a stage such that $x_i[s] = x_i$ for all $i \leq i_0$ and $s \geq s_0$. Define $g(0) = x_{i_0}$ and let $g(i+1) = x_{i+1}[s]$ for the least stage $s \geq s_0$ such that $x_{i+1}[s] > f(g(i))$ and $s > g(x_{i+1}[s])$. In this case we say that $g(i+1)$ was defined at stage s. According to the hypothesis about (x_i), the function g is total and computable. Moreover $f(g(n)) < g(n+1)$ for all n. The set E is defined recursively as follows. To compute $E \restriction_n$ find the least j, s_j such that $g(j)$ is defined at stage s_j and $g(j) > n$. For each $t < n$, if $t \notin g(\mathbb{N})$ and $t = x_k[s]$ for some $k < j$ and $s \leq s_i$ let $E(t) = 1$; otherwise let $E(t) = 0$. According to the properties of (x_i), after stage s_i there will be no additional positions $< n$ occupied by approximations to the values of (x_i). Hence the sets E, D agree on the positions in $\mathbb{N} - g(\mathbb{N})$. Hence $D = E \otimes_g X$ for some set X. $\square$

The proof of Theorem 2.1 can be modified to a construction of a hyperimmune Π_1^0 sparse set. Curiously enough, such sets have to be high.

Proposition 2.2. *Every sparse hyperimmune set is high.*

Proof. Let A be a sparse set which is not high. Let f be an A-computable function such that for each i there exist at least $2i$ numbers in $A \restriction_{f(i)}$. Since A is not high there exists a computable function h such that $h(i) > f(i)$ for infinitely many i. Since A is sparse, we may choose a computable set E and a computable function g such that $g(i) > h(i)$ for all i and $A = E \otimes_g X$ for some set X. From this presentation it follows that A is not hyperimmune. $\square$

It is, perhaps, not surprising that there are K-resolute sets that are not sparse. By applying Proposition 2.2 to a K-trivial hyperimmune set we get the following (since K-trivial sets are not high [6]).

Corollary 2.2. *There exists a K-trivial set which is not sparse.*

At this point we have justified the diagram in Figure 1. We note that all of the depicted classes are *meager*, in the sense of Baire category. Indeed, every weakly 2-generic set has effective packing dimension 1 so it is not ultracompressible.

Another way to obtain sparse and resolute sets is to use basis theorems on effectively closed sets. For this purpose, we need the following fact.

Theorem 2.2. *There exists a non-empty Π_1^0 class with no computable paths which consists entirely of sparse sequences.*

Proof. We will define a partial computable function φ with binary values such that the set of all total extensions of it, is the required Π_1^0 class. Fix a computable double sequence $(x_n[s])$ with $x_0[s] = 0$ and such that (a)-(d) of Theorem 2.1 hold. Let (φ_e) be an effective sequence of all partial computable functions. We assume the standard convention that if $\varphi_c(n)[s] \downarrow$ then e, n are less than s. The construction of φ is as follows: at stage s, if $\varphi_e(x_e)[s] \downarrow$ for some $e < s$ and $\varphi(x_e[s])$ is undefined, then define $\varphi(x_e[s]) = 1 - \varphi_e(x_e)[s]$. Moreover for each $i < s$, if $i \neq x_e[s]$ for all $e < s$ and $\varphi(i)$ is undefined, define $\varphi(i) = 0$.

For the verification, first note that since each $x_e[s]$ reaches a limit as $s \to \infty$ (and each time it is redefined it takes a value on which φ is currently undefined) the Π_1^0 class of total extensions of φ is perfect. Indeed, there are infinitely many $e \in \mathbb{N}$ for which φ_e is the empty function, and for these numbers e the function φ will be undefined on $x_e := \lim_s x_e[s]$. Second, there are no computable extensions of φ. Indeed, given $e \in \mathbb{N}$, if φ_e is total then by the construction we have $\varphi(x_e) \neq \varphi_e(x_e)$ where $x_e := \lim_s x_e[s]$. Finally, we show that every extension of φ is sparse. Let g be an increasing computable function. Then by the properties of $(x_e[s])$ there exists some e_0 such that $f(x_e) < x_{e+1}$ for all $e > e_0$. We may define a computable set E as follows. Let $E \restriction_{x_{e_0}} = \varphi \restriction_{x_{e_0}}$ and let s_0 be a stage where $x_{e_0}[s]$ has reached a limit. Also let $y_0 = x_{e_0}$. At step $e > e_0$ find a stage $s > s_0$ such that $f(x_i)[s] < x_{i+1}[s]$ for each $i \leq e$ and $x_{e+1}[s] < s$. Then define the bits of E in the interval $(y_{e-1}, x_e[s]]$ to be the bits of φ in the same interval, except where φ is (currently) undefined in which case we choose value 0. Also let $y_e = x_e[s]$.

For each n, positions between y_n and y_{n+1} in E include the values of some codes $x_e[s]$, where s is the stage found in step $n+1$ of the construction of E. By the construction, if (z_i) is the sequence of these positions we have $g(z_j) < z_{j+1}$ for each j. Therefore any extension A of φ can be written as $E \otimes_f X$ where $f(i) = z_i$ and X is some set (giving the bits of A on positions z_i). $\qquad\square$

A set is called *computably dominated* if every function computable from it is dominated by a computable function. The strong version of the computably dominated basis theorem (see [2, Theorem 1.8.44]) says that every Π_1^0 class without computable paths has a perfect subclass of computably dominated sets. By applying this basis theorem to the class of Theorem 2.2 we get more examples of sparse sets.

Corollary 2.3. *There are uncountably many computably dominated sparse sets.*

In particular, there are uncountably many resolute sets.

3. Jump inversion with K-resolute sequences

A set A is called superlow if $A' \equiv_{tt} \emptyset'$. Curiously enough, it is more involved to produce a non-trivial sparse c.e. set which does not realize the highest jump, than it is to produce one that does. A possible heuristic explanation for this is that the the notion of sparseness involves some type of domination (which is characteristic to high sets). Similar remarks apply to the construction of K-resolute sets (compare with the straightforward constructions of Section 2 that produce high sets).

Theorem 3.1. *There exists a superlow sparse c.e. set A which is not K-trivial.*

Proof. We use a priority tree construction to construct a c.e. set A with the required properties. Let (Φ_e) be an effective enumeration of all Turing functionals and let (φ_e) be an effective enumeration of all strictly increasing partial computable functions. Without loss of generality we may assume that, for all e, i, s, if $\varphi_e(i+1)[s] \downarrow$ then $\varphi_e(i)[s] \downarrow$. Moreover let $*$ denote concatenation of strings. In order to ensure that A is sparse it suffices to satisfy the following conditions.

$R_e : \varphi_e$ is total $\Rightarrow \exists E, X, g, \ (A = E \otimes_g X \ \text{ and } \ \forall i, \ \varphi_e(g(i)) < g(i+1))$

where E ranges over all computable sets, X ranges over all sets and g ranges over all computable functions. In order to ensure that A is not K-trivial it suffices to construct a prefix-free machine N (as usual, by enumerating a Kraft-Chaitin set of requests) such that the following conditions are satisfied.

$$P_e : \exists n\ K(A \restriction_n) > K_N(n) + e.$$

We will also ensure that $A' \equiv_{tt} \emptyset'$ by ensuring that the values of A' can be approximated with a computable modulus of convergence. The priority tree is the full binary tree where each level e is associated with requirements R_e, P_e. In particular, each node of length e has two branches with labels 1,0 that correspond to a guess about whether φ_e is total or not. In addition, each such node (based on the guesses about the totality of φ_i, $i < e$) will work toward the satisfaction of P_e. The construction will proceed in stages $s + 1$, where a path δ_s of length s will be defined through the tree. Given a node α, we say that stage s is an α-stage if $\alpha \subset \delta_s$. Define $\ell_\alpha(s)$ to be $\max\{i \mid \forall j < i\ \varphi_{|\alpha|}(j)[s] \downarrow\}$ if s is an α-stage and 0 otherwise. A stage s is called α-expansionary if $\ell_\alpha(s) > \ell_\alpha(t)$ for all $t < s$.

Let $\alpha \to n_\alpha$ be a one-one function from the nodes of the tree to $\mathbb{N}$ such that the sum of 2^{-n_α} for all nodes α is at most $1/2$. We fix the priority list $P_0, R_0, P_1, \ldots$. Each node carries a strategy P_α for $P_{|\alpha|}$ and a strategy R_α for $R_{|\alpha|}$. Injury of a strategy P_α means the initialization of it and all of its parameters. There will be no injury of the R_α strategies. Strategy P_α may be injured either because δ_s moves to the left of it, or because $\Phi_i^A(i)[s]$ becomes defined at some stage s for some $i < |\alpha|$ (with appropriately large use). At stage $s + 1$ the quota for weight of the N-requests that node α may enumerate is $2^{-t_\alpha[s]}$, where $t_\alpha[s] = n_\alpha + u_\alpha[s] + |\alpha|$ and $u_\alpha[s]$ is the number of times that P_α has been injured in the stages up to s. A number is called *large* at some stage of the construction if it is larger than the value of every parameter of the construction up to that stage.

Strategy R_α will define a computable sequence (q_i^α) of potential 'codes' such that $\phi_e(q_i^\alpha) < q_{i+1}^\alpha$ for all i such that $q_{i+1}^\alpha \downarrow$. If $\varphi_{|\alpha|}$ is partial, the sequence (q_i^α) will be finite. These 'codes' will be chosen inductively as a subsequence of (q_i^β), where β is the largest initial segment of α with $\beta*1 \subseteq \alpha$ (if such segment does not exist, codes are chosen as a subsequence of the identity sequence). For each α we let $p_i^\alpha = q_i^\beta$ where β is as above, and $p_i^\alpha = i$ if such β does not exist.

Strategy for P_α

(1) Pick a *large* number m_α.

(2) Let D_α be a set of 2^{t_α} terms of $(p^\alpha_{\langle\alpha,2i+1\rangle})$ with $i > m_\alpha$.

(3) Let $r_\alpha = \max D_\alpha + 1$ and enumerate an N-description of r_α of length $t_\alpha - |\alpha|$.

(4) Wait until $K(A \upharpoonright_{r_\alpha})[s] \le K_N(r_\alpha)[s] + |\alpha|$.

(5) Enumerate $\max(D_\alpha - A[s])$ into A and go to step 4.

Note that the loop between steps 4 and 5 can only be repeated at most $2^{t_\alpha} - 1$ times. Hence $K(A \upharpoonright_{r_\alpha}) > K_N(r_\alpha) + |\alpha|$. We say that node P_α *requires attention* at stage $s + 1$ if $\alpha \subseteq \delta_s$ and the strategy for α is ready to perform the next step. In other words, in the following cases:

(a) m_α is undefined;

(b) $m_\alpha \downarrow$ but D_α is undefined, and there are 2^{t_α} terms of (p^α_i) as required in step (2) of the strategy for P_α;

(c) the strategy is in step 4 and $K(A \upharpoonright_{q_\alpha})[s] \le K_N(n)[s] + e$.

The strategy R_α operates at α-expansionary stages and defines (q^α_i). Note that since $\varphi_{|\alpha|}$ is increasing, we also have $q^\alpha_i \ge i$ for all I such that q^α_i is defined.

Strategy for R_α

(1) Let q^α_0 be $p^\alpha_{\langle\alpha,0\rangle}$.

(2) Let j be the largest number such that $q^\alpha_j \downarrow$ and define q^α_{j+1} to be the least $p^\alpha_{\langle\alpha,2i\rangle}$ which is greater than $\varphi_{|\alpha|}(q^\alpha_j)$.

We say that R_α *requires attention* at stage $s + 1$ if $\alpha * 1 \subseteq \delta_s$ and the strategy is ready to perform the next step. In other words, if q^α_0 is undefined or $\varphi_{|\alpha|}(q^\alpha_j)$ is defined for some j but q^α_{j+1} is undefined. The construction includes an injury of the strategies from the implicit lowness requirement. Injury of α means injury P_α.

Construction At stage $s + 1$ define a path δ_s of length s inductively, starting from the root and from each node α choosing branch 1 if s is an α-expansionary stage and 0 otherwise. Injure all nodes to the right of δ_s. For each $e < s$ such that $\Phi^A_e(e)[s] \downarrow$ with use u_e and each α such that $|\alpha| \ge e$, $m_\alpha[s] < u_e$ injure P_α. For each $\alpha \subset \delta_s$ for which R_α requires attention, execute the next step of R_α. If some P_α with $\alpha \subset \delta_s$ requires attention, pick the least such α, execute the next step of its strategy.

Verification We first verify that $A' \leq_{tt} \emptyset'$. Each P_α may only be injured by computations $\Phi_e^A(e)[s] \downarrow$ for $e \leq |\alpha|$. Every injury of α initiates another round of the strategy of α. At any round, the maximum amount of enumerations into A that α may perform is bounded by the current value of t_α. On the other hand, the current value of t_α may be computed by the number of injuries that α has endured. It follows that there is a computable bound on the number of times that each computation $\Phi_e^A(e)[s] \downarrow$ may be 'disturbed' (by enumeration into A below the current use of the computation). Hence A' may be computably approximated with a computable modulus of convergence, which shows that $A' \leq_{tt} \emptyset'$.

Let δ be the leftmost path such that for all n we have $\delta \restriction_n \subset \delta_s$ at infinitely many stages s. By induction we show the following for each $\alpha \subset \delta$:

(i) P_α is injured or requires attention finitely many times;
(ii) if $\alpha * 1 \subset \delta$ then (q_i^α) is total.

Let α be a node and suppose that these clauses hold for all $\beta \subset \alpha$. According to the above discussion, beyond a certain stage the computations $\Phi_e^A(e)$ will either converge permanently or diverge permanently. Therefore the lowness requirements will stop injuring P_α. On the other hand (by the induction hypothesis) beyond a certain stage the strategies P_γ for $\gamma \subset \alpha$ and γ to the left of α will stop requiring attention. Therefore they will cease enumerating numbers in to A and P_α will stop being injured. After such a stage, P_α will stop requiring attention before it has completed 2^{t_α} enumerations into A. This completes the induction step for (i).

For (ii) note that by the induction hypothesis the sequence (p_i^α) is total. We may assume that $\alpha * 1 \subset \delta$ (otherwise (ii) holds trivially). Then there will be infinitely many α-expansionary stages and, by the construction, (q_i^α) will be totally defined. This completes the inductive proof of (i), (ii).

Finally we show that A meets all requirements P_e, R_e. Let α be the unique node on δ of length e. For P_e, let s_0 be a stage after which P_α is not injured. By properties (i)-(iii) that we established it follows that (p_i^α) is total. Moreover the terms $p_{\langle \alpha, 2i+1 \rangle}^\alpha$ may only be enumerated into A by P_α. Hence the strategy P_α will complete the preliminary steps (1)-(3) and will enter the loop (4)-(5) thus ensuring (as explained in the remark following the description of this strategy) that $K(A \restriction_{r_\alpha}) > K_N(r_\alpha) + |\alpha|$ for a certain number r_α. It follows that P_e is met.

For R_e, if $\alpha * 0 \subseteq \delta$ then φ_e is partial and R_e is met. Otherwise $\alpha * 1 \subseteq \delta$ and by (iii), (q_i^α) is total. Note that when a term q_i^α is defined, the strategies to the right of $\alpha * 1$ will not enumerate into A any numbers $\leq q_i^\alpha$. The same holds for the strategies $\gamma \subseteq \alpha$ or those that lie to the left of $\alpha * 1$. Moreover the only numbers enumerated in A by the strategies extending $\alpha * 1$ are terms of (q_i^α). It follows that α defines a computable set E and the function $g(i) = q_i^\alpha$ such that $A = E \otimes_g X$ for some set X. Moreover by the definition of (q_i^α) we have $\varphi_e(g(i)) < g(i+1)$ for all i. Hence R_e is met. $\qquad\square$

In some respect, the construction in the proof of Theorem 3.1 resembles the construction of a maximal set. However, the lowness requirements apparently make the use of some type of a tree argument necessary.

Corollary 3.1. *There exists a superlow K-resolute c.e. set A which is not K-trivial.*

A general jump inversion theorem for sparse sets is easy to obtain since we have already constructed a perfect Π_1^0 class of sparse sets with no computable paths.

Theorem 3.2 (Jump inversion with sparse sets). *Every jump class contains a sparse set. In particular, every degree above $\mathbf{0}'$ contains the jump of a sparse set.*

Proof. By [11] given a degree $\mathbf{a} \geq \mathbf{0}'$ and a Π_1^0 class P with no computable members, there exists $X \in P$ such that X' is a member of $\mathbf{a}$. Therefore the theorem is a consequence of Theorem 2.2. $\qquad\square$

The jump inversion for c.e. sparse sets involves a modification of the argument that we used in the proof of Theorem 3.1.

Theorem 3.3 (Jump inversion with sparse c.e. sets). *For every Σ_2^0 set $S \geq_T \emptyset'$ there exists a sparce c.e. set A such that $A' \equiv_T S$.*

Proof. The argument here is similar to the one in the proof of Theorem 3.1, so we use the same notation and terminology. Requirements R_e remain the same and the tree of strategies is also the same. We also need to satisfy $S \equiv_T A'$. The coding of S into A' will be achieved via the standard 'thickness' requirements, only that the codes that are used need to be chosen from the ones produced by the R_e strategies. Let D be a c.e. set such that if $e \in S$ then $D^{[e]} = \mathbb{N} \restriction_n$ for some n, and if $e \notin S$ then $D^{[e]} = \mathbb{N}^{[e]}$. We may fix an enumeration of D such that at each stage s, if $\langle e, n \rangle \in D^{[e]}[s]$ then $\langle e, i \rangle \in D^{[e]}[s]$ for all $i < n$. In the following construction, 'injury' of a node α is merely a way to say that δ_{2s+1} moved to the left of α. We may assume that if Φ_e^A is undefined then $\Phi_e^A[s]$ is undefined for infinitely many s.

Construction At stage $2s + 1$ define a path δ_{2s+1} of length s inductively, starting from the root and from each node α choosing branch 1 if s is an α-expansionary stage and 0 otherwise. For each $\alpha \subset \delta_{2s+1}$ for which R_α requires attention, execute the next step of R_α and injure all nodes that lie to the right of δ_{2s+1}. At stage $2s + 2$, for each α with $|\alpha| < s$ enumerate into A all $p^\alpha_{\langle \alpha, 2i+1 \rangle}$ which are defined with $i \in D^{[|\alpha|]}[2s + 2]$ and are larger than all uses of any computations $\Phi^A_i(j)[2s + 1] \downarrow$ with $j < e$ and larger than the last stage where α was injured.

Verification We first verify that $A' \leq_T S$. Since S computes $\emptyset'$, it also computes D. In order to decide if $\Phi^A_e(e) \downarrow$ we first compute a stage s_0 at which $D^{[i]}[s_0] = D^{[i]}$ for all $i < e$ such that $D^{[i]}$ is finite. Moreover inductively, we may compute whether $\Phi^A_i(i) \downarrow$ for each $i < e$, and a stage $s_1 > s_0$ such that $\Phi^A_i(j)[s] \downarrow$ for all $s \geq s_1$ and each $j < e$ such that $\Phi^A_j(j) \downarrow$. Let v_e be the maximum use of the oracle A in these computations. Next, we ask if there is a stage $2s+2 > s_1$ such that $\Phi^A_e(e)[2s+2]$ is defined with some use u_e and for all $p^\alpha_{\langle \alpha, 2i+1 \rangle}[s]$ with $|\alpha| < e$, $D^{[|\alpha|]} = \mathbb{N}^{[|\alpha|]}$, $p^\alpha_{\langle \alpha, 2i+1 \rangle}[s] > v_e$ and $p^\alpha_{\langle \alpha, 2i+1 \rangle}[s] \leq u_e$ we have $i \in D^{[|\alpha|]}[2s+2]$ or the last stage where α was injured is larger than u_e. If such a stage does not exist, then clearly $\Phi^A_e(e)$ is undefined. Otherwise, the construction will preserve the computation, hence $\Phi^A_e(e)$ is defined. Hence S computes A'.

Next, we show that $S \leq_T A'$. In order to decide if $e \in S$ we first find a stage s_0 at which all computations $\Phi^A_i(i)$, $i < e$ that eventually converge, actually converge at s_0 with correct A use. Moreover let u_e be the maximum of these uses. Then we search for some x such that one of the following holds:

(a) for all α with $|\alpha| = e$, all $i > x$ and all $s > x$ either $p^\alpha_{\langle \alpha, 2i+1 \rangle}[s]$ is undefined, or it is $\leq u_e$ or it is a member of $A[s]$;

(b) for all α with $|\alpha| = e$, all $i > x$ and all x either $p^\alpha_{\langle \alpha, 2i+1 \rangle}[x]$ is undefined, or it is $\leq u_e$ or it is not a member of $A[s]$.

If $e \in S$ we show that (a) holds for some x. Indeed, in this case $D^{[|\alpha|]} = \mathbb{N}^{[|\alpha|]}$ and all defined terms $p^\alpha_{\langle \alpha, 2i+1 \rangle}[s]$ which are not prohibited by the convergence of $\Phi^A_i(i)$, $i < e$ will eventually enter A according to the construction. By a standard use of 'true stages' in the enumeration of A (i.e. stages s where for the least number n entering A we have $A[s] \restriction_n$ is a prefix of A) we get that all of these terms that are larger than u_e will be permitted to enter A at some point of the construction.

If $e \notin S$ we show that (b) holds for some x. Indeed, in this case $D^{[|\alpha|]}$ is finite for all α with $|\alpha| = e$. Let x be larger than all the elements of this set. Then none of the codes $p^{\alpha}_{\langle \alpha, 2i+1 \rangle}$ with $i > x$ that are defined may enter A.

Furthermore, it is not possible that both (a), (b) occur for some x. Indeed, let δ be the leftmost path such that $\delta \subset \delta_{2s+1}$ for infinitely many s. If $\alpha = \delta \restriction_{|\alpha|}$ then (p^{α}_i) is total, which shows that at least one of (a), (b) must fail (for sufficiently large x). Since the search for an x such that (a) or (b) hold is computable in A', this gives a computation of whether $e \in S$ from A'.

We conclude with a proof that A meets each R_e requirement. Let α be the unique node on δ of length e and let s_0 be a stage such that δ_s is to the right of α or an extension of it, for all $s \geq s_0$. If $\alpha * 0 \subseteq \delta$ then φ_e is partial and R_e is met. Otherwise $\alpha * 1 \subseteq \delta$ and (q^{α}_i) is total. Note that when a term q^{α}_i is defined at some stage $s > s_0$, no numbers p^{β}_i for β to the right of $\alpha * 1$ will enumerated into A after s, unless they are larger than s. The same holds for the nodes β which lie to the left of $\alpha * 1$. Moreover the only numbers enumerated in A by the strategies extending $\alpha * 1$ are terms of (q^{α}_i). Finally the finitely many nodes γ that prefix α enumerate computable (possibly infinite) sets of codes $p^{\gamma}_{\langle \gamma, 2i+1 \rangle}$ into A. It follows that α can define a computable set E and the function $g(i) = q^{\alpha}_i$ such that $A = E \otimes_g X$ for some set X. Moreover by the definition of (q^{α}_i) we have $\varphi_e(g(i)) < g(i+1)$ for all i. Hence R_e is met. $\square$

4. Completely resolute and resolute-free degrees

We are interested in two extremes, namely the degrees which do not contain resolute sets and the degrees that consist entirely of resolute sets. A degree is called *completely K-resolute* if every set in it is K-resolute. Similar definitions apply to the other notions of resoluteness that we have considered. Note that every K-trivial degree is completely K-resolute, so we will be interested in nontrivial examples of such degrees. A degree is called *resolute-free* if it does not contain any resolute set (with respect to any of the definitions of resoluteness that we have discussed).

It turns out that the existence of such degrees is very related to two observations between bounded Turing reductions (i.e. weak truth table reductions) and resolute sets. Note that if f is a computable shift and $A \equiv_K B$ then $A_f \equiv_K B_f$.

Proposition 4.1. *If A is K-resolute and $A \equiv_K B$ then B is also K-resolute. The same holds for 'weakly K-resolute' in place of 'K-resolute'.*

Proof. First, we show the case for K-resolute sequences. Let f be a computable shift. Under the assumptions there are constants c_i such that for all n,

$$K(B \upharpoonright_n) \leq K(A \upharpoonright_n) + c_0 \leq K(A_f \upharpoonright_n) + c_1 \leq K(B_f \upharpoonright_n) + c_3.$$

Hence $B \equiv_K B_f$. Since f was chosen arbitrarily, B is also K-resolute.

Second, for the case of weakly K-resolute sequences let f be a computable shift. Under the assumptions there are constants d_i such that for all n,

$$K(B \upharpoonright_{f(n)}) \leq K(A \upharpoonright_{f(n)}) + d_0 \leq K(A \upharpoonright_n) + d_1 \leq K(B \upharpoonright_n) + d_3.$$

Since f was chosen arbitrarily, B is also weakly K-resolute. $\square$

Proposition 4.2. *If B is weakly K-resolute and $A \leq_{wtt} B$ then $A \leq_K B$.*

Proof. Since $A \leq_{\text{wtt}} B$ there is a computable increasing function f such that $\exists d \forall n \ (K(A \upharpoonright_n) \leq K(B \upharpoonright_{f(n)}) + d)$. Since B is weakly K-resolute, there exists a constant c such that $\forall n \ (K(B \upharpoonright_{f(n)})) \leq K(B \upharpoonright_n) + c)$. Hence $A \leq_K B$. $\square$

These observations point to the fact that in order to produce a degree which does not contain any resolute sets it suffices to produce a set that is not resolute and its Turing degree 'collapses' to (i.e. contains) a single weak truth table degree. Similarly, in order to produce a degree which consists entirely of resolute sets it suffices to produce a resolute set whose Turing degree 'collapses' to (i.e. contains) a single weak truth table degree.

Proposition 4.3. *Every computably dominated K-resolute degree is completely K-resolute. Moreover the same holds for weakly K-resolute in place of K-resolute.*

By the application of the uncountable version of the computably dominated basis theorem for Π_1^0 classes (e.g. see [2, Theorem 1.8.44]) on the class of Theorem 2.2, along with Corollary 2.1 we have the following consequence.

Corollary 4.1. *There exist uncountably many completely K-resolute degrees.*

In particular, since there are only countably many K-trivial degrees, there are completely K-resolute degrees which are not K-trivial. Similar results hold for the resolute-free degrees.

Corollary 4.2. *There exist uncountably many resolute-free degrees.*

Indeed, there are uncountably many 1-random computably dominated degrees.

Corollary 4.3. *Every (weakly) K-resolute sequence computable from a Martin-Löf random computably dominated degree is computable.*

Proof. By Demuth [12] (also see [5, Theorem 8.6.1] for a neat proof) every noncomputable set that is truth-table reducible to a Martin-Löf random set is Turing equivalent to a Martin-Löf random set. On the other hand, there is a Martin-Löf random set of computably dominated degree, and computably dominated degrees consist of a single truth-table degree. Hence the statement is a consequence of Propositions 4.1 and 4.2. □

We are interested in c.e. examples of completely K-resolute and resolute-free degrees. In [3, Theorem 4.3] it was shown that there exists a c.e. degree which contains no anti-complex sets. Since every ultracompressible set is anti-complex, this degree does not contain any ultracompressible or (by Proposition 1.1) resolute sets.

Theorem 4.1. *There exists a resolute-free c.e. degree.*

Finally, we wish to produce nontrivial examples of completely K-resolute c.e. degrees. A c.e. degree is called *contiguous* if all the c.e. sets in it are weak truth table equivalent. The existence of nontrivial contiguous degrees was first shown and exploited in [13]. A degree is called *strongly contiguous* if all the sets in it are weak truth table equivalent; in other words, it consists of a single weak truth table degree. The existence of strongly contiguous c.e. degrees was shown in [14].

Theorem 4.2. *Every strongly contiguous c.e. degree is completely K-resolute.*

Proof. We say that a c.e. degree **a** is 'wtt-bottomed' if the c.e. weak truth table degrees inside **a** have a least element. It suffices to show that every c.e. set in the least weak truth table degree inside a wtt-bottomed degree is K-resolute. Indeed, strongly contiguous c.e. degrees are clearly wtt-bottomed so the result would follow from Proposition 4.1.

Assume that **a** is a 'wtt-bottomed' c.e. degree and A is a c.e. set in the least weak truth table c.e. degree inside **a**. We show that A is K-resolute. Given a computable shift f we wish to construct a prefix-free machine M such that

$$\forall n \; K_M(A \upharpoonright_n) \leq K(A_f \upharpoonright_n). \tag{4}$$

Without loss of generality we may assume that the weight of the underlying universal prefix-free machine U (i.e. the sum of all $2^{-|\sigma|}$ such that $U(\sigma) \downarrow$) is $< 2^{-1}$. In order to define M, we construct an auxiliary c.e. set B such that $B \equiv_T A$. Let $A[s]$ be the enumeration of A with respect to a standard enumeration of all c.e. sets (and Turing functionals). The enumeration of B will be defined in these stages via a standard system of movable markers $\delta(n)[s]$ When we say 'move $\delta(n)$' at stage $s+1$ of the construction we mean

- enumerate $\delta(n)[s]$ into B;
- let $\delta(n)[s+1] = \langle n, s+1 \rangle$;
- let $\delta(i)[s+1] = \langle n, s+1 \rangle$ for each $i \in (n, s]$.

Let $\delta(0)[0] = 0$. We may assume that any number that enters A at stage s is strictly less than s. Let $g(n) = \max\{i \mid f(i) \leq n\}$. Note that $g(n) \leq n$ for all n.

Enumeration of B At stage $s+1$ define $\delta(s+1)[s+1] = s+1$. If n be the least number that enters A, move $\delta(g(n))$. If there is no such n, do nothing more.

Note that the enumeration of B is well defined. In order to show that $A \equiv_T B$, note that for each m

$$\delta(m) \text{ only moves if the approximation to } A \upharpoonright_{f(m+1)} \text{ changes.} \tag{5}$$

Hence each $\delta(m)$ reaches a limit. Moreover, since $g(n) \leq n$ for all n, every time that the approximation to $A \upharpoonright_n$ changes, the approximation to $B \upharpoonright_{\delta(n)}$ also changes. Also when $\delta(n)$ moves, its current value is enumerated in B. Hence $A \leq_T B$. On the other hand, by (5) and the fact that $\delta(m)[s] \geq m$ for all s it follows that the approximation to $B \upharpoonright_n$ does not change unless the approximation to $A \upharpoonright_{f(n+1)}$ changes. This shows that $B \leq_T A$. Hence $A \equiv_T B$.

By the hypothesis on A there exists a Turing functional Γ with a computable bound on use function γ such that $\Gamma^B = A$. Let (s_i) be an increasing computable sequence of the 'expansionary stages' in the reduction $\Gamma^B = A$, i.e. the stages s where the maximum n_s such that $\Gamma^B \upharpoonright_{n_s} = A \upharpoonright_{n_s}$ at stage

s is larger than the corresponding numbers n_t for all $t < s$. Clearly, $n_{s_i} \geq i$ for all i. We may assume that if $\gamma(n)[s]$ is defined then its value is $< s$.

Suppose that $A[s_t] \restriction_n \neq A[s_{t+1}] \restriction_n$ for some t and $n \leq t$. Then for all $k > t$, if $A[s_k] \restriction_n \neq A[s_{k+1}] \restriction_n$ we also have $\qquad$ (6) have $A[s_k] \restriction_{g(n)} \neq A[s_{k+1}] \restriction_{g(n)}$

Indeed, by stage s_t all $\gamma(i)$, $i \leq t$ are defined. In the interval of stages $[s_t, s_{t+1}]$ some number $m < n$ enters A. Hence in the construction of B (which runs on the same stages) $\delta(g(m))[s_{t+1}]$ is defined and larger than $\gamma(n)$. Hence if $A[s_k] \restriction_{g(n)} = A[s_{k+1}] \restriction_{g(n)}$ for some $k > t$ we also have $A[s_k] \restriction_{g(m)} = A[s_{k+1}] \restriction_{g(m)}$ which means that no number $\leq \gamma(n)$ will be enumerated into B in the interval of stages $(s_k, s_{k+1}]$. Since (s_i) are expansionary stages and $\Gamma^B = A$ it follows that $A[s_k] \restriction_n \neq A[s_{k+1}] \restriction_n$. This concludes the proof of (6).

Finally we may use (6) in order to construct a prefix-free machine M with the property (4). We do this dynamically during the stages (s_i) using a standard Kraft-Chaitin request set. At stage s_i, for each $n < t$ such that $K_M(A \restriction_n)[s_i] > K(A_f \restriction_n)[s_i]$ we enumerate a description of $A \restriction_n$ of length $K(A_f \restriction_n)[s_i]$. It suffices to show that the 'weight' of the requests is bounded by 1. Fix n. By (6) each description of the universal machine U of a string of length n (in particular the strings that have been current values of $A_f \restriction_n$) corresponds to at most two M-descriptions (which we enumerate in order to reduce $K_M(A \restriction_n)$). Indeed, (6) says that if we enumerate two descriptions of $A \restriction_n$ based on the same U-description of $A_f \restriction_n$ (in fact, same value of $A_f \restriction_n$) then the next description of $A \restriction_n$ will be enumerated based on a new description (and new value) of $A_f \restriction_n$. Since the weight of the domain of the universal prefix-free machine is $< 2^{-1}$, the weight of the request set for M is bounded by 1. $\qquad \Box$

We note that the proof of Theorem 4.2 is easily adaptable for the other resoluteness notions that we have considered. For example, it holds with respect to plain complexity.

A degree $\mathbf{a}$ is low if $\mathbf{a}' = \mathbf{0}'$. By [15] there exists a strongly contiguous c.e. degree which is not low. Hence we have the following consequence.

Corollary 4.4. *There exists a completely K-resolute c.e. degree which is not low.*

A number of questions regarding the relationship between the Turing degrees and the K-degrees were raised in [16] and answered in [17]. For example, in [17] it was observed that there are uncountably many sets such that all of the sets in their Turing degree are in the same K-degree (i.e. have the same initial segment prefix-free complexity). In particular, the K-trivial c.e. degrees are not the only degrees such that all of their members are in the same K-degree. Here we see this phenomenon inside the c.e. degrees.

Corollary 4.5. *There exists a c.e. Turing degree which is not low, yet all the sets in it have the same initial segment (plain or prefix-free) complexity.*

We note that by [17] the complete c.e. degree does not have this property, i.e. it contains sets with different initial segment complexity.

Acknowledgments

This research was partially done whilst the authors were visiting fellows at the Isaac Newton Institute for the Mathematical Sciences, Cambridge U.K., in the programme 'Semantics & Syntax'. Barmpalias was supported by the *Research fund for international young scientists* number 611501-10168 from the National Natural Science Foundation of China, and an *International Young Scientist Fellowship* number 2010-Y2GB03 from the Chinese Academy of Sciences; partial support was also received from the project *Network Algorithms and Digital Information* number ISCAS2010-01 from the Institute of Software, Chinese Academy of Sciences. Downey was supported by a Marsden grant of New Zealand.

References

1. J. I. Lathrop and J. H. Lutz, Recursive computational depth, *Inf. Comput.* **153**, 139 (1999).
2. A. Nies, *Computability and Randomness* (Oxford University Press, 2009).
3. J. Franklin, N. Greenberg, F. Stephan and G. Wu, Anti-complexity, lowness and highness notions, and reducibilities with tiny use, Submitted, (2012).
4. K. M. Ng, F. Stephan and G. Wu, Degrees of weakly computable reals, in *Proceedings of the Second conference on Computability in Europe: logical Approaches to Computational Barriers*, (Springer-Verlag, Berlin, Heidelberg, January 2006).
5. R. Downey and D. Hirshfeldt, *Algorithmic Randomness and Complexity* (Springer, 2010).
6. A. Nies, Lowness properties and randomness, *Adv. Math.* **197**, 274 (2005).
7. B. Kjos-Hanssen, W. Merkle and F. Stephan, Kolmogorov complexity and the recursion theorem, in *STACS*, (Berlin, 2006).
8. B. Kjos-Hanssen, W. Merkle and F. Stephan, Kolmogorov complexity and the recursion theorem, *Trans. Amer. Math. Soc.* **363** (2011).
9. R. G. Downey, D. R. Hirschfeldt, A. Nies and F. Stephan, Trivial reals, in

Proceedings of the 7th and 8th Asian Logic Conferences, (Singapore Univ. Press, Singapore, 2003).

10. D. A. Martin, Classes of recursively enumerable sets and degrees of unsolvability, *Z. Math. Logik Grundlag. Math.* **12**, 295 (1966).

11. C. G. Jockusch, Jr. and R. I. Soare, Π_1^0 classes and degrees of theories, *Trans. Amer. Math. Soc.* **173**, 33 (1972).

12. O. Demuth, Remarks on the structure of tt-degrees based on constructive measure theory, *Comment. Math. Univ. Carolin.* **29**, 233 (1988).

13. R. Ladner and L. Sasso, The weak truth-table degrees of the recursively enumerable sets, *Ann. Math. Logic* **8**, 429 (1975).

14. R. G. Downey, Δ_2^0 degrees and transfer theorems, *Illinois J. Math.* **31**, 419 (1987).

15. K. Ambos-Spies and P. A. Fejer, Degree theoretic splitting properties of recursively enumerable sets, *J. Symbolic Logic* **53**, 1110 (1988).

16. J. S. Miller and A. Nies, Randomness and computability: open questions, *Bull. Symbolic Logic* **12**, 390 (2006).

17. W. Merkle and F. Stephan, On C-degrees, H-degrees and T-degrees, in *Twenty-Second Annual IEEE Conference on Computational Complexity (CCC 2007), San Diego, USA, 12–16 June 2007*, (IEEE Computer Society, Los Alamitos, CA, USA, 2007).

Approximating functions and measuring distance on a graph

Wesley Calvert

Department of Mathematics, Mail Code 4408
1245 Lincoln Drive
Southern Illinois University
Carbondale, Illinois 62901, U.S.A.
* E-mail: wcalvert@siu.edu

Russell Miller

Department of Mathematics, Queens College – CUNY, 65-30 Kissena Blvd., Flushing,
NY 11367, U.S.A.; and Ph.D. Programs in Mathematics and Computer Science,
CUNY Graduate Center, 365 Fifth Avenue, New York, NY 10016, U.S.A.
* E-mail: Russell.Miller@qc.cuny.edu

Jennifer Chubb Reimann

Department of Mathematics
University of San Francisco
2130 Fulton Street
San Francisco, California 94117, U.S.A.
* E-mail: jcchubb@usfca.edu

We apply the techniques of computable model theory to the distance function of a graph. This task leads us to adapt the definitions of several truth-table reducibilities so that they apply to functions as well as to sets. We prove assorted theorems about the new reducibilities and especially about functions which have nonincreasing computable approximations (i.e., functions that are appoximable from above). Finally, we show that the spectrum of the distance function of a computable graph can consist of an arbitrary single 1-btt degree which is approximable from above, or of all such 1-btt degrees at once, or of the 1-btt degrees of exactly those functions approximable from above in at most n steps.

Keywords: computable approximations, distance, graphs, truth-table reducibility

1. Introduction

Every connected graph has a distance function, giving the length of the shortest path between any pair of nodes in the graph. Graphs appear in a wide variety of mathematical applications, and the computation of the distance function is usually crucial to these applications. Examples range from web search engine algorithms, to Erdős numbers and parlor games ("Six Degrees of Kevin Bacon"), to purely mathematical questions.

Therefore, the question of the difficulty of computing the distance function is of natural interest to mathematicians in many areas. This article is dedicated to exactly that enterprise, on infinite graphs. Assuming that the graph in question is symmetric, irreflexive, and computable – that is, that one can list out all its nodes and decide effectively which pairs of nodes have an edge between them – we investigate the Turing degree and other measures of the difficulty of computing the distance function. We believe that our results will convince many readers that for consideration of the distance function in particular, the notion of 1-btt reducibility (that is, bounded truth-table reducibility with norm 1) is better suited than is either the usual Turing reducibility, or m- or 1-reducibility, or the various other truth-table reducibilities that exist in the literature.

We began this study by considering the spectrum of the distance function – a standard concept in computable model theory, giving the set of the Turing degrees of distance functions on all computable graphs isomorphic to the given graph. This notion is usually used for relations on a computable structure, rather than for functions, but it is certainly the natural first question one should ask. As our studies continued, however, they led us to consider finer reducibilities than ordinary Turing reducibility, and since we were studying a function instead of a relation, we often had to adapt these reducibilities to functions. The resulting concepts are likely to be of interest to pure computability theorists, as well as to those dealing with applications, and, writing the paper in logical rather than chronological order, we spend the first sections defining and examining these reducibilities on functions. Only in the final sections do we address the original questions about the distance function on a computable graph. Therefore, right here we will offer some further intuition about the distance function, to help the reader understand why the material in the first few sections is relevant.

For a computable connected graph G, the natural first step for approximating the distance $d(x, y)$ between two nodes $x, y \in G$ is to find some path between them. By connectedness, a systematic search is guaranteed to produce such a path sooner or later, and its length is our first approximation to the distance from x to y. The next natural step is to search for a shorter path, and then a shorter one than that, and so on. Of course, these path lengths are all just approximations to the actual distance from

x to y. One of the approximations will be correct, and once we find it, its path length will never be superseded by any other approximation. That is, our (computable) approximations will *converge* to the correct answer, and so the distance function is always $\emptyset'$-computable, by the Limit Lemma (see [1, Lemma III.3.3]). Further, since the approximation are decreasing, unless $d(x, y) \leq 2$, we will never be sure that it is correct, since a shorter path could always appear.

By definition $d(x, x) = 0$, and $d(x, y) = 1$ iff x and y are adjacent, so it is computable whether either of these conditions holds. It is not in general computable whether $d(x, y) = 2$, but it is Σ_1^0, since we need only find a single node adjacent to both x and y. For each $n > 2$, however, the condition $d(x, y) = n$ is given by the conjunction of a universal formula and an existential formula, hence defines a difference of computably enumerable sets, and in general cannot be expressed in any simpler way than that. Indeed, the distance function often fails to be computable, and likewise the set of pairs (x, y) with $d(x, y) = n$ often fails to be computably enumerable when $n > 2$. In what follows, however, we will show that the distance function always has computably enumerable Turing degree – which in turn will start to suggest why Turing equivalence is not the most useful measure of complexity for these purposes. Other questions immediately arise as well. For instance, must there exist a computable graph H isomorphic to G such that H has computable distance function? Or at least, must there exist a computable $H \cong G$ such that we can approximate the distance function on H with no more than one (or n) wrong answer(s)?

The approximation algorithm described above is not of arbitrary difficulty, in the pantheon of computable approximations to functions. Our approximations to $d(x, y)$ are always at least $d(x, y)$, and decrease until (at some unknown stage) they equal $d(x, y)$. Hence $d(x, y)$ is *approximable from above*. This notion already exists in the literature; the most commonly seen function of this type is probably Kolmogorov complexity, which on input n gives the shortest length of a program outputting n. Up to the present, approximability from above has not been so common in computable model theory; the more common notion there is *approximability from below*, which arises (for instance) when one tries to find the number of predecessors of a given element in a computable linear order of order type ω. In Section 3 we give full definitions of the class of functions approximable from above, which can be classified in much the same way as the Ershov hierarchy, and compare it with the class of functions approximable from below. These two classes turn out to be more different than one would expect! First, though, in Section 2, we present the exact definitions of the reducibilities we will use on our functions, so that we may refer to these reducibilities in Section 3.

Very little background in computable model theory is actually required in order to read this article, since the distance function turned out to demand a somewhat different approach than is typical in that field of study. A background in general computability theory will be useful, however, particularly in regard to several of the so-called *truth-table reducibilities*, and for this we suggest [2] or [3], to which we will refer frequently in Section 2. We try to maintain notation from [2] as we adapt the definitions of the truth-table reducibilities to deal with functions. We do, however, choose to use an alternative notation for one of these: the weak truth-table reduction, often denoted by wtt, will be denoted bT (see Section 2), to stand for the alternative term *bounded Turing reduction*. We give the requisite definitions about graphs in Section 4, where they are first needed.

2. Reducibilities on Functions

When discussing Turing computability relative to an oracle, mathematicians have traditionally taken the oracle to be a subset of ω. To compute relative to a total or partial function from ω^n into ω, they simply substitute the graph of the function for the function itself, then apply a coding of ω^{n+1} into ω. This metonymy works admirably as far as ordinary Turing reducibility is concerned, and any alternative definition of Turing reducibility for functions should be equivalent to this one. However, *bounded* Turing reducibility (in which the use of the oracle is computably bounded; see Definition 2.4) among functions requires a new definition, and so we offer here an informal version of our notion of a function oracle.

First, to motivate this notion, consider bounded Turing (bT) reducibility with a function oracle. One would certainly assume that a total function f should be bT-reducible to itself. However, if one wishes to compute $f(x)$ for arbitrary x, using the graph of f as an oracle, and if f is not computably bounded, then there is no obvious way to compute in advance an upper bound on the codes $\langle m, n \rangle$ of pairs for which one will have to ask the oracle about membership of that pair in the graph. (This question is addressed more rigorously in Proposition 2.1 and Lemma 2.1 below.) So, when we move to reducibilities finer than $\leq_T$, there is a clear need for a notion of Turing machine having a function as the oracle.

For simplicity, we conceive of a Turing machine with a function oracle F as having three tapes: a two-way scratch tape (on which the output of the computation finally appears, should the computation halt), a one-way *question tape*, and a one-way *answer tape*. When the machine wishes to ask its function oracle for the value $F(x)$ for some specific x, it must write a sequence of exactly $(x+1)$ 1's on the question tape, and must set every cell of the answer tape to be blank. Then it executes the *oracle instruction*. In this model of computation, the oracle instruction is no longer a forking instruction. Rather, with the tapes in this state, the oracle instruction causes exactly $(1 + F(x))$ 1's to appear on the answer tape at the next step, and the machine simply proceeds to the next line in its program (which most likely will start counting the number of 1's on the answer tape, in order to use the information provided by the oracle). Notice that it is perfectly acceptable to say that a set is computable from a function oracle (using the above notion), or that a function is computable from a set oracle (using the usual notion of oracle Turing computation). If one wishes to speak only of functions, there is no harm in replacing a set by its characteristic function. We leave it to the compulsive reader to formulate the precise definition of a Turing machine with function oracle, by analogy to the standard definition for set oracles. A more immediate (and equivalent) definition uses a different approach.

Definition 2.1. The class of *partial functions on ω computable with a function oracle F*, where $F : \omega \to \omega$ is total, is the smallest class of partial functions closed under the axiom schemes I-VI from [1, § I.2] and containing F.

Of course, this is exactly the usual definition of the partial F-recursive functions, long known to be equivalent to the definition of functions computable by a Turing machine with the graph of F as its oracle. As far as Turing reducibility is concerned, nothing has changed. Only stronger reducibilities need to be considered. We start by converting the standard definitions (for sets) of $\leq_m$ and $\leq_1$ to definitions for functions.

Definition 2.2. Let φ and ψ be partial functions from ω to ω. We say that φ is *m-reducible to ψ*, written $\varphi \leq_m \psi$, if there exists a total computable function g with $\varphi = \psi \circ g$. (For strictly partial functions, this includes the requirement that $(\forall x)[\varphi(x)\!\downarrow \iff \psi(g(x))\!\downarrow].$)

If the m-reduction g is injective, we say that φ is *1-reducible to ψ*, written $\varphi \leq_1 \psi$.

This definition already exists in the literature on computability, having been presented as part of the theory of numberings studied by research groups in Novosibirsk and elsewhere. (See e.g. [4], [5] or [6, p. 477] for the notion of reducibility on numberings.) It is appropriate here as an example of our approach in generalizing reducibilities on sets to reducibilities on functions, for which reason we feel justified in calling it m-reducibility. For subsets $A, B \subseteq \omega$, it is quickly seen that $A \leq_m B$ iff $\chi_A \leq_m \chi_B$, where these are the characteristic functions of those sets; similarly for $A \leq_1 B$. The analogue of Myhill's Theorem for functions states that if $\varphi \leq_1 \psi$ and $\psi \leq_1 \varphi$, then in fact there is a computable permutation h of ω with $\varphi = \psi \circ h$ (and hence $\psi = \varphi \circ (h^{-1})$), in which case we would call φ and ψ *computably isomorphic*. The proof is exactly the same as that of the original theorem of Myhill (see [1, Thm. I.5.4]), and it makes no difference whether φ and ψ are both total or not.

Definition 2.3. A function ψ is *m-complete* for a class Γ of functions if $\psi \in \Gamma$ and, for every $\varphi \in \Gamma$, we have $\varphi \leq_m \psi$. We define 1-completeness similarly, but require that $\varphi \leq_1 \psi$.

For example, the universal Turing function $\psi(\langle e, x \rangle) = \varphi_e(x)$ is 1-complete partial computable, i.e. 1-complete for the class of all *partial* computable functions. (For each single φ_e in the class, the function $x \mapsto \langle e, x \rangle$ is a 1-reduction.) A more surprising result is that there does exist a function h which is 1-complete total computable: let $h(\langle n, m \rangle) = m$, so that, for every total computable f, the function $g(n) = \langle n, f(n) \rangle$ is a 1-reduction from f to h.

The notion of m-reducibility for sets has small irritating features, particularly the status of the sets $\emptyset$ and ω. Intuitively, the complexity of each of these is as simple as possible, yet they are m-incomparable to each other. (Also, no $S \neq \emptyset$ has $S \leq_m \emptyset$ and no $S \neq \omega$ has $S \leq_m \omega$; intuitively this is reasonable, but it is still strange to have two m-degrees containing only a single set each.) The same problem is magnified for m-reducibility on functions. Clearly, if $\varphi \leq_m \psi$, then $\mathrm{rg}(\varphi) \subseteq \mathrm{rg}(\psi)$. It follows that every total constant function forms an m-degree all by itself. Moreover the function $\varphi(x) = 2x$ is m-incomparable with $\psi(x) = 2x + 1$, even though these seem to have very similar complexity; and assorted other pathologies can be found.

We also consider reducibilities intermediate between m-reducibility and Turing reducibility, again by analogy to such reducibilities on sets.

Definition 2.4. Let α and β be total functions. We say that α is *bounded-Turing reducible* to β, or *weak truth-table reducible* to β, if there exists a Turing reduction Φ of α to β and a computable total function f such that, in computing each value $\alpha(x)$, the reduction Φ only asks the β-oracle for values $\beta(y)$ with $y < f(x)$. Thus, for each x, $\alpha(x) = \Phi^{\beta \restriction f(x)}(x)$. Bowing to the two distinct terminologies that exist for this notion on sets, we use two notations for this concept:

$$\alpha \leq_{\mathrm{bT}} \beta \quad \text{and} \quad \alpha \leq_{\mathrm{wtt}} \beta.$$

If we use D_e to denote the finite set with strong index e (i.e. the index tells the size of D_e and all of its elements, as in [1, Defn. II.2.4]), then we say that α is *truth-table reducible* to β, written $\alpha \leq_{\mathrm{tt}} \beta$, if there exist total computable functions f and g such that, for every input x to α, we have $\alpha(x) = g(x, \beta \restriction D_{f(x)})$. This is different from the related reducibilities on enumerations described by Degtev [7] — in particular, it is important for what follows in the present paper that α and β can have different ranges.

Finally, if $\alpha \leq_{\mathrm{tt}} \beta$ via f and g as above and there exists some $k \in \omega$ such that $|D_{f(x)}| \leq k$ for every $x \in \omega$, then we say that α is *bounded truth-table reducible* to β *with norm* k, and write $\alpha \leq_{k\text{-btt}} \beta$. For α to be *bounded truth-table reducible* to β (with no norm stated) simply means that such a k exists, and is written $\alpha \leq_{\mathrm{btt}} \beta$.

It should be noted that, as with sets, the relation $\leq_{k\text{-btt}}$ on function fails to be transitive, for $k > 1$. In general, if $\alpha \leq_{j\text{-btt}} \beta$ and $\beta \leq_{k\text{-btt}} \gamma$, then $\alpha \leq_{(jk)\text{-btt}} \gamma$.

As mentioned, functions and their graphs have always been conflated for purposes of Turing-reducibility. For these finer reducibilities, the conflations no longer apply.

Proposition 2.1. *A total function h is truth-table equivalent to (the characteristic function of) its own graph iff there exists a computable function b such that, for every x, we have $h(x) \leq b(x)$. (In this case, h is said to be computably bounded.)*

Proof. Let $G \subset \omega^2$ be the graph of h, and suppose first that $h(x) \leq b(x)$ for all x. Then, with a G-oracle, a Turing machine on input x can simply ask which pairs (x, n) with $n \leq b(x)$ lie in G. So we have stated in advance which oracle questions will be asked, and by assumption there will be exactly one positive answer, which will be the pair (x, y) with $y = h(x)$. Thus $h \leq_{\mathrm{tt}} G$, since we can also say in advance exactly what answer the machine will give in response to each possible set of oracle values. On the other hand, to determine whether $(x, y) \in G$, an oracle Turing machine only needs to ask an h-oracle one question: the value of $h(x)$. Thus $\chi_G \leq_{\mathrm{tt}} h$ as well.

This latter reduction is actually a bounded truth-table reduction of norm 1, under Definition 2.4, and holds even without the assumption of computable boundedness of h.

For the forwards direction, suppose $h \equiv_{\text{tt}} \chi_G$. Then the computation of h, on input x, asks for the value $\chi_G(m, n)$ only for pairs with codes $\langle m, n \rangle \in D_{f(x)}$, and outputs $g(x, \chi_G \restriction D_{f(x)})$, with f and g as in Definition 2.4. Thus $h(x)$ must be one of the finitely many values in the set $\{g(x, \sigma) : \sigma \in 2^{|D_{f(x)}|}\}$. Since f and g are computable and total, we may take $b(x)$ to be the maximum of this set, forcing $h(x) \leq b(x)$. Thus h is computably bounded. $\qquad \square$

In Proposition 2.1, tt-equivalence cannot be replaced by bT-equivalence. The following proof of this fact was devised in a conversation between E. Fokina and one of us, and completes the answer to the question asked at the beginning of this section.

Lemma 2.1 (Fokina-Miller). *There exists a total function f which is not computably bounded, yet is bT-equivalent to (the characteristic function of) its own graph G.*

Proof. Let $K = \{\langle e, x \rangle : \varphi_e(x) \downarrow\}$ be the halting set. Define $f(2x) = \chi_K(\langle x, 2x + 1 \rangle)$ on the even numbers, using the characteristic function χ_K of K, and on the odd numbers, define

$$f(2x + 1) = \begin{cases} 1 + \varphi_x(2x + 1), & \text{if } \varphi_x(2x + 1)\downarrow, \\ 0, & \text{if not.} \end{cases}$$

Of course this f is not computable, but it is total, and for each x, the input $(2x + 1)$ witnesses that φ_x is not an upper bound for f. Moreover, to determine $f(2x)$ on even numbers, we need only ask a G-oracle whether $\langle 2x, 0 \rangle \in G$. To determine $f(2x+1)$ on odd numbers, we again ask the oracle whether $\langle 2x, 0 \rangle \in G$. If so, then $\chi_K(\langle x, 2x + 1 \rangle) = f(2x) = 0$, meaning that $\varphi_x(2x + 1) \uparrow$, and so we know that $f(2x + 1) = 0$. If $\langle 2x, 0 \rangle \notin G$, then we know that $\chi_K(\langle x, 2x + 1 \rangle) = f(2x) = 1$, so $\langle x, 2x + 1 \rangle \in K$, and we simply compute $\varphi_x(2x + 1)$ (knowing that it must converge) and add 1 to get $f(2x + 1)$. In all cases, therefore, we can compute $f(y)$ by asking a single question of the G-oracle about whether a predetermined value lies in G. Thus $f \leq_{\text{bT}} G$, and of course $G \leq_{\text{1-btt}} f$. $\qquad \square$

3. Functions Approximable from Above

Having adapted several standard reducibilities on sets to serve for functions as well, we now perform the same service for the Ershov hierarchy. Traditionally this has been a hierarchy of $\emptyset'$-computable sets, determined by computable approximations to those sets and by the number of times the approximations "change their mind" about the membership of a given element in the set. In our investigations of the distance functions on computable graphs, we found that similar concepts arose, but pertaining to functions, not to sets. Therefore, the following definitions provide total Δ_2^0-functions with their own Ershov hierarchy, and then add some further structure.

Definition 3.1. Let $f(x) = \lim_s g(x, s)$ be a total function from ω to ω, with the binary function g total and computable.

- If there is a total computable function h such that

$$(\forall x) \; |\{s : g(x, s) \neq g(x, s + 1)\}| \leq h(x),$$

  then f is ω-approximable.
- If the constant function $h(x) = n$ can serve as the h in the previous item, then f is n-approximable.
- More generally, if α is a computable ordinal and there is a total computable function $h : \omega^2 \to \alpha$, nonincreasing in the second coordinate, such that

$$(\forall x \forall s) \; [g(x, s) \neq g(x, s + 1) \implies h(x, s) \neq h(x, s + 1)],$$

  then f is α-approximable.
- If, for all x and s, we have $g(x, s + 1) \leq g(x, s)$, then f is *approximable from above*. Such functions are also sometimes called *limitwise decreasing*, *semi-computable from above* or *right c.e.* functions
- If, for all x and s, we have $g(x, s + 1) \geq g(x, s)$, then f is *approximable from below*. In the literature, such functions have also been called *limitwise monotonic*, *limitwise increasing*, and *subcomputable*.
- When we combine these definitions, we assume that a single function g satisfies all of them. For instance, f is 3-*approximable from above* if f is the limit of a computable function g such that, for all x, $|\{s : g(x, s) \neq g(x, s + 1)\}| \leq 3$ *and* $(\forall s) \; g(x, s + 1) \leq g(x, s)$.
- Following [8], we define f to be *graph-α-c.e.* if α is a computable ordinal and the graph of f is an α-c.e. set in the Ershov hierarchy.

In [8], the term α-*c.e. function* was used for the functions we are calling α-approximable. We prefer our terminology, since the phrase "c.e. function" has been used elsewhere for functions approximable from below. (For such a function, the set $\{(x, y) : y \leq f(x)\}$ is c.e. This also explains the use of the term *subcomputable* for such functions.)

A characteristic function χ_A is approximable from below iff A is c.e., and approximable from above iff A is co-c.e. The definitions of approximability from below and from above may seem dual, but in fact there are significant distinctions between them. For an example, contrast the following easy lemma with the well-known fact that there exist functions which are approximable from below, but not ω-approximable.

Lemma 3.1. *Every function approximable from above is ω-approximable from above.*

Proof. Let $f = \lim_s g$ with $g(x, s + 1) \leq g(x, s)$ for all x and s. Then the computable function $h(x) = g(x, 0)$ bounds the number of changes g can make. □

On the other hand, the hierarchy of n-approximability from above does not collapse. (See also Corollary 3.3 below, which uses this lemma to show non-collapse at the ω level, as well.)

Lemma 3.2. *For every n, there is a function f which is $(n + 1)$-approximable from above but not n-approximable.*

Proof. We define a computable function $g(x, s)$ and set $f(x) = \lim_s g(x, s)$. To begin, $g(x, 0) = n + 1$ for every x. At stage $s + 1$, for each $x \leq s$, fix the greatest $t \leq s$ such that $\varphi_{x,s}(x, t)\downarrow$. If $\varphi_x(x, t) = g(x, s) > 0$, set $g(x, s+1) = g(x, s) - 1$; otherwise set $g(x, s + 1) = g(x, s)$. Thus $f(x) = \lim_s g(x, s)$ is $(n + 1)$-approximable from above, but if $f(x) = \lim_s \varphi_x(x, s)$ for some x, then $\varphi_x(x, s)$ must have assumed each of the values $(n + 1), n, \ldots, 1, 0$ as $s \to \infty$, and so φ_x is not an n-approximation to f. □

The appropriate duality pairs functions approximable from above with a subclass of the functions approximable from below, as follows.

Definition 3.2. Suppose that g is computable and total, with $g(x, s+1) \leq g(x, s)$ for all x and s, so that $f(x) = \lim_s g(x, s)$ is total and approximable from above. The *dual of g* is the function

$$h(x, s) = g(x, 0) - g(x, s).$$

Thus $j(x) = \lim_s h(x, s)$ is total, approximable from below (by h), and bounded above by $g(x, 0)$. Moreover, g is an α-approximation for f iff h is an α-approximation for j. We call j the *dual of f with respect to g*.

Conversely, let j be any function which is approximable from below via $h(x, s)$ and *computably bounded*: that is, there exists a computable total function b with $j(x) \leq b(x)$ for all x. Then the function $g(x, s) = b(x) - h(x, s)$ is the *dual of h with respect to b*, and f is the *dual of j with respect to h and b* where, again, $f(x) = \lim_s g(x, s)$. (To make the duality precise, one should assume that $h(x, 0) = 0$ for all x, so that $g(x, 0) = b(x)$. In this case, the dual of the dual is the original function.)

It is natural to call j the dual of f, but in fact j depends on the choice of the approximation g: two different approximations g and $\tilde{g}$ will often yield two different duals, though these two duals always differ by a computable function, namely $(g(x, 0) - \tilde{g}(x, 0))$. The dual of a function approximable from below also depends on the choice of computable bound. Nevertheless, it will be clear from our results below that the class of computably bounded functions approximable from below is the natural dual for the class of functions approximable from above. The computable upper bound in the former class is the obvious counterpart of the built-in computable lower bound of 0 for the latter class.

Functions approximable from below have seen wide usage in computable model theory, for example in [9–14]. Our interest in functions approximable from above arose from our investigations into the distance function on a computable graph. To our knowledge, this is the first significant use of such functions in computable model theory, although, as we will mention, they arise implicitly in the study of effectively algebraic structures and in certain other contexts. The best-known example of a function approximable from above does not come from computable model theory at all: it is the function of Kolmogorov complexity (for any fixed universal machine), mapping each finite binary string (coded as a natural number) to the shortest program which the fixed machine can use to output that string.

It was a theorem of Khoussainov, Nies, and Shore in [14] that there exists a Δ_2^0 set which is not the range of any function approximable from below. The following theorem contrasts with that result, giving a very concrete distinction between approximability from above and from below.

Theorem 3.1. *The range of every approximable function is the range of some function which is 1-approximable from above. Indeed, the ranges of the 1-approximable-from-above functions are precisely the nonempty Σ_2^0 sets.*

Proof. We prove the stronger statement. Being in the range of a 1-approximable-from-above function is clearly a Σ_2^0 condition. For the converse, let an arbitrary nonempty $S \in \Sigma_2^0$ have computable 1-reduction p to the Σ_2^0-complete set **Fin**, so that

$$(\forall x)[x \in S \iff |W_{p(x)}| < \infty].$$

We construct a computable function $g(x, s)$ that is a 1-approximation of a total function $f(x)$, the range of which is S.

We will assume that at each stage s, there is exactly one x such that $W_{p(x),s+1} \neq W_{p(x),s}$, and also that for every x, $W_{p(x)} \neq \emptyset$; both of these conditions are readily arranged. Fix the least $x_0 \in S$.

At stage 0 we define nothing.

At stage $s + 1$ we define $g(0, s + 1), g(1, s + 1), \ldots, g(s, s + 1)$, and $g(s, 0), g(s, 1), \ldots, g(s, s)$ as follows:

For the unique y such that $W_{p(y),s+1} \neq W_{p(y),s}$, let

$$g(s, 0) = g(s, 1) = \cdots = g(s, s + 1) = \begin{cases} y, & \text{if } y \geq x_0 \\ x_0, & \text{if not.} \end{cases}$$

Then, for each $x < s$, define

$$g(x, s + 1) = \begin{cases} x_0, & \text{if } g(x, s) = x_0 \\ x_0, & \text{if } W_{p(g(x,s)),s+1} \neq W_{p(g(x,s)),s} \\ g(x, s), & \text{otherwise.} \end{cases}$$

This defines g effectively on all of $\omega \times \omega$, and for every x, $g(x, s)$ is either x_0 for all s, or y for all s (where y was chosen at stage $x + 1$), or else y for $s = 0, 1, \ldots, n$ and then x_0 for all $s > n$. This last holds iff $p(y) > x_0$ and $W_{p(x)}$ received a new element at some stage $n + 1 > x + 1$. So clearly g has a limit $f(x) = \lim_s g(x, s)$ and approximates that limit from above, with at most one change. Moreover, if $y \in S$, then $p(y) \in$ **Fin**, and so when (the nonempty set) $W_{p(y)}$ receives its last element, say at stage $x + 1$, then x will have $g(x, s) = y$ for all $s > x$, making $S \subseteq \text{rg}(f)$. Conversely, if $x \notin S$, then $p(y) \notin$ **Fin**, so every x which ever had $g(x, s) = y$ will eventually get changed and will have $f(x) = x_0$; thus $\text{rg}(f) \subseteq S$. $\square$

Corollary 3.1. *There is a function 1-approximable from above, the range of which is not the range of any function approximable from below.*

Proof. This is immediate from Theorem 3.1 in conjunction with a result in [14] giving the existence of a Δ_2^0 set which is not the range of any function approximable from below. $\square$

Corollary 3.2. *There exists a function that is* 1-*approximable from above whose range is* Σ_2^0-*complete.*

Theorem 3.2. *Every ω-approximable function f is bT-reducible to some function approximable from above. (This approximation from above is known as the countdown function for f.)*

Proof. Let $f(x)$ be an ω-approximable function, approximated by $g(x, s)$, with computable function $h(x)$ bounding the number of mind changes of $f(x, s)$. Set $c(x, 0) = h(x)$. Let $c(x, s + 1) = c(x, s)$ unless $g(x, s) \neq g(x, s + 1)$, and in that case set $c(x, s + 1) = c(x, s) - 1$. Then $\lim_s c(x, s)$ is approximable from above and $f(x)$ is computable from this limit, since $f(x) = f(x, t)$ for each t with $c(x, t) = \lim_s c(x, s)$.

Our reasons for referring to this c as the *countdown function* for f (or, strictly speaking, for g with respect to h, since c does depend on the approximation and the computable bound) are clear. It is important to distinguish the countdown function $c(x, s)$, which is computable, from its limit $\lim_s c(x, s)$, which in general is not computable (and was just shown to Turing-compute f). Indeed, we have $f \leq_{bT} \lim c$, since the only value of the limit required to compute $f(x)$ is $\lim_s c(x, s)$. $\qquad\square$

On the other hand, this is not in general a truth-table reduction. For that, one would need to predict in advance what the value of $f(x)$ will be for every possible value of $\lim_s c(x, s)$ between 0 and $c(x, 0)$. Without knowing $\lim_s c(x, s)$ in advance, one cannot be sure for how many values of s we may need to compute $g(x, s)$ to determine these answers.

The limit of c is not in general Turing-reducible to f. However, if g is either an approximation from above or an ω-approximation from below, then $f \equiv_T \lim_s c(\cdot, s)$, and indeed $f \equiv_{bT} \lim_s c(\cdot, s)$, since the computation of $\lim_s c(x, s)$ only requires us to ask the oracle for the value $f(x)$. (Once an approximation from above or from below abandons a value, it cannot later return to that value, and so, once $f(x) = g(x, t)$, we know that $c(x, t) = \lim_s c(x, s)$.) However, even for approximations from above and ω-approximations from below, we have in general that $\lim c \nleq_{tt} f$, since we cannot determine the final value of the countdown without actually knowing the value $f(x)$.

Theorem 3.3. *There is a function f that is* 1-*complete within the class of all functions approximable from above.*

Proof. We construct f by constructing a computable function g approximating f from above. At stage 0, g is undefined on all inputs.

At stage $s+1$, find the least pair $k = \langle e, x \rangle$ (if any) such that $\varphi_{e,s}(x,0)\downarrow$ and n_k is undefined. Let n_k be the least element such that $g(n_k, 0)$ was undefined as of stage s, and set

$$g(n_k, 0) = g(n_k, 1) = \cdots = g(n_k, s + 1) = \varphi_{e,s}(x, 0).$$

Then (whether or not such a k existed), for each $j = \langle e, x \rangle$ such that $g(n_j, 0)$ was defined by stage s, we consider the sequence $\varphi_{e,s}(x, 0), \ldots, \varphi_{e,s}(x, t)$, for the greatest $t \leq s$ such that all these computations converge. If this finite sequence is nonincreasing, we set $g(n_j, s + 1) = \varphi_e(x, t)$. Otherwise (that is, if $\varphi_e(x, t' + 1) > \varphi_e(x, t')$ for some $t' < t$), we set $g(n_j, s + 1) = g(n_j, s)$.

This completes the construction of g. Clearly, every n is chosen at some stage to be n_k for some k, and subsequently $g(n_k, s)$ is defined for each s, so g is total and computable. Moreover, by construction, $g(n_k, s+1) \leq g(n_k, s)$ for every k and s. So the function $f(n) = \lim_s g(n, s)$ is approximable from above.

Now let h be any function which is approximable from above, say by $h(x) = \lim_s \varphi_e(x, s)$. Then, for every x, $\varphi_e(x, 0)$ converges at some finite stage, and so some n_k with $k = \langle e, x \rangle$ is eventually chosen. The function d_e mapping x to this $n_{\langle e, x \rangle}$ (for this fixed e) is computable (since we can simply wait until $n_{\langle e, x \rangle}$ is defined) and total, and also 1-1, since $n_j \neq n_k$ for $j \neq k$. Now since φ_e approximates h from above, the sequence

$$\varphi_e(x, 0), \varphi_e(x, 1), \varphi_e(x, 2), \ldots, \varphi_e(x, t), \ldots$$

is infinite and nonincreasing. So the sequence

$$g(n_k, 0), g(n_k, 1), g(n_k, 2), \ldots, g(n_k, s), \ldots$$

is exactly the same sequence, by construction, except that numbers which occur finitely often in one sequence might occur a different finite number of times in the other. This shows that

$$f(d_e(x)) = f(n_k) = \lim_s g(n_k, s) = \lim_t \varphi_e(x, t) = h(x),$$

so $f \circ d_e = h$, proving $h \leq_1 f$ via d_e. Indeed, therefore, the 1-reduction may be found uniformly in the index of a computable approximation to h from above. □

Theorem 3.3 gives an easy proof of a result which we could have shown by the method from Lemma 3.2.

Corollary 3.3. *There exists a function which is ω-approximable from above, but (for every $n \in \omega$) is not n-approximable.*

Proof. If the 1-complete function f from Theorem 3.3 were n-approximable, say via $g(x, s)$, then since every function approximable from above has a 1-reduction h to f, we would have that every such function is n-approximable (via $g(h(x), s)$). This contradicts Lemma 3.2. $\square$

Theorem 3.3 stands in contrast to the following results.

Theorem 3.4. *No function is m-complete for the class of functions approximable from below.*

Proof. Let f be any function with a computable approximation $g(x, s)$ of f from below. We build a computable approximation h from below to a function j which cannot be m-reducible to f. Define $h(\langle e, x \rangle, 0) = 0$ for every e and x.

At stage $s + 1$, if $\varphi_{e,s}(\langle e, x \rangle) \uparrow$, set $h(\langle e, x \rangle, s + 1) = 0$. Otherwise, set $h(\langle e, x \rangle, s+1) = 1 + g(\varphi_e(\langle e, x \rangle), s+1)$. By hypothesis on g, this h is clearly nondecreasing in s, and since $\lim_s g(\varphi_e(\langle e, x \rangle), s)) = f(\langle e, x \rangle)$ must exist, we see that

$$j(\langle e, x \rangle) = \lim_s h(\langle e, x \rangle, s) = 1 + \lim_s g(\varphi_e(\langle e, x \rangle), s) = 1 + f(\varphi_e(\langle e, x \rangle)).$$

However, this shows that either $j \neq f \circ \varphi_e$ or else φ_e is not total, so $j \not\leq_m f$. $\square$

Theorem 3.5. *For every function f approximable from above, there exists a function g which is 2-approximable from below and has $g \not\leq_m f$.*

Proof. Given a function f that is approximable from above, we construct an ω-approximable function, g, that f fails to m-compute. To achieve this, we assume that $f(x) = \lim_s \varphi_f(x, s)$, and consider possible witnesses for an m-reduction, diagonalizing against them. We must meet the following requirement for each $e \in \omega$:

$$R_e : \quad \varphi_e \text{ is total} \implies \exists x \; f(\varphi_e(x)) \neq g(x).$$

Set $g(x, 0) = 0$ for each $x \in \omega$. If $\varphi_e(e) \downarrow = y_e$ at stage s, set $g(e, t) = f(y_e, s) + 1$ for all $t \geq s$. For $t > s$, and any y, we have $f(y, t) < f(y, s)$, so once R_e receives attention, it is forever satisfied. Furthermore, $f(e, 0)$ serves as a computable upper bound for $g(e, s)$ so g is computably bounded, and the number of mind changes is at most 2. $\square$

The following result suggests the insufficiency of Turing reducibility, and even of bounded Turing reducibility, to classify functions approximable from above.

Prop 3.1. Every function approximable from above is bounded-Turing equivalent to (the characteristic function of) a computably enumerable set, and every c.e. set is bT-equivalent to a function approximable from above.

Proof. Let $f = \lim_s g(x, s)$ be an approximation of f from above. Define the c.e. set

$$V_f = \{\langle x, n \rangle : \exists^{>n} s \ [g(x, s+1) \neq g(x, s)]\}.$$

That is, $\langle x, n \rangle \in V_f$ if and only if the approximation to f changes its mind more than n times. Clearly $V_f \leq_T f$, and the computation deciding whether $\langle x, n \rangle \in V_f$ requires us only to ask the oracle for the value of $f(x)$. Conversely, to compute $f(x)$ from a V_f-oracle, we first compute $g(x, 0)$, and then ask the oracle which of the values $\langle x, 0 \rangle, \ldots, \langle x, g(x, 0) - 1 \rangle$ lies in V_f; the collective answer tells us exactly how many times the approximation will change its mind, and with this information we simply compute $g(x, s)$ until we have seen that many mind changes. Both of these are bounded Turing reductions. (Neither, however, is a truth-table reduction. In fact, the characteristic function of V_f is tt-equivalent to the limit of the countdown function for f, which is not in general tt-equivalent to f.)

For the second statement, note that every c.e. set is bT-equivalent (indeed tt-equivalence with norm 1) to the characteristic function of its complement. □

Corollary 3.4. *There is a 2-approximable function that is not Turing equivalent to any function approximable from above.*

Proof. Let S be a d.c.e. set which is not of c.e. degree. (See, e.g., [15] for such a construction). Then the characteristic function χ_S is 2-approximable, and Proposition 3.1 completes the result. □

4. The Distance Function in Computable Graphs

With the results of the preceding sections completed, we may now address the intended topic of this paper: the distance function on a computable graph. Computable graphs are defined by the standard computable-model-theoretic definition.

Definition 4.1. A structure $\mathcal{M}$ in a finite signature is *computable* if it has an initial segment of ω as its domain and all functions and relations on $\mathcal{M}$ are computable when viewed as functions and relations (of appropriate arities) on that domain.

Therefore, a symmetric irreflexive graph $G = \langle V; E \rangle$ is *computable* if its domain, V, the set of vertices, is either ω or a finite initial segment thereof, and there is an algorithm which decides, for arbitrary $x, y \subset G$, whether E, the edge relation, contains (x, y).

The *distance function* d on a graph G maps each pair $(x, y) \in G^2$ to the length of the shortest path from x to y. Assuming G is connected, such a path must exist. By definition $d(x, x) = 0$, and this is quickly seen to be a metric on G. Moreover, if G is a computable graph, then the distance function on G is approximable from above, since for any x and y, we can simply search for the shortest path from x to y. For the rest of this paper, we will concern ourselves only with infinite graphs. Formally, letting G_s be the induced subgraph of G on the vertices $\{0, \ldots, s\}$, we find the least t for which G_t contains x, y, and a path between them, and let $g(x, y, 0)$ be the length of the shortest such path in G_t. Then we define $g(x, y, s+1)$ to be the minimum of $g(x, y, s)$ and the length of the shortest path (if any) between x and y in G_s. This sequence is nonincreasing in s, with limit $d(x, y)$. In the language of computable model theory, we say that for every connected computable graph G, the distance function is *intrinsically approximable from above*, since it is approximable from above in every computable graph isomorphic to G. (Cf. Definition 4.3 below.) We also borrow from from computable model theory the notion of the *degree spectrum* of a relation, adapting it for functions and for 1-btt degrees.

Definition 4.2. Let $\mathcal{A}$ be a computable structure, and f a function from $\mathcal{A}^n$ to $\mathcal{A}$. The 1-*btt degree spectrum* of f on $\mathcal{A}$ is the following set of 1-btt degrees of functions:

$$\{\deg_{\text{1-btt}}(g) \ : \ (\exists \text{ a computable structure } \mathcal{B}) \ (\mathcal{B}, g) \cong (\mathcal{A}, f)\}.$$

We make the analogous definitions for 1-btt degree spectra of relations on $\mathcal{A}$, and also with degrees defined by any other reducibility in place of 1-btt degrees. It should be borne in mind that, in contrast to the case of Turing degrees, 1-btt degrees of functions are not the same as 1-btt degrees of relations.

In fact, computable graphs can encode functions approximable from above in a very strong way:

Theorem 4.1. *For any function f which is approximable from above, there is a computable graph G so that for every computable $H \cong G$, we have that*

$$f \equiv_{1\text{-}btt} d_G \equiv_1 d_H.$$

That is, the 1-btt degree spectrum of d_G is just $\{\deg(d_G)\}$.

Proof. Let f be approximable from above, and $f(x) = \lim_s f(x, s)$, with $f(\cdot, \cdot)$ computable. We construct $G = \langle V, E \rangle$ by finite approximation so that $G = \langle V, E \rangle = \lim_s \langle V_s, E_s \rangle = \lim_s G_s$.

Stage 0. Set $V_0 = \{0\}$ and $E_0 = \emptyset$. (Here 0 is labeling what will be the "center" vertex of the graph. Subsequent stages establish "spokes" eminating from this center.)

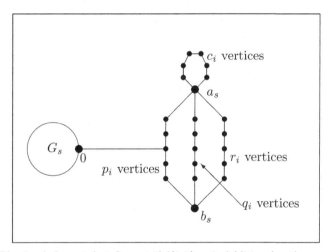

Fig. 1. *The "spoke" created at Stage 4 if $f(3,0) = 2$.* Additional paths may be added between a_s and b_s at subsequent stages.

Stage $s + 1$. First, let a_s, b_s, $\{c_1, \ldots, c_{2s}\}$, $\{p_1, \ldots, p_k\}$, $\{q_1, \ldots, q_k\}$, $\{r_1, \ldots, r_k\}$, where $k = 2f(s, 0) + 1$, be the next natural numbers not yet in use by V_s. Modify G_s by adding to it these new vertices along with new edges so that the p, q, and r vertices form three distinct, non-intersecting paths of length $2f(s, 0) + 2$ between a_s and b_s. Also add the edge connecting 0 to $p_{f(s,0)+1}$, and use the c vertices to form a loop of length $2s + 1$ at a_x. Call the resulting graph $\widetilde{G_s}$. (See Figure 1.)

Second, for each $x < s$, if $f(x, s) \neq f(x, s - 1)$, then let $\{w_1, \ldots, w_{2*f(x,s)+1}\}$ be the next natural numbers not in yet in use by $\widetilde{G_s}$, and use these to modify $\widetilde{G_s}$ by addition of a path of length $2 * f(x, s) + 2$ from a_x to b_x that is distinct from and does not intersect any other path from a_x to b_x or any other part of the graph $\widetilde{G_s}$.

Let $G_{s+1} = (V_{s+1}, E_{s+1})$, where V_{s+1} is V_s together with all new vertices used in the modifications of G_s and $\widetilde{G_s}$, and E_{s+1} is E_s extended by the new edges used in the above modifications.

Let $G = \lim_s \langle V_s, E_s \rangle$. This ends the construction.

Clearly G is a computable. Further, this graph is computably categorical, i.e., any computable graph H isomorphic to G is *computably* isomorphic to G: To build a computable isomorphism from G to H we need only distinguish (in advance) the "center" vertices of these two graphs (for G, this is 0 in the construction above), and be able to effectively find each a_x and b_x in each of them. We can do the latter by searching for the unique node that is the base of a loop of (odd) length $2x + 1$. (Note that all the other loops at a_x will pass through the corresponding b_x and hence have *even* length.) From a_x, we can find three disjoint paths of equal length that meet at some end point, which must then equal b_x. In any isomorphism, we must map the center of G to the the center of H and, for each x, a_x and b_x in G must map to their counterparts in H. Once these parts of the isomorphism are established, all other images are effectively and uniquely determined (apart from the three initial paths from a_x to b_x, which may be interchanged without harm). As a consequence of the computable isomorphism, we see that d_G is in fact 1-equivalent to d_H.

It remains to show that the distance function on G, d_G, is 1-btt equivalent to the given function f. Note that $f(x) = 1/2(d_G(a_x, b_x) - 2)$, and that a_x and b_x can be effectively found for any x, so $f \leq_{1-\text{btt}} d_G$. Conversely, if we wish to compute $d_G(u, v)$ for a given pair of vertices u and v, and in the process query f at most once, we must consider two cases. First, though, we note that the distance from the center node, 0, to a_x is computable (it is $2 + 2f(x, 0)$), and that this is also equal to $d_G(0, b_x)$. Also, $d_G(a_x, b_x) = 2 + 2f(x)$, and computation of this distance necessitates a query to the f oracle.

Case 1 *Vertices u and v are not on the same "spoke."*

Assume u is on the spoke for x and v is on the spoke for y, and note that x and y can be found effectively and that $d_G(0, a_x) = d_G(0, b_x)$ and $d_G(0, a_y) = d_G(0, b_y)$. In this case, the distance from u to v

will be

$$\min\{d_G(u, a_x), d_G(u, b_x)\} + d_G(0, a_x)$$
$$+ d_G(0, a_y) + \min\{d_G(v, a_y), d_g(v, b_y)\},$$

which is computable.

Case 2 *Vertices u and v are on the same spoke.*
Assume both are on the spoke for x. In this case, the distance from u to v is the least of the following:

- $d_G(u, a_x) + d_G(a_x, v)$
- $d_G(u, b_x) + d_G(b_x, v)$
- $d_G(u, a_x) + d_G(a_x, b_x) + d_G(b_x, v)$
- $d_G(u, b_x) + d_G(a_x, b_x) + d_G(a_x, v)$
- If u and v happen to lie on the *same path* from a_x to b_x, then the distance *along* that path may be the shortest one.

Since all of these distances are computable *except* $d_G(a_x, b_x)$, and the latter is equal to $2 + f(x)$, we see that indeed $d_G \leq_{1-\texttt{btt}} f$. $\square$

For Turing degrees, it is known that there exist relations on computable structures whose Turing degree spectra are singletons consisting of an arbitrary Turing degree. Theorem 4.1, which establishes the same result for the much finer notion of 1-$\texttt{btt}$ reducibility, can be seen as saying in a very strong way that f represents precisely the complexity of the distance function of the graph G which the theorem constructs – and not just for the particular computable presentation G, but for the isomorphism type of that G. We state the obvious (weaker) corollary.

Corollary 4.1. *For every Turing degree $\mathbf{d}$ containing a function approximable from above, there exists a computable graph G such that the Turing degree spectrum of the distance function on G is the singleton $\{\mathbf{d}\}$.*

We also note the following corollaries.

Corollary 4.2. *Every function which is approximable from above is $\texttt{btt}$-equivalent, with norm 1, to the distance function for some computable graph.*

Corollary 4.3. *There exists a computable graph G such that, for every n and for every computable graph $H \cong G$, the distance function on H is not n-approximable from above. (By analogy to Definition 4.3, we say that the distance function on this graph G intrinsically fails to be n-approximable from above.)*

Proof. By Corollary 3.3, there is a function f which is ω-approximable from above, but not n-approximable for any n. Apply Theorem 4.1 to this f, and note that if f were 1-btt reducible to a function n-approximable from above, then f itself would be n-approximable from above. □

We would also like to be able to make the n-approximability intrinsic to the graph, in the following sense (which is standard in computable model theory).

Definition 4.3. The distance function on a computable graph G is *intrinsically n-approximable from above* if, for every computable graph H isomorphic to G, the distance function on H is n-approximable from above. G is *relatively intrinsically n-approximable from above* if for every graph $H \cong G$ with domain ω, the Turing degree of the edge relation on H computes an n-approximation from above to the distance function on H.

One could give the same definition with ω in place of n, but we have already remarked that the distance function of every computable graph is approximable from above (which is to say, ω-approximable from above), so this would be trivial. In a graph satisfying Definition 4.3, there is some structural reason for which the distance function is always n-approximable from above, no matter how one presents the graph. (Relative intrinsic n-approximability from above says that this holds even when we consider noncomputable presentations.)

We now show how to use distance functions on individual computable graphs to characterize various classes of functions on ω, particularly classes of functions approximable from above. Again, the characterization will use the very strong notion of 1-btt-equivalence, and the proof uses similar techniques to those of Theorem 4.1. We shall call a *spoke of type σ*, where σ is a strictly decreasing string of natural numbers, a figure of the sort seen in the above proof. Such a spoke has nodes a_σ and b_σ which are connected by three paths of length $2\sigma(0) + 2$ (known as the *initial paths*) and also, for each $0 < i < |\sigma|$, by a single path of length $2\sigma(i) + 2$. The middle node of one of the longest paths will typically be connected to the center "hub" of a larger graph.

Theorem 4.2. *There is a computable graph G such that, for each function f which is approximable from above, there is a computable graph $H \cong G$ with $d_H \equiv_{1\text{-}btt} f$. That is, the 1-btt-degree spectrum of the distance* *function on G contains exactly those 1-btt-degrees which contain a function approximable from above.*

Proof. We use the same notion of spokes that was applied in Theorem 4.1. Also, fix a computable enumeration $\{\sigma_i\}_{i\in\omega}$ of all strictly decreasing sequences of natural numbers. (Of course, every such σ_i is finite.) The graph G consists of one center node 0 and (for each i) infinitely many spokes of type σ_i. We write a_{ij} and b_{ij} for the end nodes of the j-th spoke of type σ_i. On every one of these spokes, a single node on one of the three initial paths is adjacent to 0, with that single node lying at the middle of that path, at a distance of $(1 + \sigma_i(0))$ from a_{ij} and equally far from b_{ij}. None of these spokes has any "tag," such as was created using the nodes c_i in Theorem 4.1.

Of course, the distance function on each computable graph $H \cong G$ must be approximable from above. For the converse, let f be any function with a computable approximation $\langle f_s\rangle_{s\in\omega}$ from above. We build a computable graph $H \cong G$, starting with G itself as H_0 (except that we use only **even** numbers to build this G, with 0 as the center). Then we proceed as follows.

Stage $s > 0$. First, with $k = 2f_0(s) + 1$, let a_s, b_s, $p_1,\ldots,p_k$, $q_1,\ldots,q_k$, $r_1,\ldots,r_k$, be the next **odd** natural numbers not yet in use by V_{s-1}. Modify H_{s-1} by adding to it these new vertices along with new edges so that the p, q, and r vertices form three distinct, non-intersecting paths of length $2f_0(s) + 2$ between a_s and b_s. Thus we have begun a new spoke. Also add the edge connecting 0 to $p_{f_0(s)+1}$. Call the resulting graph $\widetilde{H}_s$, and note that again we have not used c_i vertices to "tag" a_x.

Second, for each $x < s$, if $f_{s-x}(x) \neq f_{s-x-1}(x)$, then let $w_1,\ldots,w_{2f_{s-x}(x)+1}$ be the next **odd** natural numbers not in yet in use by $\widetilde{H}_s$, and use these to modify $\widetilde{H}_s$ by addition of a path of length $2f_{s-x}(x)+2$ from a_x to b_x that is distinct from and does not intersect any other path from a_x to b_x or any other part of the graph $\widetilde{H}_s$.

Let V_s be V_{s-1} together with all new vertices used in the modifications of H_s and $\widetilde{H}_s$, and let E_s be E_{s-1} extended by the new edges used in the above steps. Set $H_s = (V_s, E_s)$.

Let $H = \lim_s\langle V_s, E_s\rangle$. This ends the construction.

Now for each spoke in H_0, we know the σ_i used to build that spoke, and so we know $d_H(a_{ij}, b_{ij})$, the distance between the end nodes of that spoke. For the spoke begun at stage s, with end nodes a_s and b_s, we have $d_H(a_s, b_s) = f(s)$. Immediately this shows that $f \leq_{1\text{-btt}} d_H$, and by the same analysis as in the proof of Theorem 4.1, we see that $d_H \leq_{1-\text{btt}} f$. Finally, to show that $H \cong G$, notice that the spoke begun at each stage s is isomorphic to a spoke of type σ, where σ is the finite sequence giving the elements of the (finite) set $\{f_0(s) \geq f_1(s) \geq \cdots\}$ in strictly decreasing order. Since G already contained infinitely many copies of this spoke, and since $H_0 \cong G$, it is clear that, for every i, both G and H contain infinitely many copies of the spoke of type σ_i, and that neither contains anything else. Thus H is a computable copy of G whose distance function has the same 1-btt degree as the function f. □

The same technique as in Theorem 4.2 allows us to make the 1-btt degree spectrum of the distance function contain precisely the degrees of functions n-approximable from above.

Theorem 4.3. *Fix any $n \in \omega$. Then there is a computable graph G, so that for each function f which is n-approximable from above, there is a computable graph $H \cong G$ with $d_H \equiv_{1-\text{btt}} f$, and such that, for each computable graph $H \cong G$, d_H is itself n-approximable from above. That is, the 1-btt degree spectrum of the distance function on G contains exactly those 1-btt degrees which contain a function m-approximable from above.*

Proof. The construction of G closely parallels the corresponding construction in Theorem 4.2. The only difference is that, instead of using all strictly decreasing sequences of natural numbers, we use only those sequences of length $\leq n$, setting $\{\sigma_i\} i \in \omega = \{\sigma \in \omega^{<n+1}\} (\forall i < |\sigma| - 1)\sigma(i) > \sigma(i+1)$. (Each value $\sigma_i(j)$ can still be any natural number; only the length of σ_i is restricted.)

Then, when we build H for a given function f, we use an n-approximation $\langle f_s \rangle_{s \in \omega}$ of f from above, and proceed exactly as in the stages in the construction from Theorem 4.2. Every spoke thus created consists of three initial paths and at most n more (shorter) paths, since the approximation to $f(x)$ changes at most n times. Therefore, the spoke for x is isomorphic to one of the spokes already appearing infinitely many times in G, and so we do achieve $H \cong G$.

There is one subtle point in the argument that $d_h \equiv_{1-\text{btt}} f$, and it also arises in the proof that every computable $H \cong G$ has a computable n-approximation from above to d_H. The subtlety corresponds to the two cases in the proof of Theorem 4.1. Case 2, determining the distance between vertices u and v on the same spoke, goes through just as before: this distance varies according to the distance between the end nodes of that spoke. However, in Case 1, u and v are on different spokes: say that the spoke of u has end nodes a_x and b_x, while the spoke of v has a_y and b_y. If $d_H(u, v)$ depended on both $d_H(a_x, b_x)$ and $d_H(a_y, b_y)$, then it could conceivably decrease as many as $2n$ times, since each of those two values could decrease n times. However, in fact $d_H(u, v)$ is computable in Case 1, with no approximation whatsoever. The distances $d_H(u, a_x)$ and $d_H(u, b_x)$ are both computable, just by following the path containing u, and the distances from each of a_x and b_x to the center node are equal and are known from the beginning; likewise on the spoke of v. Since every path from u to v must go through the center, and must go through either a_x or b_x en route to the center and through either a_y or b_y en route from there to v, $d_H(u, v)$ is exactly determined, without any queries about values of $f(x)$ or $f(y)$. Therefore d_H in general (including Case 2) is indeed n-approximable from above whenever $H \cong G$, and is indeed 1-btt-reducible to f whenever H was built from a particular f. This is all that is needed for the proof. □

By combining the preceding constructions, we may realize 1-btt degree spectra which are upper cones within the various classes of functions approximable from above.

Theorem 4.4. *Fix any two numbers $m \leq n$ in ω, and any function f which is m-approximable from above. Then there exists a computable graph G such that the 1-btt degree sprectrum of its distance function contains exactly those 1-btt degrees which are n-approximable from above and to which f is 1-btt-reducible.*

Likewise, for any function g which is approximable from above, there exists a computable graph G such that the 1-btt degree sprectrum of its distance function contains exactly those 1-btt degrees which are approximable from above and to which g is 1-btt-reducible.

Proof. For the first result, with the m-approximable f, recall the constructions from Theorems 4.1 (of a graph G_0 for f and 4.3 (of a graph G_1 from f). We combine these two constructions, with our graph G being the union of G_0 and G_1, a disjoint union except that we identify the centers of the two graphs. So G has a single center node, around which lie infinitely many spokes of type σ_i, (where $\sigma_0, \sigma_1, \ldots$ enumerates all strictly decreasing tuples of natural numbers of length $\leq n$) and also, for each $x \in \omega$, a single spoke whose type comes from the sequence $f_0(x), f_1(x), f_2(x), \ldots$, only with repetitions deleted (so that this becomes a finite, strictly decreasing sequence). As in G_0, the spoke with this type $\langle f_0(x), \ldots \rangle$ has end nodes a_x and b_x, and an additional loop is created by adding $2x$ more vertices, which along with a_x form a loop of length $2x + 1$ to identify this spoke.

Theorem 4.3 yields, for every function h which is n-approximable from above and has $f \leq_{1-\text{btt}} h$, a graph H_1 isomorphic to G_1 whose distance function is 1-btt-equivalent to h. If we repeat the preceding construction with H_1 in place of G_1 (still with G_0), then the resulting computable graph H is isomorphic to G. Its distance function d_H certainly has $d_{H_1} \leq_{1-\text{btt}} d_H$, hence $h \leq_{1-\text{btt}} d_H$. On the other hand, to compute $d_H(u, v)$, we again apply the two cases from Theorem 4.1. If u and v lie on distinct spokes in H (or if either is the center of H), then $d_H(u, v)$ can be computed directly, since the distance of a node from the center is a computable function. In Case 2, where u and v lie on the same spoke, we check first whether that spoke comes from H_1 or from G_0. If it comes from H_1, then we use the reduction $d_{H_1} \leq_{1-\text{btt}} h$, while if it comes from G_0, then we use the reductions $d_{G_0} \leq_{1-\text{btt}} f \leq_{1-\text{btt}} h$. In either case, we get a 1-btt computation of $d_H(u, v)$ from h, and since it was decidable whether the spoke comes from G_0 or from H_1 (and since 1-btt reducibility is transitive!), this shows $d_H \leq_{1-\text{btt}} h$. Thus the 1-btt degree of h lies in the 1-btt degree spectrum of the distance function of G.

For the converse, suppose that H is a computable graph isomorphic to G, with distance function d_H. First, notice that all spokes in G_0 have three initial paths and at most m additional paths, since we built G_0 using an m-approximation to f. Likewise, the spokes in G_1 all have at most m more paths after the three initial paths, and with $H \cong G$, the distance function d_H must be m-approximable from above. (As usual, for nodes on distinct spokes, we compute the distance from one to the center and then the distance from the center to the other, with no approximation at all.) Next we give a 1-btt reduction from f to d_H. Given x, we compute $f(x)$ by searching for the unique spoke in H with a loop of length $2x + 3$ at one of its two end points a_x and b_x. (We start, in each spoke, by finding the unique pair of nodes with three disjoint paths of equal length between them; these are the end nodes of the spoke, and we then search for a loop of length $2x + 3$ which starts at either end point and does not contain the other end point. This uniquely describes the identifier loops from G_0.) Having found this spoke, we know that $f(x) = d_H(a_x, b_x)$, and this is the required 1-btt reduction to d_H. Thus every 1-btt degree in the spectrum of the distance function on G contains a function which is m-approximable from above and to which f is 1-btt-reducible. This completes the proof of the first statement in the theorem.

The proof of the second statement is very much analogous, and we omit the details here. The only difference is that, instead of the graph G_1 from Theorem 4.3, we use the graph G_2 which was built for this f by Theorem 4.2. No proofs of m-approximability are needed here, and so this proof goes through for every function f which is approximable from above, even if, for every $m \in \omega$, f fails to be m-approximable from above. □

5. Related Topics

The thesis of this article, in a certain sense, is that 1-btt reducibility is the correct reducibility to use when considering distance functions on computable graphs. While the notion of "correctness" is not rigorous, we nevertheless see the results of Section 4 as a justification for this thesis: graphs with natural 1-btt degree spectra are readily constructed. We conjecture further that there are no such straightforward constructions of a graph for which the Turing degree spectrum of the distance function is an upper cone above an arbitrary c.e. degree, for example. (In this context, the spectrum must lie within the c.e. degrees, by Proposition 3.1, so by "upper cone" we really mean an upper cone of c.e. degrees.) One naturally asks whether there exist c.e. sets C with the property that the set $\{\deg(D) \ : \ D \in \Sigma_1^0 \ \& \ C \leq_T D\}$ cannot be the Turing degree spectrum of a distance function, and we would be most interested to know the answer to this question, or to its analogues for spectra of 1-degrees, bounded Turing degrees, tt-degrees, etc.

For graphs with infinitely many connected components, the distance function is $(\omega + 1)$-approximable from above, assuming we allow ∞ as the distance between any two nodes in distinct components. We approximate the distance function $d(x, y)$ at stage 0 by $g(x, y, 0) = \omega$ (or ∞), which will continue to be the value as long as x and y are not known to be in the same connected component. Meanwhile, we search systematically for a path from x to y within increasing finite subgraphs of G, and if we find one, say of length l, at some stage s, then we set $g(x, y, s) = l$, and then continue exactly as in the connected case, searching for shorter paths. It is clear that this distance function d is therefore $(\omega + 1)$-approximable from above, in the obvious definition, provided that one allows ∞ as an output of the function. (If ∞ is not allowed, then no notion of $(\omega + 1)$-approximability from above makes sense and distinguishes the concept from ω-approximability from above.) The investigation of this extension of the problem to disconnected graphs would be a worthy endeavor.

In a different context, recent work by Steiner in [16] has considered the number of realizations of various algebraic types within a computable structure, and has asked in which cases one can put a computable upper bound on the number of realizations of each algebraic type. Over the theory $\mathbf{ACF_0}$, for example, an algebraic type is generated by the formula $p(X) = 0$, where p is a polynomial irreducible over the ground field, and the degree of the polynomial is an upper bound for the number of realizations of this type in an arbitrary field (not necessarily algebraically closed). Since the minimal polynomial of the element over the prime subfield can be found effectively, one can compute such an upper bound, whereas for other computable algebraic structures considered by Steiner, no such upper bound exists. The function counting the number of realizations of each algebraic type in a computable algebraic structure is approximable from below, and when a computable upper bound exists, this function becomes the dual of a function approximable from above, exactly as described in Definition 3.2. Therefore, the theorems proven in Section 3 apply to such functions. On the other hand, when there is no computable upper bound, the standard results about functions approximable from below apply, and we saw in Section 3 that these results differ in several ways from the results when a computable bound does exist.

To close, we ask what connection, if any, there might be between distance functions on computable graphs and Kolmogorov complexity. Is it possible that Kolmogorov complexity can be presented as the distance function on some computable graph? (This is a different matter than using Kolmogorov complexity to construct structures with prescribed model-theoretic properties, as in [17].) It might be useful to fix one node $e \in G$ – call it the *Erdős node* – and to consider $d(e, x)$, a unary function on G, in place of the full distance function; in this case one could directly build a computable graph G and a computable function $f : \omega \to G$ such that $d(e, f(n))$ is exactly the Kolmogorov complexity of n. Having done so, one could then ask about other computable copies of G: does the distance function on those copies correspond to Kolmogorov complexity under some different universal (prefix-free?) machine? Right now, this question is not well-formed, and there is no obvious reason to expect to find any connections at all between these topics, except for their common use of functions approximable from above, and their common triangle inequalities. (If one knows the Kolmogorov complexity of binary strings σ and τ, one gets an upper bound on the Kolmogorov complexity of the concatenation $\sigma^\frown\tau$.) However, any connection that might arise would be a potentially fascinating link between algorithmic complexity and computable model theory.

Acknowledgments

The authors wish to acknowledge useful conversations with Denis Hirschfeldt. The work of the first author was partially supported by a Fulbright-Nehru Senior Research Scholarship, the George Washington University Research Enhancement Fund, and NSF grant DMS–1101123. That of the second author was partially supported by grant DMS–1001306 from the National Science Foundation, by the Isaac Newton Institute, the Centre de Recerca Matemática, and the European Science Foundation, and by grants numbered 62632-00 40, 63286-00 41, and 64229-00 42 from The City University of New York PSC-CUNY Research Award Program. The third author had partial support from The George Washington University Research Enhancement Fund and from NSF grant DMS–0904101.

References

1. R.I. Soare; *Recursively Enumerable Sets and Degrees* (New York: Springer-Verlag, 1987).
2. P. Odifreddi; *Classical Recursion Theory: The Theory of Functions and Sets of Natural Numbers*, vols. 1 & 2, published as vols. 125 & 143 of *Studies in Logic and the Foundations of Mathematics* (North Holland, 1992 & 1999).
3. R. Downey & D. Hirschfeldt; *Algorithmic Randomness and Complexity* (New York: Springer-Verlag, 2010)
4. Yu.L. Ershov; Theorie der Numerierungen, *Zeits. Math. Logik Grund. Math.* **23** (1977), 289–371.
5. Yu.L. Ershov; Теория Нумераций, (Moscow: Nauka, 1977).
6. Yu.L. Ershov; Theory of numberings, in *Handbook of Computability Theory*, ed. E.R. Griffor (Amsterdam: Elsevier, 1999), 473–503.
7. A. N. Degtev; On p-reducibility of numerations, *Annals of Pure and Applied Logic* **63** (1993) 57–60.
8. D. Cenzer, G. LaForte, & J. Remmel; Equivalence structures and isomorphisms in the difference hierarchy, *Journal of Symbolic Logic* **74** (2009) 2, 535–556.
9. W. Calvert, D. Cenzer, V. Harizanov, & A. Morozov; Effective categoricity of equivalence structures, *Annals of Pure and Applied Logic* **141** (2006), 61–78.
10. W. Calvert, D. Cenzer, V. Harizanov, & A. Morozov; Δ_2^0-categoricity of Abelian p-groups, *Annals of Pure and Applied Logic* **159** (2009), 187–197.
11. D. Hirschfeldt, R. Miller, & S. Podzorov; Order-computable sets, *The Notre Dame Journal of Formal Logic* **48** (2007) 3, 317–347.
12. N. G. Khisamiev; Constructive Abelian p-groups, *Siberian Advances in Mathematics* **2** (1992), 68–113.
13. N. G. Khisamiev; Constructive Abelian groups, in *Handbook of Recursive Mathematics*, (North Holland, 1998), 1177–1231.
14. B. Khoussainov, A. Nies, & R.A. Shore; Computable models of theories with few models, *Notre Dame Journal of Formal Logic* **38** (1997), 165–178.

15. R. L. Epstein, R. Haas, & R. L. Kramer; Hierarchies of Sets and Degrees Below $0'$, in *Logic Year 1979-80, The University of Connecticut, USA,* Lecture Notes in Mathematics 859, (Springer, 1981), 32–48.

16. R.M. Steiner; Effective Algebraicity, to appear in *Archive for Mathematical Logic.*

17. B. Khoussainov, P. Semukhin, & F. Stephan; Applications of Kolmogorov complexity to computable model theory, *Journal of Symbolic Logic* **72** (2007), 1041–1054.

CARNAP AND McKINSEY
Topics in the Pre-History of Possible-Worlds Semantics

M. J. Cresswell

Victoria University of Wellington

Carnap's 1946 semantics for S5 modal logic depends on the claim that $\Box\alpha$ is true iff α is L-true, i.e., valid. I investigate some problems with this definition, and then consider the relation between this early attempt, and work in the late 1950s by Kanger, Hintikka and Kripke, which led to the development of the current possible-worlds semantics for modal logic. Finally I look at J.C.C. McKinsey's 'syntactical' account of validity for a modal logic between S4 and S5.

1. The 'metalinguistic' approach to the logical modalities

In Quine 1953b [27], W.V. Quine distinguishes three grades of what he calls 'modal involvement'. The first grade he regards as innocuous. It is no more than the metalinguistic attribution of validity to a formula of non-modal logic. In the second grade we say that where α is any sentence then $\Box\alpha$ is true iff α itself is valid — or logically true. On pp. 166–169 Quine argues that while such a procedure is possible it is unilluminating and misleading.[a] The aim of this paper is to look at Carnap's attempts to produce a (propositional) modal logic on this basis.[b] I shall look at how to derive a 'metalinguistic' semantics for propositional S5, as presented in

---

[a]The principal aim of Quine 1953b [27] is to argue against the 'third grade' in which there is quantification into modal contexts. Although this is perhaps the most important aspect of Quine's objections to modal logic there are nevertheless important questions which arise even in the propositional case, which will concern us in the present work. In terms of Quine's criticism of course you would need to investigate the slightly different logic presented in Carnap 1947 [7], and try to find out whether it really can provide an adequate account of possibility and necessity, along the lines that Carnap thought it could. That is another, and much longer story.

[b]Carnap's modal logic is presented in Carnap 1946 [6], Carnap 1947 [7] pp. 173–177 and Carnap 1963 [9] pp. 889–905. Modal logic is also discussed on pp. 250–260 of Carnap 1937 [5]. It does seem that Carnap felt that you could only make sense of the logical modalities by understanding them as 'quasi–syntactical'. (1937, pp. 246 and 250.)

Carnap 1946 [6], and then look at J.C.C. McKinsey's 1945 [21] 'syntactical' account of weaker systems.[c]

2. Carnap validity

Begin with a language $\mathcal{L}$ of the classical proposition al calculus (PC) whose atomic wff are the propositional variables p, q, r,... etc., together with the 'falsum' $\bot$, and whose complex wff are made from these by the operator $\supset$.[d] $\mathcal{L}$ is interpreted by models — PC models — where each PC model $\mathcal{M}$ assigns exactly one of two truth values to every atomic wff. I write $\mathcal{M} \models p$ to mean that $\mathcal{M}$ gives the value 'true' to p and $\mathcal{M} \dashv p$ otherwise. In the rest of the paper, unless otherwise indicated, a model $\mathcal{M}$ will always be understood to be a PC model — that is to say it will assign a truth value to every propositional variable. Where $\mathcal{M}$ is a PC model, the truth relation $\models$ can be extended to an assignment of truth values to all PC-wff by means of the rules

$$\mathcal{M} \dashv \bot \tag{1}$$

$$\mathcal{M} \models \alpha \supset \beta \text{ unless } \mathcal{M} \models \alpha \text{ and } \mathcal{M} \dashv \beta. \tag{2}$$

We may extend $\mathcal{L}$ to a modal language $\mathcal{L}_M$. $\mathcal{L}_M$ is simply the modal propositional language which extends $\mathcal{L}$ by the addition of the (monadic) necessity operator $\Box$, and the formation rule that if α is a wff then so is $\Box\alpha$. $\mathcal{L}_M$ can be given a standard S5 modal interpretation in the following way. An S5 model is a pair $\langle W,V \rangle$ where W is a set of 'possible worlds' and V is an assignment to the propositional variables such that $V(p) \subseteq W$. Then, for $w \in W$, we define $\langle W,V \rangle \models_w p$ iff $w \in V(p)$, $\langle W,V \rangle \dashv_w \bot$, and $\langle W,V \rangle \models_w \alpha \supset \beta$ iff either $\langle W,V \rangle \dashv_w \alpha$ or $\langle W,V \rangle \models_w \beta$. For $\Box$ we have

$$\langle W,V \rangle \models_w \Box\alpha \text{ iff, for every } w' \in W, \langle W,V \rangle \models_{w'} \alpha. \tag{3}$$

By contrast, the idea behind Carnap's account of modality is that $\Box$ (which he writes as N) is to be interpreted in such a way that $\Box\alpha$ is to be true iff α is *valid* — or, as Carnap puts it, iff α is L-true. This can be made precise as follows.[e] (1) and (2) shew how to extend $\models$ from a truth assignment to

---

[c]In his criticisms of modal logic Quine seems to have Carnap in mind (see essay 7 of Quine 1953a [26]) and does not consider McKinsey's work.

[d]The other truth functions are all definable in terms of $\supset$ and $\bot$. in particular $\alpha \lor \beta$ may be defined as $(\alpha \supset \beta) \supset \beta$, $\sim \alpha$ as $\alpha \supset \bot$ and $\alpha \land \beta$ as $\sim(\sim \alpha \lor \sim \beta)$. Carnap takes $\sim$, $\lor$ and '.' ($\land$) as primitive, but these differences do not affect the issues discussed here.

[e]Carnap 1946 [6], p. 38, leaves the details of a model theory unspecified, but the definition in the text is clearly in accordance with his treatment of propositional logic.

the variables to an assignment to any wff of PC. To extend $\models$ to apply to all wff of $\mathcal{L}_M$ we add:

If $\mathcal{M}$ is a PC model, $\mathcal{M} \models \Box\alpha$ iff $\mathcal{M}' \models \alpha$ for every PC model $\mathcal{M}'$. (4)

We can then say that a wff α of $\mathcal{L}_M$ is C-valid (*Carnap-valid*) iff $\mathcal{M} \models \alpha$ for every PC model $\mathcal{M}$.

Theorem 2.1. *If α is S5-valid then α is C-valid.*

Proof. We prove that the axioms of S5 are all C-valid. The axioms of S5 can be stated as the following schemata, where α and β are any wff of $\mathcal{L}_M$:

PC	All instances of valid PC-wff
K	$\Box(\alpha \supset \beta) \supset (\Box\alpha \supset \Box\beta)$
T	$\Box\alpha \supset \alpha$
5	$\sim\Box\alpha \supset \Box\sim\Box\alpha$

together with the rules

MP	$\vdash \alpha, \vdash \alpha \supset \beta \rightarrow \vdash \beta$
N	$\vdash \alpha \rightarrow \vdash \Box\alpha$

Instances of valid PC-wff are trivially C-valid. For **K** assume that $\mathcal{M} \nvDash \Box\alpha \supset \Box\beta$. Then (i) $\mathcal{M} \models \Box\alpha$ and (ii) $\mathcal{M} \nvDash \Box\beta$. From (ii) there is some $\mathcal{M}'$ such that $\mathcal{M}' \nvDash \beta$. From (i) and (4) $\mathcal{M}' \models \alpha$, and so by (2) $\mathcal{M}' \nvDash \alpha \supset \beta$ and so by (4), $\mathcal{M} \nvDash \Box(\alpha \supset \beta)$. For **T**, if $\mathcal{M} \models \Box\alpha$ then for every $\mathcal{M}'$, $\mathcal{M}' \models \alpha$, and so, in particular $\mathcal{M} \models \alpha$. Thus by (2), $\mathcal{M} \models \Box\alpha \supset \alpha$. For **5**, if $\mathcal{M} \nvDash \Box\sim\Box\alpha$ there is some $\mathcal{M}'$ such that $\mathcal{M}' \nvDash \sim\Box\alpha$, and so $\mathcal{M}' \models \Box\alpha$. But then for every $\mathcal{M}''$, $\mathcal{M}'' \models \alpha$, and so $\mathcal{M} \models \Box\alpha$. So $\mathcal{M} \nvDash \sim\Box\alpha$.

The two rules of inference preserve C-validity. For MP suppose that $\mathcal{M} \models \alpha$ but $\mathcal{M} \nvDash \beta$. Then by (2) $\mathcal{M} \nvDash \alpha \supset \beta$. For N if $\Box\alpha$ is not C-valid then there is some $\mathcal{M}$ such that $\mathcal{M} \nvDash \Box\alpha$. So there is some $\mathcal{M}'$ such that $\mathcal{M}' \nvDash \alpha$, and so α is not C-valid either. So the axioms of S5 are C-valid and the rules preserve C-validity, and so every theorem of S5 is C-valid. $\quad\Box$

This is of course the basis of Carnap's defence of S5 over other modal logics. (See Carnap 1946 [6], p. 41.) So S5 is sound with respect to C-validity. However it is not complete. This is because $\sim\Box p$ is C-valid even though it is not a theorem of S5. (It obviously fails in any S5 model in which p is true in every world.)

Theorem 2.2. $\sim\Box p$ *is C-valid.*

Proof. There is at least one model $\mathcal{M}'$ such that $\mathcal{M}' \dashv p$, where p is a propositional variable, and therefore we have, for every model $\mathcal{M}$, $\mathcal{M} \dashv \Box p$, and so $\mathcal{M} \vDash \sim \Box p$. $\qquad\square$

Clearly something has gone wrong. To see what, notice that the axiomatisation of S5 used here is formulated using axiom schemata, and does not include a primitive rule of uniform substitution for propositional variables. Despite this, every S5 theorem remains a theorem if uniform substitution is made for its variables. Theorem 2.2 shews that a rule of uniform substitution for propositional variables does not preserve C-validity, since even though $\sim \Box p$ is C-valid the instance you get by substituting $\bot \supset \bot$ for p is $\sim \Box (\bot \supset \bot)$, which is not C-valid. This problem was noticed in Makinson 1966 [22] p. 333, though Makinson mentions Carnap only in a footnote. C-validity was axiomatised by Thomason 1973 [30], though Thomason doesn't mention Carnap at all. Axiomatising C-validity is straightforward. Assume S5 axiomatised by schemata. Then add the rule

For any PC wff α, if α is not PC-valid, then $\sim \Box \alpha$ is a theorem. $\qquad(5)$

Call S5 with the addition of (5) S5$^+$. (Thomason's axiomatisation, though equivalent, is slightly different.) The completeness proof which follows relies on the fact that all modal wff are equivalent in S5 to truth functions of propositional variables and wff of the from $\Box \alpha$ where α is a PC-wff.[f]

For any first degree wff α let α' be a PC wff in which every sub-wff β of α of the form $\Box \gamma$ is replaced by $\bot \supset \bot$ if γ is PC-valid and by $\bot$ if γ is not valid. (Since α is first degree γ will be a PC- wff.)

Theorem 2.3. $\vdash_{S5+} \alpha \equiv \alpha'$

Proof. The proof is by induction on the construction of α. If β is a propositional variable or $\bot$ then β' is β and the result holds by PC. If β is $\Box \gamma$ then we have two cases. If γ is PC-valid then $\vdash_{S5+} \Box \gamma$. But in that case β' is $\bot \supset \bot$, and $\vdash_{S5+} \Box \gamma \equiv (\bot \supset \bot)$. If γ is not PC-valid then, by (5), $\vdash_{S5+} \sim \Box \gamma$. But in that case β' is $\bot$, and $\vdash_{S5+} \Box \gamma \equiv \bot$. If β is $\gamma \supset \delta$, and

---

[f]Note that if (5) were changed to read

(5a) If α is not a theorem then $\sim \Box \alpha$ is a theorem

you would get the same theorems, but this new rule would make the logic non-monotonic in the sense that adding new axioms could make (5a) no longer validity-preserving. In his study of Carnap's modal logic Schurz 2001 [28] defends the various non-standard features of the logic of C-validity. Cocchiarella 1975b [11] also welcomes the various unusual features of Carnap's logic. (See for instance pp. 45 and 54f., *op. cit.*)

we have $\vdash_{S5+} \gamma \equiv \gamma'$ and $\vdash_{S5+} \delta \equiv \delta'$, then by PC we have $\vdash_{S5+} (\gamma \supset \delta)$ $\equiv (\gamma' \supset \delta')$, i.e., $\vdash_{S5+} (\gamma \supset \delta) \equiv (\gamma \supset \delta)'$. $\qquad\square$

Theorem 2.4. α *is C-valid iff* α' *is PC-valid.*

Proof. It is sufficient to prove that for any model $\mathcal{M}$, $\mathcal{M} \vDash \alpha \equiv \alpha'$. The proof is by induction on the construction of α. If a sub-wff β of α is a propositional variable or $\bot$ then β' is β and we are done. If β is $\gamma \supset \delta$ and $\mathcal{M} \vDash \gamma \equiv \gamma'$ and $\mathcal{M} \vDash \delta \equiv \delta'$ then $\mathcal{M} \vDash (\gamma \supset \delta) \equiv (\gamma \supset \delta)'$. If β is $\Box\gamma$ then let $\mathcal{M}$ be any Carnap model. If γ is PC-valid then $\mathcal{M} \vDash \Box\gamma$, and so $\mathcal{M} \vDash \Box\gamma \equiv (\bot \supset \bot)$. If γ is not PC-valid then $\mathcal{M} \vDash \sim\Box\gamma$, and so $\mathcal{M} \vDash \Box\gamma \equiv \bot$. $\qquad\square$

From theorem 2.3 we have:

Lemma 2.5. $\vdash_{S5+} \alpha$ *iff* α' *is PC-valid.* $\qquad\square$

Lemma 2.5 and theorem 2.4, together with the fact that every wff is equivalent in S5 to a first-degree wff give us

Theorem 2.6. $\vdash_{S5+} \alpha$ *iff* α *is C-valid.* $\qquad\square$

3. Quine/Carnap validity

Makinson considers it to be a defining feature of logical validity, that the rule of uniform substitution be preserved. If so then C-validity is problematic as an adequate semantics for modal logic. One way of handling this problem is to follow some ideas developed by Quine as early as Quine 1934 [25]. Quine suggests that the theorems of logic are always schemata. So that rather than speak of $p \supset p$ where p is a propositional variable, we speak of the schema $\alpha \supset \alpha$, whose instances are sentences like

> Wellington is the capital of New Zealand $\supset$ Wellington is the capital of New Zealand.

If necessity is to be connected with logical validity then we might want to say that $\sim\Box p$ is indeed valid for every simple sentence. But it will not be a logical theorem since the *schema* $\sim\Box\alpha$ will not be logically valid, and will in fact fail if α is 'Wellington is the capital of New Zealand $\supset$ Wellington is the capital of New Zealand'. Carnap explicitly follows Quine in this:

> We here make use of 'p', 'q', etc. as *auxiliary variables*; that is to say they are merely used (following Quine) for the description of certain forms of sentences. (1946 [6], p.41)

On p. 38 he has already told us that "our systems will not contain propositional variables", and on p. 46 he contrasts his axiomatic system (which he calls a 'calculus') with a version of S5 (MPC_v) closer to that of Lewis, which contains propositional variables and a rule of uniform substitution for them. So let us assume, following Quine and Carnap, that the p, q etc that we spoke of before are not really 'variables', but are just simple sentences. It may indeed be true that $\sim \Box p$ will still be valid, but it will not count as a truth of logic, since the schema $\sim \Box \alpha$ is not valid. We shall call this approach the *Quine/Carnap semantics*, and it is this semantics that Carnap seems to have in mind when it comes to proving completeness.[g]

Definition 3.1. A wff α is QC-valid iff, for every wff β obtained from α by uniform substitution for the variables of α, β is C-valid.

This definition immediately rules out $\sim \Box p$ as QC-valid, since, as noted above, an instance of it, $\sim \Box(\bot \supset \bot)$, is not C-valid, and is not an S5-theorem. Of course for non-modal wff C-validity and QC-validity coincide, and in the case of non-modal wff we shall frequently simply speak of validity.

Corollary 3.2. *S5 is QC-sound.* □

This corollary is immediate because, as established in theorem 2.1, every instance of the S5 axioms is C-valid, and the rules clearly preserve QC-validity. Given this, we may make use of standard principles of S5, which ensure that we need only consider S5 wff of first-degree, in which no modal operator occurs in the scope of any other. Despite this, the completeness of S5 with respect to QC-validity is not entirely trivial.[h]

Theorem 3.3. *If α is QC-valid then α is S5-valid.*

Proof. What we have to shew is that for every non-theorem of S5 you can find a substitution instance which is not C-valid. Standard results about S5, together with facts about the ordinary classical propositional calculus,

---

[g]This is true in the propositional case, where Carnap seems unaware of the $\sim \Box p$ problem. The case is different in predicate logic, where, on p. 54f, he includes axioms which are only valid with respect to the first-order version of what I have called C-validity.

[h]The proof which follows should be compared to that given in Thomason 1973 [30], p. 283f (theorem 2) which does not use normal forms. Gottlob 1999 [15], pp. 246 and 251, mentions a proof in Makinson's PhD thesis. (Gottlob also has a fairly full history of work done on Carnap's propositional modal logic up to 1999, noting links with computation theory.) A proof along these lines is indicated by Carnap himself on pp. 46-48 of 1946 [6], and also on p. 87 of Burgess 1999 [4].

make it sufficient to consider solely wff in what may be called *perfect modal conjunctive normal form*[i], where a wff is in PMCNF iff it is a conjunction of wff, each having the form

$$\alpha_0 \vee \Box\alpha_1 \vee \ldots \vee \Box\alpha_n \vee \Diamond\beta \tag{6}$$

where there is a single collection of propositional variables $p_1, \ldots, p_k$ such that

(i) Each α_i $(0 \leqslant i \leqslant n)$ is a disjunction containing either p_j or $\sim p_j$ for every $1 \leqslant j \leqslant k$; and

(ii) β is a conjunction containing either p_j or $\sim p_j$ for every $1 \leqslant j \leqslant k$.

(6) will be S5-valid iff at least one of $\alpha_i \vee \beta$ is PC-valid, and this means that if (6) is not S5-valid then no variable in any α_i will be the negation of any variable in β or vice versa. As a consequence no α_i will contain a variable both negated and unnegated. So we may use the following notation. Let $\alpha_i(j)$ for $1 \leqslant j \leqslant k$ denote p_j if p_j is a disjunct of α_i, and let $\alpha_i(j)$ denote $\sim p_j$ if $\sim p_j$ is a disjunct of α_i. (So if (6) is not S5-valid α_i will contain exactly one of p_j or $\sim p_j$ for every $1 \leqslant j \leqslant k$.) Similarly let $\beta(j)$ denote p_j or $\sim p_j$ according to which of them is a conjunct of β. (If β should contain some p_j and its negation then β will be contradictory and $\alpha_i \vee \beta$ will only be valid if α_i is, and if α is valid $\Box\alpha_i$ will be true on the Quine/Carnap account. So without loss we may consider only cases in which β contains exactly one of p_j and $\sim p_j$.) To shew that (6) is not QC-valid it is sufficient to produce a substitution instance of (6) which is not C-valid. Such an instance is

$$\alpha_0' \vee \Box\alpha_1' \vee \ldots \vee \Box\alpha_n' \vee \Diamond\beta' \tag{7}$$

where β' is defined as follows. For the case where $k = 1$ every $\alpha_i = \beta$ and we can let $\beta' = \bot$ if β is p_1 and $\bot \supset \bot$ if β is $\sim p_1$. For $1 < k$ consider one of the conjuncts of β — for definiteness say $\beta(1)$:

(i) If $\beta(1)$ is p_1 then, β' is $\beta[p_1 \wedge \sim(\beta(2) \wedge \ldots \wedge \beta(k))/p_1]$.

(ii) If $\beta(1)$ is $\sim p_1$ then, β' is $\beta[p_1 \vee (\beta(2) \wedge \ldots \wedge \beta(k))/p_1]$.

Obviously β' will be contradictory. What we have to shew is that where each α_i' is α_i with the same substitutions as obtained β' from β, then α_i' is not PC-valid.

---

[i]This procedure is due to Wajsberg 1933 [31], and is followed by Carnap 1946 [6] pp. 43–46, who calls a wff reduced to PMCNF the 'MP-reductum' of the original wff.

Take case (i). If (6) is not S5-valid then there is at least one h ($1 \leqslant$ h $\leqslant$ k) such that β(h) = α_i(h). There are two possibilities. If $\beta(1) = \alpha_i(1)$ then if p_1 is made false, $\alpha_i'(1)$ will be false, whatever values are given to the other variables, and since the other disjuncts never include a variable both negated and unnegated, α_i' is not PC-valid and so $\square \alpha_i'$ is false. If on the other hand $\beta(1) \neq \alpha_i(1)$, then, since $\beta(1)$ is p_1, $\alpha_i(1)$ is $\sim p_1$, and then $\alpha_i'(1)$ will be

(iii) $\sim(p_1 \wedge \sim(\beta(1) \wedge \ldots \wedge \beta(k)))$

i.e., it will be equivalent to $\sim p_1 \vee (\beta(2) \wedge \ldots \wedge \beta(k))$. For every p_j, make p_j false if $\alpha_i(j)$ is p_j and true if $\alpha_i(j)$ is $\sim p_h$. Then, in particular, $\alpha_i(h)$ will be false, and since $\beta(1) \neq \alpha_i(1)$, then $\beta(h) = \alpha_i(h)$ for some $2 \leqslant h \leqslant k$, and so $\beta(2) \wedge \ldots \wedge \beta(k)$ will be false independently of the values to the other variables, and therefore α_i' will be false. So α_i' is not PC-valid in case (i).

Now take case (ii). As before first suppose that $\beta(1) = \alpha_i(1)$. In that case if p_1 is made true $\alpha_i'(1)$ is false, whatever values are given to the other variables, and since the other disjuncts never include a variable both negated and unnegated, α_i' is not PC-valid and so $\square \alpha_i'$ is false. If on the other hand $\beta(1) \neq \alpha_i(1)$, then, since $\beta(1)$ is $\sim p_1$, $\alpha_i(1)$ is p_1, and then $\alpha_i'(1)$ will be

(iv) $p_1 \vee (\beta(2) \wedge \ldots \wedge \beta(k))$.

So for every p_j, including p_1, make p_j false if $\alpha_i(j)$ is p_j and true if $\alpha_i(j)$ is $\sim p_j$. Then, in particular, $\alpha_i(h)$ will be false and since $\beta(1) \neq \alpha_i(1)$ then $\beta(2) \wedge \ldots \wedge \beta(k)$ will be false independently of the values to the other variables, so α_i' is not PC-valid in case (ii).

So in each of these cases a single substitution makes β' contradictory and, for $0 \leqslant i \leqslant n$, makes no α_i' PC-valid. So, first, α_0' can be made false, and second, $\square \alpha_i$ ($1 \leqslant i \leqslant h$) is false, and third, $\lozenge \beta'$ is false. So that, (7), an instance of (6), turns out to be false in this model, and so is not C-valid. So (6) is not QC-valid. $\square$

It is not clear that Carnap was aware of the complexity required in a completeness proof for QC-validity using normal forms. On p. 45 he says that if a wff is valid (L-true) then 'An examination of the forms listed ... shows that all of them except [the ones which satisfy the tests for validity] are impossible in this case.' So it is not obvious how clearly he saw what was required to carry the completeness proof through successfully.

4. Meaning postulates

The previous section used QC-validity to solve the $\sim \Box p$ problem. But there is another way, suggested by Carnap's device of 'meaning postulates', as set out in Carnap 1952 [8].[j] In that paper (p. 222) Carnap is concerned to allow a logic in which sentences like

$$\text{If Jack is a bachelor, then Jack is unmarried} \qquad (8)$$

can be displayed as non-contingent. Since (8) will not come out as L-true according to Carnap's usual definition — it will not be true in all models — a theorist who wants it to be analytic needs to add it as a *meaning postulate*. Carnap is concerned to argue that it is up to the theorist to stipulate which sentences are to be regarded as analytic. In a sense this is simply the theorist stipulating how the language is to be understood. In a formal language a meaning postulate will be a wff of that language. In his paper Carnap is not explicit about just which formal languages he has in mind, though none of his examples of meaning postulates involve modal formulae. Indeed his aim of explaining modality in non-modal terms would make this plausible, and I shall follow him. Given a set of meaning postulates Carnap (p. 225) describes a wff as L-true with respect to that set of meaning postulates iff it is L-implied by those postulates, which means that it is true in all models which satisfy the meaning postulates.

I will set out a version of Carnap's theory for the propositional case, where the meaning postulates can be expressed as PC wff. Let Π be a consistent set of PC-wff. Where $\mathcal{M}$ is a PC model giving truth values to the atomic wff, say that $\mathcal{M}$ is a Π-model iff it extends to all PC wff in accordance with (1) and (2) in section 2 above, and is such that $\mathcal{M} \models \beta$ for every $\beta \in \Pi$. Let Π_M be the set of all PC models $\mathcal{M}$ such that $\mathcal{M} \models \beta$ for every $\beta \in \Pi$. Note that since Π contains only PC wff (1) and (2) determine truth values for all wff in Π in any PC model. $\models$ can then be extended to provide truth values to all wff of $\mathcal{L}_M$ by replacing (4) above with

$$\text{If } \alpha \text{ is a wff of } \mathcal{L}_M, \text{ and } \mathcal{M} \text{ is any PC model, then } \mathcal{M} \models \Box\alpha \qquad (9)$$
$$\text{iff } \mathcal{M}' \models \alpha \text{ for every } \mathcal{M}' \in \Pi_M.$$

Say that α is CΠ-valid iff $\mathcal{M} \models \alpha$ for every $\mathcal{M} \in \Pi_M$. Say that α is CM-valid iff α is CΠ-valid for every consistent set Π of meaning postulates. It is easy

---

[j]Meaning postulates do not appear in Carnap 1946 [6] or 1947 [7], and I know of no suggestion in his published work that he ever considered applying them in his modal logic.

to see that $\sim\Box p$ is not CM-valid, since Π can be taken to be $\{p\}$, and then $\Box p$ will be CΠ-valid, so that $\sim\Box p$ will not be CΠ-valid for every Π, and so not CM-valid.

Theorem 4.1. *If $\vdash_{S5} \alpha$ then α is CM-valid.*

Proof. It is sufficient to prove that for any consistent set Π of PC-wff, if $\vdash_{S5} \alpha$ then α is CΠ-valid. The proof proceeds as in theorem 2.1, but relativised to Π. PC is trivial. For **K** assume a model $\mathcal{M} \in \Pi_M$ such that $\mathcal{M} \dashv \Box\alpha \supset \Box\beta$. Then (i) $\mathcal{M} \vDash \Box\alpha$ and (ii) $\mathcal{M} \dashv \Box\beta$. From (ii) there is some $\mathcal{M}' \in \Pi_M$ such that $\mathcal{M}\dashv \beta$. From (i) $\mathcal{M}' \vDash \alpha$, and so by (2) $\mathcal{M}'\dashv \alpha \supset \beta$ and so by (9), $\mathcal{M} \dashv \Box(\alpha \supset \beta)$. For **T**, if $\mathcal{M} \in \Pi_M$, and $\mathcal{M} \vDash \Box\alpha$ then for every $\mathcal{M}' \in \Pi_M$, $\mathcal{M}' \vDash \alpha$, and so, in particular, since $\mathcal{M} \in \Pi_M$, $\mathcal{M} \vDash \alpha$. Thus by (2), $\mathcal{M} \vDash \Box\alpha \supset \alpha$. For **5**, if $\mathcal{M} \dashv \Box\sim\Box\alpha$ there is some $\mathcal{M}' \in \Pi_M$ such that $\mathcal{M}'\dashv \sim\Box\alpha$, and so $\mathcal{M}' \vDash \Box\alpha$. But then for every $\mathcal{M}'' \in \Pi_M$, $\mathcal{M}'' \vDash \alpha$. So $\mathcal{M} \vDash \Box\alpha$, and therefore $\mathcal{M} \dashv \sim\Box\alpha$.

The two rules of inference preserve CΠ-validity. For MP suppose that, for $\mathcal{M} \in \Pi_M$, $\mathcal{M} \vDash \alpha$ but $\mathcal{M} \dashv \beta$. Then by (2) $\mathcal{M} \dashv \alpha \supset \beta$. For N if $\Box\alpha$ is not CΠ-valid then there is some $\mathcal{M} \in \Pi_M$ such that $\mathcal{M} \dashv \Box\alpha$. So there is some $\mathcal{M}' \in \Pi_M$ such that $\mathcal{M}'\dashv \alpha$, and so α is not CΠ-valid either. So the axioms of S5 are CΠ-valid and the rules preserve CΠ-validity, and so every theorem of S5 is CΠ-valid. $\qquad\Box$

It is important here to see why Π cannot contain modal wff. Intuitively the rule would then be circular. In (9) circularity is ruled out because whether or not $\mathcal{M}$ is a Π model depends only on whether or not $\mathcal{M} \vDash \beta$ for every β in Π, and these are all PC wff. If modal wff were allowed in Π, then the following situation could easily arise. Suppose that Π is $\{\Box p\}$. Let $\mathcal{M}$ be a PC model in which $\mathcal{M} \dashv p$. Then (9) would state that

$$\mathcal{M} \vDash \Box p \text{ iff for every } \mathcal{M}' \text{ such that } \mathcal{M}' \vDash \Box p, \mathcal{M}' \vDash p. \qquad (10)$$

So let $\mathcal{M}'$ be any model for which $\mathcal{M}' \vDash \Box p$. Then, from (9)

$$\text{If } \mathcal{M}' \vDash \Box p \text{ then, for every } \mathcal{M}'' \text{ such that } \mathcal{M}'' \vDash \Box p, \mathcal{M}' \vDash p. \qquad (11)$$

Since $\mathcal{M}'$ is an $\mathcal{M}''$ such that $\mathcal{M}'' \vDash \Box p$, then $\mathcal{M}' \vDash p$, i.e. we have the situation that $\mathcal{M}'\vDash p$ for every $\mathcal{M}'$ such that $\mathcal{M}' \vDash \Box p$, and therefore, by (9), $\mathcal{M} \vDash p$, which is a contradiction. Thus, if Π is $\{\Box p\}$ then *no* PC model could make p false, which is absurd. By insisting that Π contain no modal operators, we guarantee that whether or not a wff β in Π is true according to a given model $\mathcal{M}$ does not depend on the rule (9), and so does not assume a previously given value to β.

It is possible to prove completeness in the following way. Suppose that α is not a theorem of S5. Then there is a finite S5 model $\langle W, V \rangle$ with W $= \{w_1,...,w_n\}$ such that, for some $w^* \in W$, $\langle W,V \rangle \dashv_{w*} \alpha$. Since the value of α is unaffected by values of variables not in α, we may consider a PC model $\mathcal{M}_i$ (for $1 \leqslant i \leqslant n$) to be specified so that, for any variable p in α,

$$\mathcal{M}_i \vDash p \text{ iff } \langle W,V \rangle \vDash_{w_i} p. \tag{12}$$

For $1 \leqslant i \leqslant n$, let β_i be the conjunction of p or $\sim p$, for every variable p in α, according as $\mathcal{M}_i \vDash p$ or $\mathcal{M}_i \dashv p$. Let β be the disjunction $\beta_1 \vee ... \vee \beta_n$, and let $\Pi = \{\beta\}$. Then

Lemma 4.2. $\mathcal{M}_i \vDash \beta$ *for every* $1 \leqslant i \leqslant n$. $\qquad\square$

In other words, every $\mathcal{M}_i$ is a Π model.

Lemma 4.3. *For any* $\mathcal{M} \in \Pi_M$, $\mathcal{M}$ *coincides with some* $\mathcal{M}_i$ *on all variables in* α.

Proof. If $\mathcal{M} \in \Pi_M$ then $\mathcal{M} \vDash \beta$, and so, for some β_i, $\mathcal{M} \vDash \beta_i$. Since β_i is a conjunction of p or $\sim p$ for every p in α then $\mathcal{M} \vDash p$ iff $\mathcal{M}_i \vDash p$, and so $\mathcal{M}$ coincides with $\mathcal{M}_i$ on variables in α. $\qquad\square$

Write $\mathcal{M} \approx \mathcal{M}'$ to mean that $\mathcal{M}$ coincides with $\mathcal{M}'$ on all the variables in α.

Lemma 4.4. *If* $\mathcal{M} \approx \mathcal{M}'$ *and* γ *is a subformula of* α *then* $\mathcal{M} \vDash \gamma$ *iff* $\mathcal{M}'$ $\vDash \gamma$. $\qquad\square$

Theorem 4.5. α *is not* $C\Pi$-*valid.*

Proof. The proof is by induction on the construction of α. For γ a subformula of α take the induction hypothesis to be that, for $\mathcal{M} \in \Pi_M$, $\mathcal{M} \vDash \gamma$ iff $\langle W,V \rangle \vDash_{w_i} \gamma$ for any $\mathcal{M} \approx M_i$. Since $\mathcal{M} \approx \mathcal{M}_i$ for some $1 \leqslant i \leqslant n$, where $\mathcal{M} \in \Pi_M$ that will ensure that $\mathcal{M} \dashv \alpha$ for some $\mathcal{M} \in \Pi_M$, and so α is not $C\Pi$-valid (and therefore not CM-valid). Without loss of generality we may state the induction for $\mathcal{M}_i$, since if γ is a subwff of α and $\mathcal{M} \approx \mathcal{M}_i$ then $\mathcal{M} \vDash \gamma$ iff $\mathcal{M}_i \vDash \gamma$, for all $\mathcal{M} \in \Pi_M$. The induction for the variables in α is by definition of $\mathcal{M}_i$, and the induction is trivial for $\bot$ and $\supset$. Suppose $\langle W,V \rangle \dashv_{w_i} \Box\gamma$. Then, for some $w_j \in W$, $\langle W,V \rangle \dashv_{w_j}\gamma$, and so, by the induction hypothesis, $\mathcal{M}_j \dashv \gamma$, and so $\mathcal{M}_i \dashv \Box\gamma$. Suppose $\mathcal{M}_i \dashv \Box\gamma$. Then for some $\mathcal{M} \in \Pi_M$, $\mathcal{M} \models \gamma$. By lemma 4.3 $\mathcal{M} \approx M_j$ for some $1 \leqslant j \leqslant n$, and so, by the induction hypothesis, $\langle W, V \rangle \models_{w_j} \gamma$, and so $\langle W, V \rangle \models_{w_i} \Box \gamma$.

$\Box$

5. Classes of models

In this section we look at a more flexible way of solving the $\sim \Box p$ problem, which does not require a notion like QC-validity. This approach has quite wide-ranging implications, in that it leads fairly easily to the current indexical semantics for modal logic.

Let $\mathcal{M}^+$ be a class of PC models, and let $\models$ satisfy (1) and (2). (4) is replaced by

$$\mathcal{M} \models \Box \alpha \text{ iff } \mathcal{M}' \models \alpha \text{ for every } \mathcal{M}' \in \mathcal{M}^+. \tag{13}$$

Say that a wff α is valid in $\mathcal{M}^+$ iff $\mathcal{M} \models \alpha$ for every $\mathcal{M} \in \mathcal{M}^+$, and say that α is $\mathcal{M}^+$-valid iff α is valid in every $\mathcal{M}^+$. With this change it is easy to see that $\sim \Box p$ is not $\mathcal{M}^+$-valid, since one need only consider, for instance, a class $\mathcal{M}^+$ whose only member is a model $\mathcal{M}$, where $\mathcal{M} \models p$. Moreover, it is also easy to establish that every class $\mathcal{M}^+$ of models satisfying (13) corresponds to an S5 model $\langle W, V \rangle$, and vice versa:

(a) Suppose that $\mathcal{M}^+$ is a class of models. Assume it is indexed by some set I, which of course could be $\mathcal{M}^+$ itself. Then let $\langle W, V \rangle$ be an S5 model in which $W = I$, and for every variable p, $i \in V(p)$ iff $\mathcal{M}_i \models p$. An easy induction establishes the following:

Theorem 5.1. *For any wff α, and any $\mathcal{M}_i \in \mathcal{M}^+$, $\langle W, V \rangle \models_i \alpha$ iff $\mathcal{M}_i \models \alpha$.* $\Box$

(b) For the converse, given $\langle W, V \rangle$, let $\mathcal{M}^+$ be the set of PC models, indexed by W, in which, for any variable p, $\mathcal{M}_w \models p$ iff $w \in V(p)$. As before, a straightforward induction establishes:

Theorem 5.2. *For any wff α, and any $\mathcal{M}_w \in \mathcal{M}^+$, $\langle W, V \rangle \models_w \alpha$ iff $\mathcal{M}_w \models \alpha$.* $\Box$

$\mathcal{M}^+$-validity has an historical importance. In the late 1950s there developed an approach to modal logic which was what led to the current indexical semantics. In this approach the indices were in fact models, and the intention appears to have been to connect the validity of a wff α with the truth of $\Box \alpha$, just as Carnap had suggested. Consider the work found in Kanger 1957 [18], Hintikka 1957 [16], Bayart 1958 [1], 1959 [2], Kripke

1959 [19] or Montague 1960 [23]. All of these authors, except Bayart, produced a semantics in terms of a restricted class of models.[k] What of course also emerged is that a more flexible approach involves a relation R between models. So that in addition to a set $\mathcal{M}^+$ of models it is stipulated whether or not, for $\mathcal{M} \in \mathcal{M}^+$ and for $\mathcal{M}' \in \mathcal{M}^+$, we have $\mathcal{M} \, R \, \mathcal{M}'$, and in place of (13) we have.

$$\mathcal{M} \models \Box\alpha \text{ iff } \mathcal{M} \models \alpha \text{ for every } \mathcal{M}' \in \mathcal{M}^+ \text{ such that } \mathcal{M} \, R \, \mathcal{M}'. \qquad (14)$$

Models in this sense correspond to ordinary relational models, and of course characterise every normal modal system. While the importance of the relation R in the development of modal logic has been well appreciated, it is perhaps less appreciated what a change the introduction of an arbitrary class of models effects.

Kanger's discussion on pp. 33–35 does not explicitly refer to a class of models, although such a class emerges from his classification of the various properties of the relation R.[l] Kripke's 1959 completeness proof for S5, where there is no relation between models, does make explicit reference, on p.2, to a class of models.[m] What theorems 5.1 and 5.2 do is shew that any class of models can be amalgamated into a single model by indexing the class, and stipulating that the truth of a wff at an index i in the enlarged model, is simply its truth at $\mathcal{M}_i$. When α is a complex wff involving modal operators its truth at $\mathcal{M}_i$ will depend on just what models are in $\mathcal{M}^+$. This is in contrast to QC-validity, which was intended to be independent of the choice of any particular subset of other models. The importance can be seen by recalling how $\mathcal{M}^+$-validity solves the $\sim \Box p$ problem. For when $\mathcal{M}^+$ is an arbitrary class of models there is nothing to prevent it from containing only models which make p true. And once you have taken this step, then, as we have seen, it is trivial to amalgamate the class of models into a single indexical model in which the nature of the indices is immaterial.

---

[k]Of these early works Bayart stands out as perhaps being the first to see that the indices in a model can be anything you please, and in his semantics states explicitly that worlds are not models (Bayart 1958, p. 28.) A translation and commentary of Bayart 1958 and 1959 will appear in a forthcoming issue of *Logique et Analyse*. (Cresswell, forthcoming). In a footnote to Montague 1960 (p. 73 of the 1974 reprint) Montague says that his paper was delivered at a 1955 conference at the University of California.

[l]It is perhaps worthy of note that Kanger's description of the semantics of 'logical necessity' on p. 35 appears to have the consequence that when $\Box$ ('L' in his terminology) represents logical necessity then $\sim \Box p$ is logically valid.

[m]In a 1978 paper [29] Brian Skyrms explains how to give a metalinguistic account of the semantics of S5 modal logic. Skyrms considers this important as a commentary on Quine's attitude to modal logic.

As far as I can tell this further step was not taken by Carnap, and it is not clear, to me at least, whether it would have been acceptable to him. The closest he comes appears to be the use of meaning postulates discussed in section 4 above, and indeed the method of meaning postulates is one way of restricting the class of models. But, as remarked in footnote j, meaning postulates do not appear in Carnap 1946 [6] — nor even in Carnap 1947 [7]. In any case, Carnap wanted modal logic to provide an analysis of *necessity*, and he was concerned to provide an analysis of logical necessity in terms of L-truth, or validity. So he needed an account which enabled him to solve the question of which of Lewis's modal logics was the 'correct' one. It would not help answer this question to be told that it all depends on which class of models you choose. Look at it this way. In Prior 1957 [24], Prior shews how you can get S5 in a temporal logic in which L means 'it is, was and always will be that'. What makes this a *temporal* logic — a logic of time — rather than a strictly *modal* logic — a logic of necessity and possibility — is that the indices are moments of time, whereas in the modal case they are possible worlds. So that even if restricted models get the logic right they cannot give an explanation of modality without assuming at least some metaphysics — an assumption that Carnap was attempting to avoid. In a sense Carnap was as much a sceptic about primitive necessity as Quine, and it is not clear whether he would have been sympathetic to the subsequent development of modal logic.

One author who is not sympathetic to the development of the possible worlds semantics for modal logic, or at least was not in 1975, is Cocchiarella 1975a [10]. In order to assess Cocchiarella's objections I shall first consider them in the context of the current possible worlds semantics. In so doing we must bear in mind that in Carnap's modal logic the accessibility relation, so important in current modal logic, plays no role, and that we are solely concerned with S5. For S5 we may assume a fixed set W of 'worlds' (whose nature, at least for the purposes of logic, may be anything we please). This set can be denumerably infinite.[n] With respect to this fixed set we have a model which gives each propositional variable a truth value at each world, and then is extended in the usual way to give truth values to every wff in each world. For $\Box$ the rule is simply that $\Box\alpha$ is true at a world w iff α is true at every world. The modal logic S5 is then the logic whose theorems are

---

[n]In the 'canonical model' of S5 it will in fact have the cardinality $2^{\aleph_0}$, but in fact a denumerable infinity is all that is ever required. (The canonical model of S5 splits into a number of completely disconnected sub-models, where $\Box\alpha$ is true at a world iff α is true at all points in that same sub-model. In that way it is not subject to the $\neg\Box p$ problem.)

those wff true at every world in every model. Notice that in this semantics $\Box\alpha$ is true, at any world, provided α itself is true at every world without exception — not just at every 'accessible' world.[°] This is indeed as it should be if we are talking about *logical* necessity and possibility.

Yet Cocchiarella claims that this semantics is not appropriate for the logical modalities, since it does not analyse truth as truth in *all* worlds, but only in truth at an arbitrary non-empty subset of them. Why does he claim this? Firstly Cocchiarella identifies worlds with models. He then notes that the interpretation to $\Box\alpha$ on this semantics always takes place within a particular (S5) model. Indeed this is necessary to solve the $\sim\Box p$ problem. In a particular S5 model nothing prevents a variable p from being assigned the set of all worlds as its truth range, and in such a model $\sim\Box p$ will be false. So if worlds are identified with models, as Cocchiarella supposes it might seem that we are excluding some possible worlds.

The intuition behind an S5 model $\langle W,V \rangle$ is the Wittgensteinian intuition that the meaning of a sentence is the conditions under which it is true.[P] In terms of possible worlds theory W can be a fixed set of worlds, where the assignment function V is unconstrained in terms of the values it is permitted to assign to variables. In particular nothing prevents $V(p)$ being W itself. The reason is that nothing should dictate what a simple sentence should mean, and so nothing should prevent V from assigning to a variable a value which makes it true in all worlds.[q] In specifying validity in S5 we do this in terms of a class of S5 models where there are enough models in the class to represent all possible meanings that a variable might have, each meaning being a set of worlds. When p is given the value true in every world it's not as if we are ignoring worlds in which p is false. We are ignoring the fact that in other S5 models p can be given other values *with*

[°]You can of course obtain S5 by means of an accessibility relation which is required to be an equivalence relation. The connection between that and the present semantics is technically interesting, but need not detain us now.

[P]Wittgenstein 1921 [32], 4.024. "To understand a proposition means to know what is the case if it is true. (One can understand it, therefore, without knowing whether it is true.) It is understood by anyone who understands its constituents.)" Carnap certainly appears to have taken Wittgenstein's remark as endorsing the truth-conditional theory of meaning. (See for instance Carnap 1947 [7] p. 9.)

[q]There is nothing wrong in principle with treating the worlds in an S5 model as themselves PC models, since what an S5 model does in each world is provide a truth value for each variable in that world. For that reason the only relevant feature of a world is as indexing an assignment of truth values. But it is important to see that in this case models enter in two ways. With respect to each model, a propositional variable has a truth value in every possible world (whether that world is a PC model or anything else); but S5 validity must be defined with respect to a class of S5 models.

respect to the very same fixed collection of worlds — i.e., we are ignoring the possibility that p might have been given a different meaning. To construe the difference between asking of the proposition which is the value of p in a given model, whether it is true in all worlds, and asking whether the linguistic item (propositional variable p) could have been given a different value seems to me to be guilty of precisely the kind of use/mention confusion that Quine thought infected modal logic generally.

6. McKinsey's 'syntactical' interpretation

Unlike the mechanism discussed in the last section, QC-validity does not generalise to other modal systems. Carnap 1946 [6] (p. 41) is clear that it leads at least to S5, and uses that fact as part of an argument in defence of S5. In this section we look at the slightly earlier 'syntactic' definition of possibility found in McKinsey 1945 [21]. McKinsey (p. 83, f/n 3) claims that his view is based on that found in Carnap 1937 [5]. We shall first present McKinsey's semantics in a form which mirrors his motivation, and, in the next section, look at his formal account in a way which has closer links with the more recent possible worlds semantics.[r] On p. 83 McKinsey says:

> As the intuitive basis for the syntactical definition of possibility, I take the position that to say a sentence is possible means that there exists a true sentence of the same form. Thus, for example, it would be said that the sentence, 'Lions are indigenous to Alaska,' is possible, because of the fact that the sentence, 'Lions are indigenous to Africa' has the same form and is true.

McKinsey immediately recognises that this needs to be made precise, and does so in a way we shall set out in the next section. But for now we will look at what happens if we change (4) to read

$$\mathcal{M} \models \Box\alpha \text{ iff } \mathcal{M} \models \alpha' \text{ for every wff } \alpha' \text{ which can be}$$
$$\text{obtained from } \alpha \text{ by uniform substitution of PC-wff} \quad (15)$$
$$\text{for the propositional variables of } \alpha.\text{[s]}$$

Notice that this definition does not involve any model other than $\mathcal{M}$. $\mathcal{M}$ is as before simply a PC model. McKinsey is explicit on p. 85 that his

---

[r]'Semantics' here is our word. McKinsey would not describe it so, since it does not refer to any entities outside the language. Thus the importance of his title. (See his comments on p. 94.) In many respects Carnap also, at least in the 1946 paper, seems to have thought of himself as doing syntax rather than semantics.

semantics applies to languages which contain constant sentences. He then says

> The provable formulas of the usual systems are, presumably, to be taken as those which become true when the sentential variables are replaced by arbitrary constant sentences.

In effect McKinsey is adopting a semantics analogous to our QC-validity rather than C-validity, though of course importantly different from QC-validity.[t] For that reason we shall distinguish between $\mathcal{M}$-validity and QM-validity. Say that a wff α is $\mathcal{M}$-valid iff $\mathcal{M} \vDash \alpha$ for every PC model $\mathcal{M}$, when evaluated in accordance with (15), and say that a wff α is QM-valid iff α' is $\mathcal{M}$-valid for every instance α' obtained from α by uniform substitution for its variables. Notice that $\sim \Box p$ is not QM-valid, since you don't have $\mathcal{M} \vDash \sim \Box(\bot \supset \bot)$.

The principal result of McKinsey's paper, theorem 1 on p. 90, is that (15) gives you a system at least as strong as S4, and it is not difficult to see that this is so:

Theorem 6.1. *Every theorem of S4 is true in every PC model when $\Box$ is evaluated in accordance with (15). I.e., all theorems of S4 are QM-valid.*

Proof. First take **K**. Suppose $\mathcal{M} \vDash \Box\alpha$ but $\mathcal{M} \nvDash \Box\beta$. Then there is an instance β' of β such that $\mathcal{M} \nvDash \beta'$. Let α' be α with the same substitutions as obtained β' from β. Then by (15), $\mathcal{M} \vDash \alpha'$. So $\mathcal{M} \nvDash \alpha' \supset \beta'$, and so, again by (15), $\mathcal{M} \nvDash \Box(\alpha \supset \beta)$. For **T** suppose that $\mathcal{M} \vDash \Box\alpha$. Then by (15) we have that $\mathcal{M} \vDash \alpha'$ for every instance α' of α. Since α is a substitution instance of itself we therefore have that $\mathcal{M} \vDash \alpha$. For **4** ($\Box\alpha \supset \Box\Box\alpha$) suppose that $\mathcal{M} \nvDash \Box\Box\alpha$. Then for some instance α' of α we have $\mathcal{M} \nvDash \Box\alpha'$, and so for some instance α'' of α' we have $\mathcal{M} \nvDash \alpha''$. But since α'' is an instance of α' and α' is an instance of α, α'' is already an instance of α, and so $\mathcal{M} \nvDash \Box\alpha$.

Clearly MP is M-validity-preserving, and therefore QM-validity-preserving. For N suppose that for some model $\mathcal{M}$, $\mathcal{M} \nvDash \Box\alpha$. Then there is some instance α' of α such that $\mathcal{M} \nvDash \alpha'$. But α is QM-valid, and α' is an instance of α, and so $\mathcal{M} \nvDash \alpha'$. $\square$

---

[t]Makinson 1966 [22], p. 335n, says, "In the present author's knowledge, the only writer who makes the schematic role of modal formulae explicit in his modelling is McKinsey (*op cit*)"

McKinsey points out on p. 90 that his account will not give S5, and gives a counterexample on p. 91f. It is simpler to shew that we don't have the validity of the Brouwerian axiom.[u] Take the following instance

B $\sim p \supset \Box \sim \Box p$

and let $\mathcal{M} \dashv p$, so that $\mathcal{M} \models \sim p$. Now $\mathcal{M} \dashv \sim \Box(\bot \supset \bot)$, and so $\mathcal{M} \dashv \Box \sim \Box p$.

Although McKinsey was able to shew that his semantics gives at least S4 but does not give S5, he did not attempt to prove that it gives only S4. This is understandable in 1945 when the choice was between S4 and S5. McKinsey does consider a system he calls S4.1, and our next task will be to shew that the system you get if you use (15) in place of (4) includes this system. Because McKinsey's system is not contained in S5 it is customary nowadays to call it S4M, and to refer to its proper axiom as the 'McKinsey' axiom:

M $\Box \Diamond \alpha \supset \Diamond \Box \alpha$

S4M is simply S4 with the addition of **M**. In current terms S4M is characterised by frames in which the relation R is reflexive and transitive, and in addition has the property that every world can see an end world — a world that can see only itself.

Theorem 6.2. *M is QM-valid.*

Proof. It is simplest to consider **M** in the form

M1 $\Diamond(\Diamond \alpha \supset \Box \alpha)$

since this is trivially equivalent to **M**. If **M1** is false in a model $\mathcal{M}$ then for every instance α' of α we have $\mathcal{M} \models \Diamond \alpha'$ but $\mathcal{M} \dashv \Box \alpha'$. So let α' be the instance of α in which every propositional variable is replaced by $\bot$.[v]

---

[u]That the addition of (an equivalent of) B is sufficient to turn S4 into S5 was known to Lewis and Langford 1932 [20] (see p. 498, paragraph (13)) though they do not suggest it as the proper axiom of a separate system, probably because the system T was not considered until Feys 1937 [14]. (The name B is given to what Becker 1930 called the 'Brouwersche Axiom'. For some further remarks on this see Hughes and Cresswell 1996 [17], p. 70, note 5.) So that although McKinsey does not consider this axiom it was certainly available to him, and known to give S5 when added to S4.

[v]McKinsey uses $\sim$ and $\wedge$ as his primitive truth-functional operators. One would then define $\bot$ as some arbitrary contradiction as for instance $p \wedge \sim p$. So there would in fact be many distinct contradictions. However they would all be equivalent and would have exactly the same truth value — false — in any model.

The only instance of α' is α' itself, and so if $\mathcal{M} \models \Diamond\alpha'$ then $\mathcal{M} \models \Box\alpha'$. But $\mathcal{M} \dashv \Box\alpha'$. $\qquad\Box$

Corollary 6.1. *Every theorem of S4M is QM-valid.* $\qquad\Box$

Yet there is a proof in Drake 1962 [13] that McKinsey's semantics gives precisely S4. How can this be? The answer is to be found by looking more carefully at McKinsey's paper.

7. Restricted substitution functions

We can apply what we learned in section 5 to Drake's completeness proofs of McKinsey's semantics. (15) in fact relies on a number of important facts about substitution instances, which McKinsey imposes by special conditions. He treats a *substitution* as simply a function from wff to wff satisfying certain structure-preserving conditions.[w] A substitution function s must satisfy the following conditions which preserve logical form. Adapting McKinsey's definitions from p. 84 supplemented by an addition for the modal operator on p. 85, to a language based on $\bot$, $\supset$ and $\Box$ we have, for every substitution s,

$\quad$ **S$\bot$** $\quad s(\bot) = \bot$
$\quad$ **S$\supset$** $\quad s(\alpha \supset \beta) = (s(\alpha) \supset s(\beta))$
$\quad$ **S$\Box$** $\quad s(\Box\alpha) = \Box s(\alpha)$

S$\bot$–S$\Box$ ensure that if a substitution is defined for the variables then it is defined for all wff.

Lemma 7.1. *If $s(p) = s'(p)$ for every variable p, then $s(\alpha) = s'(\alpha)$ for every wff α.*

Proof. A straightforward induction on wff. $\qquad\Box$

S$\bot$–S$\Box$ are common to all sets of substitutions, but McKinsey imposes a number of other conditions, which distinguish between different modal

---

[w]More accurately McKinsey defines a substitution function from a language with propositional variables to one in which the atomic formulae are ordinary natural language sentences which have no occurrences of logical words like 'and' or 'not'. However, *pace* Quine and McKinsey, there is no bar to stating axioms as schemata and eschewing a rule of uniform substitution for variables, so that we do not need to consider two separate languages.

systems. Where S is a set of substitutions, all satisfying S⊥–S□, we may impose the following

Ref There is some $s \in$ S such that for every wff α, $s(\alpha) = \alpha$ (McKinsey calls this s, s_1).

Trans For any s and s' in S there is some $s'' \in$ S such that for all wff α, $s''(\alpha) = s(s'(\alpha))$.

In place of (15) McKinsey's semantics then reads

$$\mathcal{M} \vDash \Box\alpha \text{ iff } \mathcal{M} \vDash s(\alpha) \text{ for every } s \in \text{S}. \tag{16}$$

Call a wff MkS-valid iff for every set S of substitutions and every model $\mathcal{M}$, $\mathcal{M} \vDash s(\alpha)$ for every $s \in$ S, when $\Box$ is evaluated according to (16). All the axioms of K are MkS-valid and the rules preserve MkS-validity. (Since wff like $\sim \Box p$, though $\mathcal{M}$-valid, are not MkS-valid, we do not need an additional notion of QMkS-validity.) For T, S must satisfy **Ref**, and for S4, S must in addition satisfy **Trans**. McKinsey proves that (16) gives a system at least as strong as S4, and Drake 1962 [13] proves that it is exactly S4. To obtain S5 it is sufficient to add the following condition:

Sym For any $s \in$ S there is some $s' \in$ S such that $s'(s(\alpha)) = \alpha$ for all wff α.

It is easy to see that **B** above is MkS-valid with this restriction, and therefore that it gives S5. An example of an S which satisfies this condition would be one which restricts permitted substitutions so that you could only substitute variables for variables, and that distinct variables must be replaced by distinct variables. It is easy to see that this construal validates **B**:

Theorem 7.2. *B is MkS-valid when substitutions are restricted to distinct variables for distinct variables.*

Proof. Suppose that $\mathcal{M} \dashv \Box \sim \Box\alpha$. Then there will be a wff α' obtainable from α by replacement of distinct $p_1,...,p_n$ in α by $q_1,...,q_n$ (where, for $1 \leqslant i \leqslant n$ and $1 \leqslant j \leqslant n$, p_i may be the same as q_j but $p_i \neq p_j$ and $q_i \neq q_j$ for $i \neq j$) such that $\mathcal{M} \dashv \sim \Box\alpha'$. So $\mathcal{M} \vDash \Box\alpha$ and so $\mathcal{M} \vDash \beta$, where β is any wff obtained from α' by any replacement of distinct variables. One of these replacements will be that of q_i by p_i, and in that case β is α, and so $\mathcal{M} \vDash \alpha$, and therefore $\mathcal{M} \dashv \sim \alpha$. □

Drake's completeness proofs for MkS-semantics are surprisingly messy, and it is not at all clear whether and how they might be extended to obtain a completeness result for QM-semantics, though, as established in section 5, we know that the system contains S4M.

Did McKinsey appreciate the difference between MkS-validity and QM-validity? There is no clear evidence from the paper that he did. The formal development makes use of a set S of substitutions, and McKinsey is well aware that different conditions on S give rise to different modal systems. On the other hand the motivational remarks at the beginning of his paper strongly suggest a desire to *analyse* necessity in a way which might lead to the 'correct' modal system. In addition, the distinction between the languages L_1 and L_2 on p. 85f suggests that McKinsey wants to solve the $\sim \Box p$ problem using QM-validity, rather than MkS-validity. Or perhaps better, McKinsey's paper gives little reason to suppose that he was aware of the importance of the difference.

Acknowledgement: The research involved in this paper was begun between September and December of 2010, when I held a residential Fellowship at the Flemish Institute for Advanced Studies (VLAC) of the Royal Flemish Academy of Belgium for Science and the Arts, on a project with Dr A. A. Rini, called 'Flight from Intension', investigating early attempts to produce a semantics for modal logic without using any intensional entities. The research is also supported by a Marsden grant with E. D. Mares and A. A. Rini on a natural history of necessity. I would like particularly to thank the anonymous referee for this paper whose comments enabled me to correct many errors and infelicities.

References

1. A. Bayart, La correction de la logique modale du premier et second ordre S5, *Logique et Analyse* **1**, 28–44 (1958).
2. A. Bayart, Quasi-adéquation de la logique modale de second ordre S5 et adéquation de la logique modale de premier ordre S5, *Logique et Analyse* **2**, 99–121(1959)
3. O. Becker, , Zur Logik der Modalitäten, *Jahrbuch für Philosophie und Phänomenologische Forschung* **11**, 497-548 (1930).
4. J. P. Burgess, Which modal logic is the right one? *Notre Dame Journal of Formal Logic* **40**, 81–93 (1999).
5. R. Carnap, *The Logical Syntax of Language* (London, Kegan Paul, Trench Truber 1937).
6. R. Carnap, Modalities and quantification, *The Journal of Symbolic Logic* **11**, 33–64 (1946).

74

7. R. Carnap, *Meaning and Necessity* (Chicago, University of Chicago Press, 1947). Second edition 1956, references are to the second edition.
8. R. Carnap, Meaning postulates, *Philosophical Studies* **3**, 65–73 (1952). Reprinted in the second edition of [7, pp. 222–229]. Page references are to this reprint.
9. R. Carnap, Intellectual Autobiography, in *The Philosophy of Rudolf Carnap*, ed. P. A. Schilpp (La Salle, Ill., Open Court 1963), pp. 3–84.
10. N. B. Cocchiarella, On the primary and secondary semantics of logical necessity, *Journal of Philosophical Logic* **4**, 13–27 (1975).
11. N. B. Cocchiarella, Logical atomism, nominalism, and modal logic, *Synthese* **31**, 23–67 (1975).
12. M. J. Cresswell (forthcoming), Arnould Bayart's completeness theorems: translated with an introduction and commentary, *Logique et Analyse* (in an issue in memory of Paul Gochet).
13. F. R. Drake, On McKinsey's syntactical characterizations of systems of modal logic, *The Journal of Symbolic Logic* **27**, 400–406 (1962).
14. R. Feys, Les logiques nouvelles des modalités, *Revue Néoscholastique de Philosophie* **40**, 517–53, and **41**, 217–52 (1937).
15. G. Gottlob, Remarks on a Carnapian Extension of S5, in J. Wolenski, E. Köhler (eds.), *Alfred Tarski and the Vienna Circle* (Kluwer, Dordrecht 1999), 243–259.
16. K. J. J. Hintikka, *Quantifiers in Deontic Logic* (Helsingfors, Societas Scientiarum Fennica 1957). Commentationes Humanarum Litterarum 23.4.
17. G. E. Hughes and M .J. Cresswell, *A New Introduction to Modal Logic* (London, Routledge 1996).
18. S. G. Kanger, *Provability in Logic*, (Stockholm, Almqvist & Wiksell 1957).
19. S. A. Kripke, A completeness theorem in modal logic, *The Journal of Symbolic Logic*, **24**, 1–14 (1959).
20. C. I. Lewis and C. H. Langford, *Symbolic Logic* (New York, Dover Publications 1932).
21. J. C. C. McKinsey, On the syntactical construction of systems of modal logic, *The Journal of Symbolic Logic* **10**, 83–94 (1945).
22. D. Makinson, How Meaningful are Modal Operators? *Australasian Journal of Philosophy* **44**, 331–337 (1966).
23. R. M. Montague, Logical necessity, physical necessity, ethics and quantifiers. *Inquiry* **4**, 259–269 (1960). Reprinted in *Formal Philosophy*, (New Haven, Yale University Press 1974), 71–83.
24. A. N. Prior, *Time and modality*, (Oxford, Clarendon Press 1957).
25. W. V. O. Quine, Ontological remarks on the propositional calculus, *Mind* **433**, 473–476 (1934).
26. W. V. O. Quine, Reference and modality. In *From a Logical Point of View* (Cambridge, Mass., Harvard University Press, second edition 1961), 139–59.
27. W. V. O. Quine, Three grades of modal involvement, in *The Ways of Paradox* (Cambridge Mass., Harvard University Press 1976), 158–176.
28. G. Schurz, Carnap's modal logic. In W. Stelzner and M. Stockler (eds.), *Zwischen traditioneller und moderner Logik.* (Paderborn, Mentis 2001), 365–

380.

29. B. Skyrms, An immaculate conception of modality, *The Journal of Philosophy* **75**, 368–87 (1978).

30. S. K. Thomason, A new representation of S5, *Notre Dame Journal of Formal Logic* **14**, 281–284 (1973).

31. M. Wajsberg, Ein erweiteter Klassenkalkül, *Monatshefte für Mathematik und Physik* **40**, 113–26 (1933).

32. L. Wittgenstein, *Tractatus Logico–Philosophicus* (London, Routledge and Kegan Paul, 2nd printing 1963). Translated by D. F. Pears and B. F. McGuinness.

33. G. H. von Wright, *An Essay in Modal Logic*, (Amsterdam, North Holland Publishing Co. 1975).

Limits to joining with generics and randoms

Adam R. Day

Department of Mathematics
University of California, Berkeley
Berkeley, California 94720 U.S.A.
E-mail: adam.day@math.berkeley.edu

Damir D. Dzhafarov

Department of Mathematics
University of California, Berkeley
Berkeley, California 94720 U.S.A.
E-mail: damir@math.berkeley.edu

Posner and Robinson [4] proved that if $S \subseteq \omega$ is non-computable, then there exists a $G \subseteq \omega$ such that $S \oplus G \geq_T G'$. Shore and Slaman [7] extended this result to all $n \in \omega$, by showing that if $S \not\leq_T \emptyset^{(n-1)}$ then there exists a G such that $S \oplus G \geq_T G^{(n)}$. Their argument employs Kumabe-Slaman forcing, and so the set they obtain, unlike that of the Posner-Robinson theorem, is not generic for Cohen forcing in any way. We answer the question of whether this is a necessary complication by showing that for all $n \geq 1$, the set G of the Shore-Slaman theorem cannot be chosen to be even weakly 2-generic. Our result applies to several other effective forcing notions commonly used in computability theory, and we also prove that the set G cannot be chosen to be 2-random.

1. Introduction

Our starting point is the following well-known theorem of computability theory.

Theorem 1.1 (Posner and Robinson [4], Theorem 1). *For all $S \subseteq \omega$, if S is non-computable there exists $G \subseteq \omega$ such that $S \oplus G, S \oplus \emptyset' \geq_T G'$. (In particular, when $S \leq_T \emptyset'$ then $S \oplus G \equiv_T G' \equiv_T \emptyset'$.)*

In order to generalize the Posner-Robinson theorem to higher numbers of jumps, we need an additional condition on S. If we want $S \oplus G \geq_T G^{(n)}$, then we need S to be not computable in $\emptyset^{(n-1)}$, since otherwise $S \oplus G \leq_T \emptyset^{(n-1)} \oplus G \leq_T G^{(n-1)}$. In connection with their work on the definability of the jump, Shore and Slaman proved this is the only condition needed for the generalization.

Theorem 1.2 (Shore and Slaman [7], Theorem 2.1). *For all $n \geq 1$ and all $S \subseteq \omega$, if $S \nleq_T \emptyset^{(n-1)}$ there exists $G \subseteq \omega$ such that $S \oplus G, S \oplus \emptyset^{(n)} \geq_T G^{(n)}$.*

The proof of the Shore-Slaman theorem is not a simple generalization of Posner and Robinson's proof. While both theorems are proved using forcing constructions, they differ in the underlying forcing notion used. Posner and Robinson use Cohen forcing, and show that the set G in Theorem 1.1 can actually be chosen to be Cohen 1-generic. Shore and Slaman use a considerably more intricate notion of forcing, due to Kumabe and Slaman (discussed further in Section 3). This raises the question of whether Theorem 1.2 *can* be proved using Cohen forcing. More generally, one can ask whether Theorem 1.2 can be proved using some other commonly used forcing notion, such as Jockusch-Soare forcing or Mathias forcing.

In this article, we give a negative answer to the above questions, in the following sense. We prove a general result that applies to most of the forcing notions $\mathbb{P}$ used in computability theory, and which roughly states that if n is sufficiently large and $G \subseteq \omega$ is sufficiently generic for $\mathbb{P}$, then G does not satisfy Theorem 1.2. For Cohen forcing, $n = 1$ and weak 2-genericity suffice. We then prove a similar result for randomness, showing that for each $n \geq 2$, the set G in Theorem 1.2 cannot be chosen to be n-random.

We refer the reader to Soare [8] and Downey and Hirschfeldt [3] for background on computability theory and algorithmic randomness, respectively. A brief account of forcing in arithmetic, as we shall use it, is included in Section 3 below.

2. A non-joining theorem for generics

The purpose of this section is to prove the following theorem. We shall prove in the next section a theorem applying to forcing notions in general, of which this will be the special case for Cohen forcing. We present this argument separately in order to make the basic idea easier to understand.

Theorem 2.1. *There exists a $\emptyset'$-computable perfect tree $T \subseteq 2^{<\omega}$ such that for all $S \in [T]$ and all $G \subseteq \omega$, if G is weakly 2-generic then $S \oplus G \ngeq_T \emptyset'$.*

The theorem establishes that for all $n \geq 2$, the set G in Theorem 1.2 cannot be chosen to be n-generic.

Corollary 2.1. *For all $n \geq 2$, there exists an $S \not\leq_T \emptyset^{(n-1)}$ such that for all $G \subseteq \omega$, if G is weakly 2-generic then $S \oplus G \not\geq_T \emptyset'$. In particular, $S \oplus G \not\geq_T G^{(n)}$.*

Proof. Let $T \subseteq 2^{<\omega}$ be the tree obtained from Theorem 2.1. As T is perfect, we may choose an $S \in [T]$ such that $S \not\leq_T \emptyset^{(n-1)}$. Then for every weakly 2-generic set G, and so certainly for every n-generic G, we have $S \oplus G \not\geq_T \emptyset'$. In particular, $S \oplus G \not\geq_T G^{(n)}$. $\square$

We proceed with the proof of the theorem.

Proof of Theorem 2.1. Computably in $\emptyset'$, we construct the tree T, an auxiliary set $B \subseteq \omega$, and a sequence $\{D_e\}_{e \in \omega}$ of dense subsets of $2^{<\omega}$. Our objective is to meet the following requirements for all $e \in \omega$.

$\mathcal{R}_e$: If $S \in [T]$ and $G \subseteq \omega$ meets D_e, then $\Phi_e(S \oplus G) \neq B$.

Of course, every weakly 2-generic set meets each set D_e. And as B will be $\emptyset'$-computable, meeting these requirements suffices.

Construction. We obtain T, B, and each D_e as $\bigcup_s T_s$, $\bigcup_s B_s$, and $\bigcup_s D_{e,s}$, where T_s, B_s, and $D_{e,s}$ denote the portions of each of these sets built by the beginning of stage s. Initially, let $T_0 = \{\lambda\}$, where λ is the empty string, and let $B_0 = D_{e,0} = \emptyset$ for all e.

At stage $s = \langle e, t \rangle$, assume T_s, B_s, and $D_{e,s}$ have been defined. Let $\langle \sigma_i : i < n \rangle$ enumerate all strings of the form τb, where τ is a maximal string in T_s and $b \in \{0, 1\}$. This step adds a split above each element of T_s. Let $\langle \tau_j : j < m \rangle$ enumerate all binary strings smaller than s. For all $j < m$, we will enumerate an extension τ_j^* of τ_j into D_e, thus ensuring that D_e is dense.

For all $i < n$ and $j < m$, we will choose a unique number $x_{i,j} > s$ not in B_s, and define sequences

$$\sigma_i \preceq \sigma_{i,0} \preceq \cdots \preceq \sigma_{i,m-1}$$

and

$$\tau_j \preceq \tau_{j,0} \preceq \cdots \preceq \tau_{j,n-1}$$

with $|\sigma_{i,j}| = |\tau_{j,i}|$, such that one of the following applies:

(1) there is a $b \in \{0, 1\}$ such that $\Phi_e^{\sigma_{i,j} \oplus \tau_{j,i}}(x_{i,j}) \downarrow = b$;
(2) $\Phi_e^{\sigma \oplus \tau}(x_{i,j}) \uparrow$ for all $\sigma \succeq \sigma_{i,j}$ and $\tau \succeq \tau_{j,i}$ with $|\sigma| = |\tau|$.

We define $\sigma_{i,j}$ and $\tau_{j,i}$ simultaneously. For convenience, let $\sigma_{i,-1} = \sigma_i$ and $\tau_{j,-1} = \tau_j$. Let (i,j) be the lexicographically least pair in $\omega \times \omega$ such that $\sigma_{i,j}$ and $\tau_{j,i}$ are not defined (this implies that $\sigma_{i,j-1}$ and $\tau_{j,i-1}$ are already defined). Using $\emptyset'$, we can find a $\sigma \succeq \sigma_{i,j-1}$ and a $\tau \succeq \tau_{j,i-1}$ that satisfy (1) or (2) above, and we let $\sigma_{i,j}$ and $\tau_{j,i}$ be the least such σ and τ, respectively.

To complete the construction, add $x_{i,j}$ to B_{s+1} for all i, j such that case (1) above applies with $b = 0$. Then, for all $j < m$, let τ_j^* be the least extension of $\tau_{j,n-1}$ greater than s, and add τ_j^* to $D_{e,s+1}$. Finally, for all $i < n$, let σ_i^* be an extension of $\sigma_{i,m-1}$ of length $|\tau_j^*|$, and add all initial segments of σ_i^* to T_{s+1}.

Verification. The construction is $\emptyset'$-computable, and so, since at each stage s, only elements greater than s are added to B and D_e, it follows that B and the sequence $\{D_e\}_{e \in \omega}$ are computable in A. It is clear that the D_e are dense: given any $\tau \in 2^{<\omega}$, an extension of τ is added to D_e at each sufficiently large stage $s = \langle e, t \rangle$.

Now fix e, let G be any weakly 2-generic set, and let S be any element of $[T]$. Then we may choose a $\tau \prec G$ in D_e. Let s be the least stage such that $\tau \in D_{e,s}$, and let σ be a maximal initial segment of S in T_s, so that $|\sigma| = |\tau|$. By construction, there is an x such that one of the following cases applies:

(1) $\Phi_e^{\sigma \oplus \tau}(x) \downarrow = 1 - B(x)$;

(2) $\Phi_e^{\sigma^* \oplus \tau^*}(x) \uparrow$ for all $\sigma^* \succeq \sigma$ and $\tau^* \succeq \tau$.

Now if case (1) holds, then $\Phi_e^{S \oplus G}(x) \downarrow = 1 - B(x)$ since $\sigma \preceq S$ and $\tau \preceq G$. And if case (2) holds, then it cannot be that $\Phi_e^{S \oplus G}(x)$ converges, else some initial segments of S and G would witness a contradiction.

We conclude that $\Phi_e(S \oplus G) \neq B$, and hence that requirement $\mathcal{R}_e$ is satisfied. This completes the verification and the proof. $\square$

The proof given establishes that $S \oplus G \not\geq_T \emptyset'$ by constructing an auxiliary set B below $\emptyset'$ and showing that $S \oplus G \not\geq_T B$. In fact this proof can be easily modified to show that for any non-computable set $A \leq_T \emptyset'$, there is a perfect tree T computable in $\emptyset'$ such that for all $S \in [T]$ and all $G \subseteq \omega$, if G is weakly 2-generic then $S \oplus G \not\geq_T A$. However, this argument does not generalize directly to other forcing notions.

3. Extensions to other forcing notions

We now extend Theorem 2.1 to other forcing notions. We assume familiarity with the basics of forcing in arithmetic, but as these formulations are often dependent on the nuances of the definitions, we include a brief review of some of the particulars of our treatment. Our approach is close to that of Shore [6, Chapter 3], with some variations. The goal is to define forcing in a way that is general enough to cover the forcing notions most commonly used in computability theory.

Definition 3.1. A *notion of forcing* is a triple $\mathbb{P} = (P, \leq, V)$, where:

(1) P is an infinite subset of ω;
(2) $\leq$ is a partial ordering of P;
(3) V is a monotone map from $(P, \leq)$ to $(2^{<\omega}, \succeq)$ such that for each $n \in \omega$, the set of $p \in P$ with $|V(p)| \geq n$ is dense.

As is customary, we refer to the elements of P as the *conditions* of $\mathbb{P}$, and say $q \in P$ *extends* $p \in P$ if $q \leq p$. We call the map V a *valuation*. For each $F \subseteq P$, we let $|V(F)|$ denote $\sup_{p \in F} |V(p)|$.

Definition 3.2. Let $\mathbb{P} = (P, \leq, V)$ be a forcing notion, and F a filter on $(P, \leq)$. If $\mathcal{C}$ is a set of subsets of P, then F is $\mathcal{C}$-*generic for* $\mathbb{P}$ if

(1) $|V(F)| = \infty$;
(2) for every $C \in \mathcal{C}$, either $F \cap C \neq \emptyset$ or $F \cap \{p \in P : (\forall q \leq p)[q \notin C]\} \neq \emptyset$.

Condition (1) above ensures that if F is any generic filter then $V(F)$ uniquely determines a subset of ω. If F is $\mathcal{C}$-generic for $\mathbb{P}$ and $G = V(F)$, then we also say that G is $\mathcal{C}$-*generic for* $\mathbb{P}$.

The following definition is standard in effective applications of forcing.

Definition 3.3. Let $A \subseteq \omega$ be given, and let $\mathbb{P} = (P, \leq, V)$ be a notion of forcing.

(1) $\mathbb{P}$ is A-*computable* if P, $\leq$, and V are A-computable.
(2) A set $D \subseteq P$ is A-*effectively dense* if there is an A-computable function that takes each $p \in P$ to some $q \leq p$ in D.

We work in the usual forcing language, consisting of the language of second-order arithmetic, augmented by a new set constant $\dot{G}$ intended to denote the generic real. The (strong) forcing relation $\Vdash_{\mathbb{P}}$ is defined recursively in the standard way, starting by putting $p \Vdash_{\mathbb{P}} \varphi$ for $p \in P$ if φ is a true atomic sentence of first-order arithmetic, or if φ is $\dot{n} \in \dot{G}$ (respectively, $\dot{n} \notin \dot{G}$) for some $n < |V(p)|$ such that $V(p)(n) = 1$ (respectively, $V(p)(n) = 0$). For conjunctions we write $p \Vdash_{\mathbb{P}} \varphi \wedge \psi$ if $p \Vdash_{\mathbb{P}} \varphi$ and $p \Vdash_{\mathbb{P}} \psi$. For existential formulas we write $p \Vdash_{\mathbb{P}} \exists x \varphi(x)$ if $p \Vdash_{\mathbb{P}} \varphi(\dot{n})$ for some $n \in \omega$. Finally for negations we write $p \Vdash_{\mathbb{P}} \neg\varphi$ if for all $q \leq p$ it is not true that $q \Vdash_{\mathbb{P}} \varphi$.

It is easy to see that if $G = V(F)$ for some filter F on $(P, \leq)$ with $|V(F)| = \infty$, and if $p \Vdash \varphi(\dot{G})$ for some $p \in F$ and some Σ_1^0 sentence φ of the forcing language, then $\varphi(G)$ holds. (Of course, this is just a consequence of the more powerful fact that forcing implies truth for any sufficiently generic real, but we shall not need that here.)

We introduce the following concept.

Definition 3.4. Let $A \subseteq \omega$ be given, let $\mathbb{P} = (P, \leq, V)$ be a notion of forcing, and let $\{\varphi_i\}_{i \in \omega}$ be an enumeration of all Σ_1^0 formulas in the forcing language. We say that $\mathbb{P}$ is 1-*decidable in* A if $\mathbb{P}$ is A-computable, and there is an A-computable function $f : P \times \omega \to P \times \{0,1\}$ such that if $f(p,i) = (q,t)$ then

(1) $q \leq p$;
(2) if $t = 1$ then $q \Vdash_{\mathbb{P}} \varphi_i$;
(3) if $t = 0$ and $q \Vdash_{\mathbb{P}} \neg\varphi_i$.

Note that if $A \geq_T \emptyset'$, then being 1-decidable in A reduces to the set of conditions in $\mathbb{P}$ that decide φ_i being A-effectively dense, uniformly in i. This is because p forcing a Σ_1^0 fact corresponds a Σ_1^0 fact holding of $V(p)$, and $V(p)$ is computable in A.

Let $\mathbb{C}$ denote Cohen forcing. The *product notion* $\mathbb{C} \times \mathbb{P}$ is the notion of forcing whose elements are pairs $(\sigma, p) \in 2^{<\omega} \times P$ such that $|\sigma| = |V(p)|$, with extension defined componentwise. The induced valuation, $V_\times$, on the product, is defined by

$$V_\times(\sigma, p) = \sigma \oplus V(p).$$

Cohen forcing is easily seen to be 1-decidable in $\emptyset'$. We wish also to calibrate the level of 1-decidability for some other common notions of forcing, to which to apply Theorem 3.1 below. These include the following:

(1) Jockusch-Soare forcing;
(2) Sacks forcing with computable perfect binary trees;
(3) Mathias forcing with conditions restricted to pairs (D, E) such that E is an infinite computable set.

Each of these notions can be formalized in a manner fitting Definition 3.1 above. For (1), conditions are pairs (T, n) such that $n \in \omega$ and T is (an index for) an infinite subtree of $2^{<\omega}$ containing a unique string of length n, and extension is defined by having $(T', n') \leq (T, n)$ if $n' \geq n$ and $T' \subseteq T$. The valuation map takes (T, n) to $T \upharpoonright n$, i.e., to the unique string in T of length n. For (2), conditions are (indices for) computable perfect subtrees of $2^{<\omega}$, ordered by inclusion, with the valuation taking each such tree to its longest string before a splitting. Finally, for (3), conditions are pairs (D, E) such that D is (the canonical index of) a finite set, E is (a Δ_1^0 index for the characteristic function of) an infinite computable set, and $\max D < \min E$; extensions is defined by having $(D', E') \leq (D, E)$ if $D \subseteq D' \subseteq D \cup E$ and $E' \subseteq E$; and the valuation takes (D, E) to the string $\sigma \in 2^{<\omega}$ of length $\min E$ such that $\sigma(i) = 1$ if and only if $i \in D$. (For (3), see also Section 2 of [1], where alternatives to, and simplifications of, the coding presented here are discussed.)

In addition, we wish to consider Kumabe-Slaman forcing, which is especially important in the present discussion because of its use in proving Theorem 1.2. As this forcing may be less familiar, we include its definition. (See [7, Definition 2.5], for complete details.) Adopting the terminology of Shore and Slaman, we call a subset Φ of $2^{<\omega} \times \omega \times \omega$ a *use-monotone Turing functional* if the following conditions hold:

(1) if $\langle \sigma, x, y \rangle, \langle \sigma', x, y' \rangle \in \Phi$ and σ and σ' are comparable then $y = y'$ (the idea being that a triple $\langle \sigma, x, y \rangle$ in Φ codes that $\Phi^\sigma(x) \downarrow = y$);
(2) if $\langle \sigma, x, y \rangle$ and $\langle \sigma', x', y' \rangle$ belong to Φ and $\sigma' \prec \sigma$ then $x' < x$;
(3) if $\langle \sigma, x, y \rangle \in \Phi$ and $x' < x$, then $\langle \sigma', x', y' \rangle \in \Phi$ for some y' and $\sigma' \preceq \sigma$.

Note that under this definition, a use-monotone Turing functional is not necessarily computable and is perhaps best thought of as a Turing functional relative to an oracle. The generic real produced by Kumabe-Slaman forcing is a use-monotone Turing functional. The partial order used in this notion of forcing is $(P, \leq)$ where

(1) the elements of P are pairs $(\Phi, \vec{X})$, such that Φ is a finite $\{0, 1\}$-valued use-monotone Turing functional, and $\vec{X}$ is a finite set of subsets of ω;
(2) $(\Psi, \vec{Y}) \leq (\Phi, \vec{X})$ if

(a) $\Phi \subseteq \Psi$, and if $\langle \sigma, x, y \rangle \in \Psi - \Phi$ and $\langle \sigma', x', y' \rangle \in \Phi$ then $|\sigma| > |\sigma'|$;
(b) $\vec{X} \subseteq \vec{Y}$, and if σ is an initial segment of some $X \in \vec{X}$ and $\langle \sigma, x, y \rangle \in \Psi$ then $\langle \sigma, x, y \rangle \in \Phi$.

We can code the finite functionals Φ here simply by their canonical indices, but of course $\vec{X}$ can consist of arbitrary subsets of ω, and so cannot necessarily be coded by a number. However, our interest will be in a restriction of this forcing that can be so coded, namely, when the conditions are pairs $(\Phi, \vec{X})$ such that $\vec{X}$ consists of $\emptyset^{(n)}$-computable sets. (In this case, $\vec{X}$ can be coded simply by a Δ_{n+1}^0 index for the characteristic function of $\vec{X}$.) The valuation map then takes a condition $(\Phi, \vec{X})$ to the characteristic function of Φ truncated at its largest element.

We have the following bounds on the complexity of each of these notions.

Proposition 3.1.

(1) Cohen forcing is 1-decidable in $\emptyset'$.
(2) Jockusch-Soare forcing is 1-decidable in $\emptyset'$.
(3) Sacks forcing with computable perfect trees is 1-decidable in $\emptyset''$.
(4) Mathias forcing with computable sets is 1-decidable in $\emptyset''$.
(5) Kumabe-Slaman forcing with $\emptyset^{(n)}$-computable sets is 1-decidable in $\emptyset^{(n+2)}$.

Proof. Part (1) is clear. Part (2) is essentially by the proof of the low basis theorem, as forcing the jump involves asking whether a particular Π_1^0 subclass is non-empty. Part (3) is similar, but we gain an extra quantifier as the set of conditions is only computable in $\emptyset''$. Part (4) follows by Lemma 3.3 of [1], and (5) is implicit in Lemmas 2.10 and 2.11 of [7]. $\square$

Note that for any of the notions $\mathbb{P}$ in the previous proposition, the same bounds on 1-decidability apply also to $\mathbb{C} \times \mathbb{P}$.

We can now state and prove our extension of Theorem 2.1.

Theorem 3.1. *Let $A \subseteq \omega$ be given, and let $\mathbb{P} = (P, \leq, V)$ be a notion of forcing such that $\mathbb{C} \times \mathbb{P}$ is 1-decidable in A. There exists an A-computable perfect tree $T \subseteq 2^{<\omega}$, and an A-computable class $\mathcal{C}$ of dense subsets of P, such that for all $S \in [T]$ and all $G \subseteq \omega$, if G is $\mathcal{C}$-generic then $S \oplus G \not\geq_T A$.*

Proof. The proof is similar to that of Theorem 2.1, so we just highlight the differences. The construction is now computable in A, the sets D_e are subsets of P, and we let $\mathcal{C} = \{D_e\}_{e \in \omega}$. The requirements take the following form.

$\mathcal{R}_e$: If $S \in [T]$ and F is a filter on $(P, \leq)$ with $|V(F)| = \infty$ that meets D_e, then $\Phi_e(S \oplus V(F)) \neq B$.

The construction is modified in that at stage $s = \langle e, t \rangle$ we fix an enumeration $\langle p_j : j < m \rangle$ of all conditions smaller than s, and instead of building two sequences of strings, $\sigma_i \preceq \sigma_{i,0} \preceq \cdots \preceq \sigma_{i,m-1}$ and $\tau_j \preceq \tau_{j,0} \preceq \cdots \preceq \tau_{j,n-1}$, the latter is replaced by a sequence $p_j \geq p_{j,0} \geq \cdots \geq p_{j,n-1}$ of conditions. For all i, j, we ensure that $|\sigma_{i,j}| = |V(p_{j,i})|$ and that one of the following applies:

(1) $(\sigma_{i,j}, p_{i,j}) \Vdash_{\mathbb{C} \times \mathbb{P}} \Phi_e(\dot{G})(x_{i,j}) \downarrow = 1$;
(2) $(\sigma_{i,j}, p_{i,j}) \Vdash_{\mathbb{C} \times \mathbb{P}} \neg(\Phi_e(\dot{G})(x_{i,j}) \downarrow = 1)$.

This can be done A-computably, using the fact that $\mathbb{C} \times \mathbb{P}$ is 1-decidable in A to decide the Σ_1^0 sentence $\Phi_e(\dot{G})(x_{i,j}) \downarrow = 1$. The definitions of T_{s+1}, and $D_{e,s+1}$, are then entirely analogous to their definitions in the proof of Theorem 2.1, and B_{s+1} is obtained from B_s by adding $x_{i,j}$ for all i, j for which case (2) applies.

For the verification, fix e, let $G \subseteq \omega$ be $\mathcal{C}$-generic, and let S be any element of $[T]$. Let F be a filter on $(P, \leq)$ such that $G = V(F)$, so that $|V(F)| = \infty$ and F meets D_e. Choose $p \in F \cap D_e$, let s be the least stage such that $p \in D_{e,s}$, and let σ be a maximal initial segment of S in T_s, so that $|\sigma| = |V(p)|$. By construction, there is an x such that one of the following cases applies:

(1) $B(x) = 0$ and $(\sigma, p) \Vdash_{\mathbb{C} \times \mathbb{P}} \Phi_e(\dot{G})(x) \downarrow = 1$;
(2) $B(x) = 1$ and $(\sigma, p) \Vdash_{\mathbb{C} \times \mathbb{P}} \neg(\Phi_e(\dot{G})(x) \downarrow = 1)$.

If case (1) holds, then by the general definition of forcing, $\Phi_e(V_\times(\sigma, p))(x) \downarrow = 1$, and hence $\Phi_e(\sigma \oplus V(p))(x) \downarrow = 1 \neq B(x)$. It follows that $\Phi_e(S \oplus V(F))(x) \downarrow = 1 - B(x)$ since $\sigma \prec S$ and $V(p) \prec V(F)$. Now suppose case (2) holds. If $\Phi_e(S \oplus V(F))(x) \downarrow = 1$ then there exists some $n \geq |\sigma| = |V(p)|$ such that $\Phi_e(S \upharpoonright n \oplus V(F) \upharpoonright n)(x) \downarrow = 1$. As $|V(F)| = \infty$, F necessarily contains some $q \in P$ with $|V(q)| \geq n$, and as F is a filter, we may assume $q \leq p$. Let τ be any initial segment of S of length $|V(q)|$. Then (τ, q) is an extension of (σ, p) in $\mathbb{C} \times \mathbb{P}$ that forces $\Phi_e(\dot{G})(x) \downarrow = 1$, which is impossible. Hence, it must be that $\Phi_e(S \oplus V(F))(x) \uparrow$ or $\Phi_e(S \oplus V(F))(x) \downarrow = 0$. $\square$

Now suppose $\mathbb{P}$ is one of the notions of forcing listed above. We obtain the following consequences.

Corollary 3.1. *Let $\mathbb{P} = (P, \leq, V)$ be any of the notions of forcing discussed above, and let $A \subseteq \omega$ be such that $\mathbb{P}$ is 1-decidable in A, as calibrated in Proposition 3.1. (Hence $\mathbb{C} \times \mathbb{P}$ is 1-decidable in A, as remarked above.) There exists an A-computable class $\mathcal{C}$ of dense subsets of P, and for each*

*non-computable set $B \subseteq \omega$ an $S \not\leq_T B$, such that for all $G \subseteq \omega$, if G is
$\mathcal{C}$-generic for $\mathbb{P}$ then $S \oplus G \not\geq_T A$.*

Proof. The proof is similar to that of Corollary 2.1, which is just the special case when $\mathbb{P} = \mathbb{C}$, $A = \emptyset'$, and $B = \emptyset^{(n-1)}$. Theorem 3.1 produces the class $\mathcal{C}$, as well as a perfect tree T through which we choose an infinite path $S \not\leq_T B$. By construction, $S \oplus G \not\geq_T A$ for all $\mathcal{C}$-generic G. □

We conclude this section by noting an interesting consequence of the corollary to Kumabe-Slaman forcing. As noted above, general Kumabe-Slaman forcing, i.e., forcing with pairs $(\Phi, \vec{X})$, where $\vec{X}$ ranges over arbitrary finite collections of sets, cannot be readily coded as a subset of ω. One could nonetheless ask whether some such coding is possible, and in particular, whether the notion can be made 1-decidable in some arithmetical set A. The corollary implies that this cannot be so. Indeed, the proof of Theorem 1.2 in [7] for n only uses Kumabe-Slaman with $\emptyset^{(n)}$-computable sets, and produces a set $G = V(F)$ for a filter F that decides every Σ_n^0 sentence. It is not difficult to see that if $n \geq m + 1$ then any such G is generic for every $\emptyset^{(m)}$-computable class of subsets of conditions. So, if Kumabe-Slaman forcing were 1-decidable in $\emptyset^{(m)}$, we could apply the corollary to find an $S \not\leq_T \emptyset^{(n-1)}$ such that for all G as above, $S \oplus G \not\geq_T \emptyset^{(m)}$, and hence certainly $S \oplus G \not\geq_T G^{(n)}$, giving a contradiction.

4. A non-joining theorem for randoms

We now ask whether the set G of the Shore-Slaman theorem, Theorem 1.2, can be chosen to be n-random. Randomness and genericity are each a notion of typicality, so a negative answer would suitably complement Corollary 3.1. In this section, we show that the answer is indeed negative.

Although randomness and genericity are orthogonal notions in most respects, it is possible to think of randomness as a notion of genericity in a limited way. Namely, by a result of Kautz, the weakly n-random sets can be characterized in terms of genericity for Solovay forcing with Π_n^0 classes of positive measure (see [3, Section 7.2.5]). While it may at first seem possible to obtain the result for n-randomness as just another application of Theorem 3.1, it is worth pointing out that this is not the case. The reason is that forcing with Π_n^0 classes of positive measure is only 1-decidable in $\emptyset^{(n+2)}$, which means that Theorem 3.1 requires a higher level of genericity than is used in Kautz's characterization. This level of genericity no longer corresponds to n-randomness.

Our goal, then, is to give a direct argument for the following result, which is analogous to Theorem 2.1.

Theorem 4.1. *For any non-computable set $A \leq_T \emptyset'$, there exists a $\emptyset'$-computable perfect tree $T \subseteq 2^{<\omega}$ such that for all $S \in [T]$ and all $R \subseteq \omega$, if R is 2-random then $S \oplus R \not\geq_T A$.*

The following corollary is then obtained exactly as Corollary 2.1 was above.

Corollary 4.1. *For all $n \geq 2$, there exists an $S \not\leq_T \emptyset^{(n-1)}$ such that for all $G \subseteq \omega$, if G is 2-random then $S \oplus G \not\geq_T \emptyset'$. In particular, $S \oplus G \not\geq_T G^{(n)}$.*

To prove the theorem, we need the following lemma. Fix a Turing reduction Φ, and define $m(\tau, \rho) = \mu(\{X : \Phi(\tau \oplus X)[2 \cdot |\tau|] \succeq \rho\})$ for all $\tau, \rho \in 2^{<\omega}$, where μ is the uniform measure on Cantor space.

Lemma 4.1. *Let $A \subseteq \omega$ be non-computable. For all rational $q > 0$ and all $\sigma \in 2^{<\omega}$, there is an $n \in \omega$ such that the set of τ with $m(\tau, A \restriction n) \geq q$ is not dense above σ.*

Proof. Assume not, and let q and σ witness this fact. Let $G \subseteq \omega$ be any set which is 1-generic relative to A and extends σ. Then G meets $\{\tau \succeq \sigma : m(\tau, A \restriction n) \geq q\}$ for all n, and hence if we let $E_n = \{X \subseteq \omega : \Phi(G \oplus X) \succeq A \restriction n\}$, then $\mu(E_n) \geq q$ for all n. Let $E = \bigcap_n E_n$. It follows that $\mu(E) \geq q$, and $\Phi(G \oplus X) = A$ for all $X \in E$. By a theorem of Sacks [5], this implies that $G \geq_T A$, which is impossible since anything 1-generic relative to a non-computable set cannot compute that set. $\square$

Proof of Theorem 4.1. Because the 2-random sets are closed under the addition of prefixes, we need only consider a single functional Φ such that $\Phi(X \oplus 0^e 1Y) = \Phi_e(X \oplus Y)$ for all sets X and Y and all $e \in \omega$. Assume the function m from above is defined with respect to this Φ.

Computably in $\emptyset'$, we construct the tree T by stages, along with a sequence $\langle n_i : i \in \omega \rangle$ of numbers. Our test will be defined by

$$U_i = \{\tau \in 2^{<\omega} : (\exists \sigma \in T)[\Phi(\sigma \oplus \tau) \succeq A \restriction n_i]\}.$$

Clearly, then, $\{U_i\}_{i \in \omega}$ will be a $\emptyset'$-computable sequence of open sets, and if $S \in [T]$ and $R \subseteq \omega$ is such that $\Phi(S \oplus R) = A$, then some initial segment of R will belong to every U_i. Thus, to verify the construction below, it will suffice to ensure that $\mu(U_i) \leq 2^{-i}$ for all i.

Construction. Let T_s denote the approximation to T at stage s, with $T_0 = \{\lambda\}$. At stage s, we define T_{s+1} and $n_s \in \omega$. Let $\langle \sigma_i : i < m \rangle$ be a listing of the maximal strings in T_s. We need to add a split above each of these strings, but before can do so, we will make a series of extensions to each σ_i.

First, we initialize level s of our test. To do this we find $n_s \in \omega$, and for all i, a string σ_i^* extending σ_i such that $m(\sigma, A \upharpoonright n_s) \leq 2^{-2(s+1)}$ for all $\sigma \succeq \sigma_i^*$. Lemma 4.1 establishes that there exists an n_s and sequence $\langle \sigma_i^* : i < m \rangle$ meeting this condition, and an instance can be found using $\emptyset'$.

Now for any path extending σ_i^* we have bounded the size of the strings that will be enumerated into U_s due to this path. However, there will be continuum many paths above σ_i^* in T so we need to do better than this. For all $t \leq s$, we extend each σ_i^* in order to force a large number of elements into U_t. We determine a sequence of extensions

$$\sigma_i^* \preceq \sigma_{i,0} \preceq \ldots \preceq \sigma_{i,s}$$

inductively as follows. Suppose we have defined $\sigma_{i,t-1}$ for some $t \leq s$, where for convenience we write $\sigma_{i,-1} = \sigma_i^*$. Then we find $\sigma_{i,t} \succeq \sigma_{i,t-1}$ such that

$$\sup_{\sigma \succeq \sigma_{i,t-1}} m(\sigma, A \upharpoonright n_t) - m(\sigma_{i,t}, A \upharpoonright n_t) \leq 2^{-2(s+2)}. \tag{1}$$

We define T_{s+1} to be downward closure of $\{\sigma_{i,s} b : i < m \wedge b \in \{0,1\}\}$.

Verification. For each i and s, define

$$U_{i,s} = \{\tau \in 2^{<\omega} : (\exists \sigma \in T_{i+s+1})[\Phi(\sigma \oplus \tau) \succeq A \upharpoonright n_i]\}.$$

As $T = \bigcup_s T_s$, and $T_s \subseteq T_{s+1}$, we have $U_i = \bigcup_s U_{i,s}$. Note that for all s, the measure of $U_{i,s}$ is equal to the sum over the maximal $\sigma \in T_{s+1}$ of $m(\sigma, A \upharpoonright n_i)$. We prove by induction on $j \in \omega$ that $\mu(U_{i,j}) \leq \sum_{k \leq j} 2^{-(i+k+1)}$, and hence that $\mu(U_i) \leq 2^{-i}$, as desired.

For the base case $j = 0$, the claim holds because there are 2^{i+1} many maximal strings in T_{i+1}, and for any such string σ we have $m(\sigma, \emptyset' \upharpoonright n_i) < 2^{-2(i+1)}$ by choice of n_i. Assume, then, that $j > 0$ and that the claim holds for $j - 1$. Let σ be a maximal string in T_{i+j+1}, and let $\tau \preceq \sigma$ be a maximal string in T_{i+j}. By (1), we have that $m(\tau, A \upharpoonright n_i)$ is within $2^{-2(i+j+1)}$ of $\sup_{\rho \succeq \tau} m(\rho, A \upharpoonright n_i)$. Thus, the total new contribution of $m(\sigma, A \upharpoonright n_i)$ to the measure of $U_{i,j}$ can be at most $2^{-2(i+j+1)}$. Since there are 2^{i+j+1} many maximal strings in T_{i+j+1}, the claim follows. $\square$

For the readers familiar with the basic concepts of algorithmic randomness, we remark that if T is the tree constructed in Theorem 4.1, then it is easy to see that $\emptyset'$ can construct a set $S \in [T]$ such that S is not K-trivial.

This gives an example of a set that is not K-trivial and cannot be joined above $\emptyset'$ with a 2-random set. This contrasts with a recent result of Day and Miller [2] who show that any set S that is not K-trivial can be joined above $\emptyset'$ with an incomplete 1-random set, and indeed even with a weakly 2-random set.

Acknowledgements

The authors are grateful to Theodore Slaman for posing the question that motivated this article as well as for insightful discussions. They also thank the anonymous referee for helpful comments. The first author was supported by a Miller Research Fellowship in the Department of Mathematics at the University of California, Berkeley. The second author was supported by an NSF Postdoctoral Fellowship.

References

1. Peter A. Cholak, Damir D. Dzhafarov, Jeffry L. Hirst, and Theodore A. Slaman Generics for computable Mathias forcing. To appear.
2. Adam R. Day and Joseph S. Miller. Cupping with random sets *Proc. Amer. Math. Soc.*, to appear.
3. Rodney G. Downey and Denis R. Hirschfeldt. *Algorithmic randomness and complexity*. Springer-Verlag, 2010.
4. David B. Posner and Robert W. Robinson. Degrees joining to $0'$. *J. Symbolic Logic*, 46(4):714–722, 1981.
5. Gerald E. Sacks. *Degrees of Unsolvability*. Princeton University Press, 1963.
6. Richard A. Shore. Lecture notes on Turing degrees. In *Computational Prospects of Infinity II: AII Graduate Summer School*, Lect. Notes Ser. Inst. Math. Sci. Natl. Univ. Singap. World Sci. Publ., Hackensack, NJ, to appear.
7. Richard A. Shore and Theodore A. Slaman. Defining the Turing jump. *Math. Res. Lett.*, 6(5-6):711–722, 1999.
8. Robert I. Soare. *Computability theory and applications*. Theory and Applications of Computability. Springer, New York, to appear.

Freedom & Consistency

Michael Detlefsen

Department of Philosophy, University of Notre Dame
Notre Dame, IN 46556, USA
mdetlef1@nd.edu
www.nd.edu

Claims emphasizing the supposed freedom of mathematics (e.g. those by Cantor and Hilbert) are considered. A challenge by Frege to such claims is also analyzed and sustained.

Keywords: Cantor, Hilbert, Frege, Conceptual Freedom, Futility Argument.

1. Introduction

"The essence of mathematics", Cantor famously wrote, "lies precisely in its freedom" ([11], §8).[a] The language is dramatic, but for Cantor and other foundational thinkers of his time, it expressed a serious truth concerning the legitimate formation and use of concepts in mathematics.[b]

This was that formation and use of concepts in mathematics (or, more accurately, in pure mathematics) need not be held to the standards of concept formation common and proper outside mathematics. Specifically, Cantor maintained, it does not have to meet a traditional constraint requiring that the contents of a justifiedly deployed concept be derivable by abstraction from the contents of an apprehension of an instance(s) of the concept. At the same time he emphasized that adopting these less demanding standards should not pose a threat to the distinctive security of mathematical reasoning.

Similar claims were made by a number of other nineteenth and early

---

[a] The German was: "Das Wesen der Mathematik liegt gerade in ihrer Freiheit."

[b] The German phrase was "freie Begriffsbildung." This has commonly been translated as "free concept formation."

Sometimes phrases such as "Einführung eines Begriffes" were used instead of "Begriffsbildung" (cf. [40], 163) in claims concerning conceptual freedom. Such phrases seem better rendered as "concept introduction" than as "concept formation."

twentieth century foundational writers.[c] In addition to conceptual freedom there were other freedoms commonly believed to be characteristic of mathematics as well.[d] My focus here will be freedom in the formation/introduction of concepts or, more accurately, freedom in the axiomatic introduction of formal expressions corresponding to said concepts. For convenience, I'll refer to this type of freedom generically as *conceptual freedom*.

Cantor defended conceptual freedom by identifying certain safeguards on its legitimate exercise. Adherence to these safeguards, he argued, is enough to secure the distinctive reliability of mathematical reasoning while at the same time securing the benefits of conceptual freedom. These benefits were for the most part taken to be benefits of simplicity and/or efficiency.[e]

By a "liberal" standard of concept introduction we will therefore mean one which does not require satisfaction of a constraint on concept introduction that was widely recognized and accepted in Cantor's day and before. This was the following traditional constraint on the justified use of concepts.

Empirico-constructive concept introduction (ECCI): A concept C may be justifiedly employed in a reasoning practice only if the content of C is derivable from the content of an experience (or, to use another common term, the content of an *intuition*) of an instance of C by application of a legitimate process of abstraction.

The clause which comes after the "only if" in this statement is what, for convenience, I will refer to as the Instantiation Constraint (IC, for short). The empirico-constructive view of concept introduction took the IC to be the pivotal constraint on justified concept introduction, a constraint which was widely (though by no means universally) accepted by foundational thinkers both of the modern era and of the nineteenth and twentieth centuries.

The ECCI (or some similar view) seems to have been behind Gauss' so-called *geometrical* interpretation of the complex numbers. He described

---

[c]Cf. [57], §78; [34], 20; [15], 335, 338; [18], 8–10; [10], 2; [69], 52–53; [49], 679–680; [39], 18–20; [53], 333–334; [14], 88–89.

[d]The two most prominent were:

I. *Logical Freedom* ($\approx$ freedom to determine the principles of logical reasoning appropriate to use in mathematical proof)

and

II. *Axiomatic Freedom* ($\approx$ freedom to adopt other than evident propositions as investigative starting points).

[e]For more on this, see [17], §1 and §§5.2, 5.6 and 5.7.

this interpretation as providing an "intuitiono-sensual" ([29], 174) representation of the complex numbers that "leaves nothing to be desired" (*op. cit.*, 175), and he suggested this was sufficient to justify the use of concepts for imaginary quantities in analysis.

On this way of thinking, the key fact concerning the introduction of concepts of imaginary quantities in mathematics is that it can be justified in the same way that use of concepts is generally justified in science—namely, by showing that they have contents that can be derived from contents given in intuitiono-sensory experience.[f]

Many nineteenth century thinkers saw the IC as a welcome tightening of what they believed were improperly liberal standards of concept introduction.[g]

These thinkers were concerned, among other things, to distinguish the use of concepts in *genuine reasoning* from something they believed had been but ought not to be confused with it—namely, the symbolic or syntactical manipulation (i.e. non-contentual use) of linguistic expressions.[h] In their view, knowledge consists of genuine judgments, genuine judgments are attitudes taken towards genuine propositional contents, propositional contents are generally speaking relations between conceptual contents and conceptual contents are only secured when they are derived from contents given in experience.

Concern to provide for justified, contentual use of concepts of imaginary quantities in our thinking was thus a prime motive for late nineteenth and early twentieth century support of the IC. It was also a traditional motive.

Commitment to a standard of concept introduction like the ECCI has

---

[f]Among philosophers of the relevant period, the empirico-constructive view was perhaps most influentially advocated by Kant.

> "One ... demands that a ... concept be made sensible (*sinnlich*), that is, that an object corresponding to it be presented (*darzulegen*) in intuition. Without this the concept would ... remain (*bleiben*) without sense (*Sinn*), that is, without meaning (*Bedeutung*). Mathematics fulfills this demand by the construction of a figure (*Gestalt*), which, although produced a priori, is an appearance present to the senses."

> [47], A240, B299

[g]Schopenhauer gave a memorable statement of such a view (cf. [61], vol. 2, ch. 7), directed towards what he took to be the lax standards of Fichte, Schelling and Hegel.

[h]Schopenhauer's phrase for this latter was "mere words in our heads" (*blosse Worte im Kopfe*), *loc. cit.*. He stressed that manipulation of mere words or symbols ought not to be confused with genuine operations of thought (e.g. genuine judgment or genuine reasoning) on *real contents* (*realen Gehalt*).

perhaps been most closely associated with constructivist views (e.g. Kronecker's finitism and Brouwer's intuitionism).[i] Brouwer in fact urged not only a strictly contentual view of the nature of mathematical reasoning, but a strict discipline in keeping content and language separate in our thinking. He thus advocated a kind of *pure contentuality*—a language-less contentuality if you will—in mathematical reasoning. He codified this in his so-called *first act of intuitionism*, which he described as mandating a complete separation of mathematics from mathematical language in recognition of the fact that mathematics is an "essentially languageless activity of the mind" (cf. [6], 140–141).

There were non-constructivists too, though, who emphasized the essentially contentual character of genuine reasoning. Frege is a prime example. His contentualist view of reasoning was in fact the pivotal point of his disagreement with the formalists.[j]

Frege was not, however, unusual among non-constructivists in his adherence to a broadly contentualist understanding of the nature of proof. As Weyl noted, the contentualist conception of reasoning was the standard conception prior to the development of Hilbert's proof-theoretic ideas.[k]

Concern for the contentuality of concepts, judgments and proofs was thus a powerful motive for the IC as a condition on concept introduction. If concepts (judgments, proofs, etc.) are fundamentally contentual in nature, there must be an account of how they get their contents. The IC was instrumental in providing such an account. In the view of many, it was the only or at least the best way to do so.

Another motive was objectivity—or at least inter-subjectivity—in

---

[i]As an aside I would note that constructivist adoption of the IC should be borne in mind in understanding the so-called *existence requirement* of the constructivists. The existence requirement is generally presented as a constraint on the acceptability of existence claims—namely, that a ϕ cannot justifiably be claimed to exist unless and until a ϕ has been constructed or exhibited in intuition. It is possible to see the IC as adding in certain ways to this. It implies that the concept of a ϕ cannot legitimately be used in reasoning unless an instance of a ϕ can be exhibited by appropriate means. It should also be noted, though, that the IC applies to use of concepts in settings other than those involving existential claims.

[j]Cf. [26], 387.

[k]Compare Weyl's remark: "Before Hilbert constructed his proof theory everyone thought of mathematics as a system of contentual (*inhaltliche*), meaningful (*sinnerfüllte*), and evident (*einsichtige*) truths; this point of view was the common platform of all discussions. ... Brouwer, like everyone else, required of mathematics that its theorems be (in Hilbert's terminology) "real propositions", meaningful truths." ([67], 22).

See [68], 640, [3], 336, [5], 490–492 and [7], 2–5 for similar confirmation of the prevalence of this view.

mathematical judgment. Here the IC was seen as supporting the use of concepts whose contents, at least in their pre-abstractive origins, are *impressed on* the conceiver by external sources and thus proceed from sources that are free of the subjectivity associated with voluntary human influence or control.

Conceptual libertarians[1], on the other hand, have not generally been moved by such considerations. In their view, justified concept introduction in mathematics does not require the type or degree of objectivity the ECCI is intended to enforce—namely, transient impression of experiential contents of the type suggested by the IC.

As the famous remark quoted at the beginning of this paper indicates, Cantor was emphatic on this point. He regarded it as the key to understanding (what he took to be) the cardinal difference between mathematics (more exactly, *pure* mathematics) and the other sciences. In his view, though the IC may be an appropriate standard for concept introduction in the natural sciences, it is not a proper standard for concept introduction in (pure) mathematics. In mathematics, Cantor (and others) maintained, the only proper constraints are consistency and "fruitfulness" (cf. [11], §8).

Consistency provides a properly secure standard for concept introduction while also being more liberal than the IC as a standard of concept introduction (or formal expression introduction).[m] That it is properly secure was taken to follow from the supposed fact that pure mathematics is intended to provide only for the "intrasubjective or immanent reality of its concepts" ([11], §8), and not for that "extra-subjective or transient reality" (*loc. cit.*) that is generally required in the empirical sciences.

In what way(s) it might be more liberal is more difficult to say and is the central question of this paper. Frege gave what I think is now a largely forgotten argument in this connection. This is the so-called Futility Argument, presented in §3 and discussed thereafter.

It is a direct challenge to the libertarian view of justified introduction of concepts or formal expressions associated with them. Cantor and other

---

[1]I understand the term 'libertarian' to broadly signify views which promote liberty in one or another sphere of human activity. I intend no additional connection to current political views which go by that name and which generally emphasize reasons for minimizing the role(s) of government in civil life.

[m]By formal expression introduction, I mean the introduction of linguistic expressions by formal axiomatic means. The same arguments given here can also be given for concept introduction. By framing my argument in terms of formal expression introduction I hope to make it more accessible and familiar to the average reader of this collection.

libertarians maintained that replacing the IC or the CIC^n with a consistency constraint results in an easing of the standards required for their justified introduction. Frege denied this and argued that, judged from a practical vantage, establishing consistency is not only as difficult as establishing the CIC, it imposes an essentially equivalent restriction.

Now for the statement of the CIC. It is intended to be a condition on the justified introduction or use of formal expressions that constrains it in a way parallel to that in which the IC constrains the justified use of concepts in mathematical reasoning. It thus stipulates that the introduction of an expression (or family of expressions) by a set of formal axioms is justified just in case there is an expressly known interpretation (of the language of those axioms) which manifestly satisfies those axioms.

The focal concern of the remainder of this paper will be the benefits of freedom, if any, that may plausibly be ascribed to replacement of the CIC by a consistency constraint.[o]

2. Freedom & Consistency

Is it reasonable to think that replacing the IC or the CIC by a consistency condition would make for freer introduction of formal expressions? The view that it would seems to require a further belief that consistency is a condition whose verified satisfaction is within our powers in some way or sense in which verified satisfaction of the IC or the CIC is not.

---

[n]CIC is short for 'Confirming Interpretation Constraint'. It is stated in the next paragraph of the text.

[o]It should be noted at least in passing that not everyone accepted the idea that consistency is a legitimate condition on the "admission" or concepts. Frege, in particular, denied this.

> "A concept is still admissible (*zulässig*) even though its defining characteristics (*Merkmale*) do contain a contradiction: all that we are forbidden to do, is to presuppose that something falls under it. From the assumption that a concept contains no contradiction, however, we cannot infer that for that reason something falls under it. If such concepts were not admissible, how could we generally prove that a concept does not contain a contradiction?"

> [24], §94

In his final question, Frege's point seems to be that the need to prove the consistency of the characteristic properties of a concept could not properly be regarded as a sensible requirement were it not in some sense possible for inconsistent concepts (or sets of concepts) to be in some meaningful sense legitimately "introduced". In saying this, he seems to have been suggesting that one might contemplate an inconsistent concept for a consistency proof, and that such contemplation would itself require that the concept be somehow introduced into our thinking.

What, more exactly, might this mean? And what if any reason is there to accept it?

The IC demands not only that a legitimately introduced concept be consistent, but that an instance of it be intuited (or apprehended by some other justificatively privileged means). Without intuition of an instance, the thinking has generally been, there will not guaranteedly be an externally impressed (as contrasted with subjectively generated) content from which the content of a given concept might appropriately be derived by abstraction. And without an externally impressed content from which to derive the contents of a concept, those contents will not have the objectivity we generally desire in scientific concepts.

To put the point another way, the contents of legitimately introduced concepts or formal expressions must be witnessed or found, not invented or created. Consistency is a less demanding condition. Every witnessing of a formal expression establishes its consistency. Or, more accurately, every witnessing of a set of introducing axioms for a formal expression establishes their consistency.[P]

On the other hand, consistency does not necessarily guarantee witnessing. Even more clearly, it does not guarantee witnessing of any epistemically privileged type (e.g. witnessing by some type of intuition or experience). This being so, it ought at least in principle to be easier to satisfy the demands of consistency than to satisfy the demands of passive witnessing, or so the thinking seems to have been.

Frege offered a proposal somewhat along these lines. In particular, he urged a distinction between the consistency of a concept and what he termed its *fulfillment* (*Erfülltheit*), which he characterized as the condition that something falls under it.

In his view, the formalists had mistakenly taken freedom from contradiction in a concept to be "sufficient guarantee" ([24], §109) of its fulfillment.[q]

[P]Not everyone has agreed. Paul Weiss, for one, considered this a dogma. He wrote: "Is it possible that the only way we can determine whether a set is consistent is by seeing all the postulates actually exemplified in some one object? If so, we must arbitrarily assume that the object is self-consistent, so that the proof of consistency must ultimately rest on a dogma." ([66], 468).

[q]Kant famously urged a distinction similar to Frege's.

"I can think (*denken*) whatever I want, provided only that I do not contradict myself. This suffices for the possibility of a concept, even though I may not be able to answer for there being, in the sum of all possibilities, an object corresponding to it."

[48], Bxxvi, note a

This was indeed, as he saw it, the cardinal failing of the formalists.[r]

As we'll see in the next section, though, while Frege affirmed the logical distinctness of consistency from instantiation or fulfillment, he seems to have seen little practical difference between what is required to establish the one and what is required to establish the other. Aside from seeing no truth in the libertarian idea that, in mathematics, concepts may be freely introduced or *created* (i.e. introduced simply on the strength of their consistency), he also believed it to be of little practical significance.

Beliefs which separated the (mere) consistency of a concept from its fulfillment were thus a main source of views which took the replacement of the IC or the CIC by a simple consistency requirement to represent a move in the direction of increased conceptual freedom.[s] As we will now see, though, not everyone believed that the logical distinction between consistency and fulfillment provides meaningful support for conceptual freedom. Frege, in particular, argued that it provided no practical support for it.

3. The Futility Argument

That consistency is not to be seen as a practically less demanding condition on formal expression introduction than the CIC was not only Frege's view but one shared by many foundational thinkers of the late nineteenth and early twentieth centuries.

What set Frege apart was his articulation of the idea.

"[T]he power to create (*schöpferische Macht*) is ... restricted by the proviso that the properties must not be mutually inconsistent. But how does one know that the properties do not contradict one another? There seems to be no criterion for this except the occurrence of the properties in question in one and the same object. But the creative power with which many mathematicians credit themselves thus becomes practically worthless. For as it is they must certainly prove, before they perform a creative act, that there is no inconsistency between the properties they want to assign to the objects that are to be, or have already been, constructed; and apparently they can do this only by proving that there is an object

---

[r]He believed, too, that they had exhibited unwholesome tendencies to infer consistency from absence of revealed inconsistency.

[s]Later, of course, there was also development of a radical alternative to model construction as a means of proving consistency—namely, Hilbert's proof-theoretic approach to consistency. More on this later.

with all these properties together. But if they can do that, they need not first create (*schaffen*) such an object."

[25], §143[t]

Reduced to essentials, and transposed from the setting of concept introduction to formal expression introduction, Frege's argument seems to be as follows:

The Futility Argument

Premise 1: The formal axiomatic introduction of an expression $\mathcal{E}$ is justified only if the consistency of the axioms (call them $\mathcal{A}^{\mathcal{E}}$) proposed as the rules of usage of $\mathcal{E}$ (what I have been calling the *introducing axioms* for $\mathcal{E}$) is established.

Premise 2: Practically speaking, to establish the consistency of $\mathcal{A}^{\mathcal{E}}$ requires finding an interpretation of $\mathcal{A}^{\mathcal{E}}$ which is recognized by appropriate means to be a model of it.

Premise 3: But to find an interpretation of $\mathcal{A}^{\mathcal{E}}$ that recognizedly models it is essentially to satisfy the CIC for $\mathcal{E}$ (or is essentially equivalent to doing so).

Premise 4: But if the introduction of $\mathcal{E}$ satisfies the CIC, it is not free.

We thus obtain the following A̲nti-L̲ibertarian T̲hesis:

ALT: Free, justified formal expression introduction is not generally a practical possibility.

How successful a challenge to conceptual freedom this argument is is the question that will occupy us for the remainder of this paper. My focus will be on premises 2 and 3.

4. Premise 2

That finding or constructing models is the only generally applicable practical means of proving consistency has been an express view since at least the middle of the nineteenth century.[u] During this period, of course, the term 'model' has often been used in a somewhat looser sense(s) than that of contemporary mathematical logic.

---

[t] Other statements by Frege to similar effect can be found in the *Grundlagen* (cf. §95, [24]), and in his correspondence with Hilbert (cf. [28], pp. 43, 47, (letter of January 6, 1900) and pp. 49–50 (letter of September 16, 1900)).
[u] Cf. [35].

'Model construction' has also not been the only terminology employed.[v] Whatever the terminology and its understanding, though, the basic idea has been that the only generally applicable and practically feasible way to establish the consistency of a set of sentences is to provide an interpretation of them on which they are evidently true.[w]

This was Frege's view, and the generally prevailing view, at the time he (Frege) formulated the Futility Argument. He maintained this view despite an earlier discussion with Hilbert concerning alternative methods of proving consistency (cf. [28], 43, [28], 49).

Frege had apparently read the text of Hilbert's Paris address and parts of it had suggested to him that Hilbert believed himself to have conceived of such an alternative. Frege rightly observed that development of such an alternative would be a matter of the highest importance ("ungeheurer Bedeutung", *loc. cit.*). He also expressed skepticism concerning whether it could be a genuine alternative to model construction and have wider scope than it (*op. cit.*, 50). Finally, he asked Hilbert for a statement of its 'principle' and an example (*loc. cit.*).

Hilbert replied a few days later that the pressures of preparing a new set of lectures would not permit him to give Frege the information he had requested. He said only that he found himself "forced" (*op. cit.*, 51) to his alternative approach by (i) "the requirements of strictness in logical inference and in the logical construction of a theory" (*loc. cit.*) and by (ii) the need to avoid circularity (or to achieve groundedness) in the larger scheme of consistency proofs for classical mathematics (cf. *loc. cit.*).

This effectively ended the correspondence between Frege and Hilbert. Hilbert sent a final letter in July of 1903, but it contained no further information concerning either the underlying ideas or an illustration of Hilbert's "direct" (cf. [36], 265) approach to consistency.

In time, Hilbert's approach was developed into a coherent alternative to model construction as a means of proving consistency. This development

---

[v]Other designations include: "method of suitable specialization" (cf. [37], 135); "construction of examples" (*loc. cit.*); "concrete representation" ([65], 3; [69], 43–44); "exhibition of entities" ([55], 484–485) and proving "an *existence theorem*" [56], 11.

[w]A brief remark concerning the term 'construction' may be in order. Consistency proofs were viewed as existence proofs, and such proofs had two distinguished varieties. There were those which provided witnessing instances. These were commonly called *constructions*. There were also those which proceeded more abstractly. These were called *abstract* existence proofs.

Constructions in the aforementioned generic sense were not necessarily assumed to meet some more particular set of standards (e.g. those of the intuitionists) concerning the means by which they provide their witnesses.

revealed it to be as radically different an alternative as Frege expected it would have to be.

At the core of Hilbert's direct approach was the so-called "fundamental idea" of his proof theory. Roughly, this was that the operations of our mathematical reasoning either are or are intimately related to operations on linguistic expressions regarded from the point of view of their elementary outward appearances.

This being so, Hilbert reasoned, one should be able to determine at least some of the properties of a body of mathematical reasoning (e.g. consistency) by attending to the outward appearances of signs and their combinations. As he put it:

> "The fundamental idea (*Grundidee*) of my proof theory is none other than to describe the activity (*Tätigkeit*) of our understanding (*Verstandes*), to make a protocol of the rules according to which our thinking (*Denken*) actually proceeds (*verfährt*). Thinking, it so happens, parallels speaking and writing: we form statements and place them one behind another. If any totality of observations and phenomena deserves to be made the object of a serious and thorough investigation, it is this one ... "

> [41], 16–17; English translation in [63], 475

In short, instead of *interpreting* formulae and their combinations, as in model construction, Hilbert's proposal was to *analyze* them and their combinations from the point of view of their outward appearances and to investigate their interrelationships (e.g. their consistency) in terms of these appearances.^x

The approach is thus "direct" in the sense of being based on *examination of* arithmetical reasoning rather than on *interpretation of* it. Hilbert put the basic point this way:

> "The axioms of arithmetic are essentially nothing but the known rules of calculation, with the addition of the axioms of continuity. ... I am convinced that it must be possible to find a direct proof (*einen direkten Beweis*) for the consistency of the arithmetical axioms by means of a careful working through ... of the known methods of reasoning in the theory of irrational numbers."

> [36], 265–266

---

^xMore accurately, his aim was to analyze and investigate the visualizable relationships between the outward appearances of actual and imaginary signs and combinations thereof.

Analysis rather than interpretation of the methods of reasoning with real numbers was thus the core idea of Hilbert's "direct" approach to the consistency problem for the arithmetic of the reals. He combined this strategic idea with a general view of the requirements of rigor in solving problems. According to this view, rigorous reasoning in mathematics requires satisfaction of a finiteness condition, an exactness condition and a type of purity condition.[y]

The exactness condition included a disclosure requirement—that is, a requirement that all premises and steps of inference used in a piece of mathematical reasoning be explicitly declared. Such declaration was in turn taken to require a syntactical specification of the premises and inferences used. A specification in terms of content could not accomplish the intended explicitness because, though a content may be syntactically expressed, it is not explicitly given or exhibited by a syntactical expression in the way that the syntactical appearance of that expression is.

Full disclosure of the elements used in a piece of reasoning was thus taken to require that it be syntactically specified and that it be conducted exclusively by reference to this specification. Anything less would allow the use of contents that are not fully disclosed, and any such use would constitute a failure of rigor.

In Hilbert's view, axiomatic reasoning, properly understood, is reasoning which embodies this type of disclosure. It is often, perhaps even typically, the result of deliberate efforts. Sometimes though, Hilbert suggested, it is not.[z] In either case, however, the essential point is the same: mathematical reasoning is not only syntactically representable (e.g. for purposes of meta-mathematical investigation), but syntactically conductable. That is, when rigorously conducted, its proper conduct (resp. improper conduct) can be established by appeal to the properties of and relations between syntactical objects considered as such.[aa]

Rigorous reasoning was thus taken to be reasoning that is in some

---

[y]As Hilbert put it: "[I]t should be possible to establish the correctness (*Richtigkeit*) of the solution (*Antwort*) of a problem by means of a finite number of steps based on a finite number of hypotheses which must always be exactly formulated and which are contained in the presentation of the problem (*in der Problemstellung liegen*). This requirement ... is simply the requirement of rigor (*Strenge*) in proving (*Beweisführung*)." ([36], 258).
[z]Cf. [42], 380.
[aa]The properties of and relationships between what I am calling syntactical objects can of course include idealizations of various types.

sense conductable by application of syntactically stated rules.[ab] Consistency which applies directly to such reasoning was then also syntactically conceived, and a properly pure proof of such consistency was correspondingly regarded as one whose premises and inferences remain within this syntactically oriented *Problemstellung*.

On this way of seeing things, rigor demanded a direct approach to the consistency problem for arithmetic. Relative consistency proofs could be admitted but, in the end, they had to be rooted in a direct proof of consistency of the base theory (i.e. the theory on whose consistency their consistency ultimately rested).

Digression. Hilbert had an additional, somewhat more special reason for regarding a direct approach to the consistency problem for arithmetic as necessary. This was that proofs by model construction were not, in his view, even in-principle possible there.

The "method of suitable specialization or construction of examples",[ac] he wrote, "necessarily breaks down" ([37], 252) in the case of arithmetic. This is because (a) the only promisingly non-circle-engendering place to explore for a model of arithmetic is the realm of logic and (b) construction of a model for logic will inevitably make use of what is essentially arithmetical evidence and judgment.

As Hilbert saw it, this implied the inevitable failure of logicism. Arithmetic and logic are so intimately intertwined as to require "simultaneous" development of their foundations (cf. [37], 245–246), and the only way to meet this demand is to provide a direct analysis of arithmetico-logical reasoning, and a corresponding direct definition and proof of its consistency.
End Digression

Hilbert's *direct* approach to consistency seems to raise an in-principle alternative to model construction as a means of proving consistency. The question for us, though, is whether it also raises an effective challenge to premise 2.

The answer, I think, is that it does not. Premise 2 does not assert that model construction is the only in-principle possible means of giving consistency proofs. Rather, it asserts that it is the only generally practicable

---

[ab]See [46], 67–68 for a statement of the idea that rigor requires syntactic conduct of inference. Similarly for the use of axioms. Full disclosure requires that they be both specified and applied syntactically (cf. [41], 7–10, English trans. in [63], 469–471; [43], §1)

[ac]This was Hilbert's terminology for what I have been calling 'model construction.'

means of doing so.[ad]

History seems to support this view. It indicates, in particular, that the main historical alternative to model construction—Hilbert's direct approach—is not generally viable. Its lack of successful application became a concern among foundational thinkers soon after the Paris address (cf. [54], 405). The early attempt to apply the direct approach to a system of arithmetic in [37] didn't change that significantly, and the later discovery of Gödel's incompleteness theorems only heightened concerns regarding the viability of the direct approach.

There were a few modest successes in the 1920s (e.g. for fragments of elementary arithmetic by Ackermann (cf. [1]) and von Neumann (cf. [64])).[ae] Gentzen's later proof of the consistency of full elementary arithmetic (cf. [31]) was also an important application of the direct approach or a modification of it.[af]

These cases notwithstanding, the generally prevailing view has been and remains that model construction is the only generally practicable means of approaching consistency problems.[ag,ah] This view, or a slight weakening of it (to allow for the minor class of cases where successful application of the direct approach is a realistic option), has been borne out by experience.[ai]

[ad]Even this is more than what is seemingly necessary to sustain the anti-libertarian conclusion of the Futility Argument, or some slight weakening of it (e.g. to a claim which admits a minor class of possible exceptions to the general pattern asserted in the conclusion of the Futility Argument).

[ae]Von Neumann's work is said (cf. [63], 129) to have been influenced by [37] via Julius König's [50].

[af]It seems worth noting, though, that Gentzen did not himself see direct proof as a generally available alternative to model construction. In particular, he suggested that use of model construction methods are likely indispensable to the justification of set theoretic axioms (cf. [32] and [33]).

[ag]H. C. Brown made a statement to this effect, while adding that the conditions of effective model construction proofs of consistency are difficult to satisfy (cf. [8], 530). He also cited [37] as an exception to the general pattern of using model construction to prove consistency.

See [56], 11 for another, later statement of this claim.

[ah]Hilbert doesn't seem to have seen a strategic need for broad application of the direct approach. He thought of it instead as desgined especially for arithmetic—the theory whose consistency in his view lay at the base of a hierarchy of relative consistency proofs encompassing the various sub-areas of mathematics. Whether and/or how well this coheres with his motivation of the direct approach in [36] is a question worth considering, though I lack the space to consider it here.

[ai]This is perhaps the place to add that not all who questioned Hilbert's direct approach questioned its possibility (practical or theoretical). There were those who instead questioned its value and, in some cases, the value of consistency proofs generally.

Brouwer, for example, said he saw no reason to doubt that a direct consistency proof

5. Premise 3

If construction of a model $\mathcal{M}$ for a theory T is to establish T as consistent, $\mathcal{M}$'s modelling of T must be evident. According to premise 3, to satisfy this condition is tantamount to satisfying the CIC.

Whether this is plausible, though, may seem to depend on one's understanding of what in premise 2 are referred to as "appropriate" means of making clear that a proposed model for a set of introducing axioms is in fact a model of them.

Such means have often been described in experiential or quasi-experiential terms. Model construction has, for example, been said to

(i) provide for "an observed fact of [the] coexistence [of the axioms of the system] as literal truths in a particular case produced" ([35], 220),

to

(ii) "show ... something real, or at least imaginable, with respect to which we can interpret the definitions, or do all the things indicated by the postulates" ([9], 629),

to

(iii) "exhibit within some accepted region independent of the defining postulates, an entity with respect to which all the postulates permit of interpretation together" ([8], 530),

to

(iv) "point to familiar sets of objects which ... actually fulfill [the] fundamental laws" ([13], 72),

and to

(v) give a "concrete representation" of the systems whose consistency it is intended to prove ([65], 3; [69], 43–44; [20], 185–186).

Model construction has thus been described as providing special knowledge of the models constructed—specifically, knowledge that consists in somehow "exhibiting" them as being true or making them "real" or "concrete."[aj]

---

for arithmetic could be given. "[T]he unjustified application of the principle of excluded middle to properties of well-constructed mathematical systems", he wrote, "can never lead to contradiction" ([4], 336, footnote 2). Still, he claimed, finding such a proof of consistency would yield nothing of much value because "an incorrect theory is none the less incorrect" even if "it cannot be inhibited by any contradiction" (*loc. cit.*).

[aj]Here I should perhaps note that not everyone considered model construction to be a satisfactory general means of settling consistency problems. As noted earlier, the American philosopher Paul Weiss described such proofs as dogmatic. There were also those

More basic, perhaps, is the element of "finding" that figures in premise 3. This suggests that the model construction with which it is concerned is in some sense and/or in certain respects a matter of discovering a model rather than of creating it or devising it. Let's consider more carefully what this might mean.

For some modellable theories, it seems, we have a capacity to in some sense and to some extent devise structures that model them. In particular, we seem to have some discretion or freedom in the choice of domains and the choice of interpretations of basic expressions when constructing models.

It does not follow, of course, that the resulting interpretation's being a model of the intended theory is likewise a matter of discretion. Our intending or desiring that a devised interpretation be a model of a given theory does not guarantee that it will be.

That we may in some meaningful sense choose domains and/or interpretations, then, and even that we may create them with an intention that they be models of certain theories, is compatible with our finding or discovering, and even with our having to find or to discover, that the interpretations devised are in fact models of the theories they are intended to model.

Since, however, construction of a structure $\mathcal{M}$ for T is of use in establishing the consistency of T only to the extent that it is evidently a model for T, model construction that is capable of establishing consistency seems not to be sheerly a matter of constructor discretion.

There are other considerations concerning the 'finding' of models that seem also to be worth noting here. One of these concerns matters of evidensory independence such as that mentioned by H. C. Brown in the remark in quotation (iii) above.

Some of the theories for which we may seek consistency proofs are theories that come with an interpretation. For convenience, let's call such intepretations *received* interpretations. Let T be such a theory, and let $\mathcal{M}$ be a structure constructed for the purpose of establishing T's consistency.

If construction of a new model $\mathcal{M}$ for T is to make strategic sense as a means of establishing T's consistency, it must seemingly support increased rational confidence concerning T's consistency beyond what it might reasonably be taken to be were it to be based solely on knowledge of the received interpretation of T.

Under what conditions can this be expected to happen? One possibil-

---

(cf. [55], 484–485; [19], 474n; [20], 186) who suggested that only long, thorough experience can provide proper assurance of consistency.

ity is that $\mathcal{M}$'s modelling of T should be more evident than the so-called *received* interpretation's modelling of T. Were this so, construction of $\mathcal{M}$ could conceivably provide for increased confidence in the consistency of T without thereby similarly increasing confidence that T is true under the *received* interpretation.

Another possibility, though, is that (a) the evidence which bears on the question whether $\mathcal{M}$ is a model of T should on balance be affirmative and that (b) it should be *independent of* (in an appropriate sense) the evidence that the received interpretation of T is a model of T. Without attempting a full characterization of the relevant notion(s) of independence, which I would not know how to give, it seems clear that it should at a minimum guarantee that not every reason to doubt that the received interpretation of T is a model of T is equally a reason to doubt that $\mathcal{M}$ is.

Some such independence between the evidence that a received interpretation is a model of a given theory and the evidence that a constructed interpretation is would thus seem to be necessary if the latter were to be capable of adding significantly to the evidence for a theory's consistency.

This leads to a more general question related to premise 3, namely, that concerning the conditions under which a consistency proof might generally be expected to be gainful. To put it another way: "Why shouldn't received interpretations themselves (as distinguished from *constructed* interpretations) be enough to provide all necessary assurances concerning the consistency of introducing axioms?"[ak]

Past writers have sometimes taken the view that consistency proofs are not always, or perhaps even not generally useful. In particular, they have claimed, they do not add anything of value in cases where the axioms of an axiomatic system are *typical* axioms—that is, where they are evident in the ways in which axioms have traditionally been required to be evident.

Where axioms are truly axioms, the thinking goes, there is no need for and no meaningful possibility of gainful proof of their consistency. Where axioms are truly axioms, they are evident, and this evidentness provides a guarantee of their consistency that is both sufficient and as good as any we may reasonably hope to obtain through a proof of consistency.

Peano believed that this was essentially the situation for both elementary arithmetic and elementary geometry.

"A consistency proof for arithmetic, or for geometry, is in my opin-

---

[ak] Here by 'introducing axioms' I mean formal axiomatic systems used to introduce formal expressions into a system(s) of mathematical reasoning.

ion not necessary. In fact, we do not create the axioms arbitrarily, but assume instead as axioms very simple propositions which appear explicitly or implicitly in every book of arithmetic or geometry. The axiom systems of arithmetic and geometry are satisfied by the ideas of number and point that every author in arithmetic or geometry knows. We think of numbers, and therefore numbers exist. A consistency proof can be useful if the axioms are intended as hypotheses which do not necessarily correspond to real facts."

[58], 343[a1]

Frege took what may have been an even stronger view. He maintained not only that consistency proofs for elementary arithmetic and/or geometry were not necessary or helpful, they were, in a serious sense, not even "possible."

"Axioms do not contradict one another because they are true; this admits of no proof."

[25], 321

As Peano and Frege saw it, then, a search for a consistency proof only makes sense when the propositions in question are not themselves genuine axioms, but only *hypotheses*, to use the common if also somewhat imprecise term.

Broadly speaking, then, construction of a model for a set of introducing axioms may generally speaking accomplish one of two things:

(1) it may increase evidence for the consistency of a typical (i.e. an evident) set of introducing *axioms*,

or

(2) it may provide evidence for the consistency of a set of *hypotheses* (as distinct from typical axioms) which are to serve as introducing conditions for a given formal expression (or set of formal expressions).

It is perhaps worth noting that increase of the type mentioned in 1. may generally be expected from model constructions in which the evidentness of the constructed model as a model of the given theory is in an appropriate sense *independent* of that provided by the so-called received interpretation.

[a1]English translation in [2], 38.

Contrary to the suggestions of Peano and Frege, then, there might be reason to construct models even for sets of introducing axioms that are evident under their received interpretations.

Reliable model construction seems to involve two types of tests. One concerns the intrinsic or immediate evidentness of the axioms under the constructed interpretation. Let's call this the *test of axiomatic evidentness.* The other is less direct and harder to describe, and its significance is harder to assess. It concerns the testing of identified consequences of the axioms of the target theory for evidentness under the constructed interpretation. Let's call this the *test of consequences.*

How significant the test of consequences is as a source of rational support for a model construction[am] will depend on at least two things: (A) the extent to which consequences of the axioms of the target theory are identifiable and (B) the extent to which sentences identified as consequences are evidently true on the constructed interpretation.

Requiring that an interpretation score well on both the test of axiomatic evidentness and the test of consequences can be expected to limit the class of interpretations qualified to serve as consistency-establishing models for a given theory. It seems moreover to limit it in ways comparable to those in which enforcement of the CIC might be expected to limit it.

Separation might be attainable if one were to interpret the CIC as requiring not only evident modeling of the theory whose consistency is to be established, but empirical modeling of it (cf. [35], 220), or modeling which is in some special way 'familiar' (cf. [13], 72), or 'concrete' (cf. [65], 3) or 'exhibitable' (cf. [8], 530).

If, however, such special constraints are not included in the requirements of the CIC, constructing a recognized model of a set of introducing axioms and satisfying the CIC with respect to them would appear to come to substantially the same thing. Both impose essentially the same restriction on our freedom to make use of such axioms in our mathematical reasoning. Specifically, both require that we know an interpretation of the introducing axioms which we further know to be a model of them.

This being so, premise 3 of the Futility Argument appears to be plausible, at least on those understandings of the CIC which do not require of a model of a set of introducing axioms that, in addition to its being recognized as such, this recognition should have some further special characteristics.

[am] As a source of rational support, that is, for the belief that the interpretation constructed is a model of the theory it is intended to model.

How should we understand such characteristics and their role(s)? Specifically, should we regard them (or some of them) as indicating further reasonable constraints on model construction?

The answer is not altogether clear. Allow me to briefly indicate why it isn't.

Recent psychological research has revealed evidence of the operation of *dual processes* in at least large tracts of human deductive reasoning.[an] Dual process views see human deductive reasoning as at least often combining two very different types of processes. Roughly speaking, these are:

(1) processes that are unconscious, automatic, evolutionarily old, nonverbal, domain specific, contextualized, pragmatic, parallel and relatively rapid,

and

(2) processes that are explicit or conscious, that involve deliberate application of rules, that are evolutionarily recent, linked to language, domain general, abstract, logical, sequential and relatively slow.

These processes are sometimes (cf. [51]) referred to as "reflexive" and "reflective" processes, respectively. It will be convenient for me to use this terminology here.

Evidence for dual processing has centered on so-called *belief-bias* experiments. These have revealed certain tendencies among human reasoners (including instructed, educated adult reasoners) to misclassify invalid arguments as valid when their conclusions are believable and to misclassify valid arguments as invalid when their conclusions are not believable, or when they are presented abstractly (cf. [21]).

Roughly stated, the overall findings are that human reasoners find it difficult to disregard or abstract away from the contents of propositions when judging their logical relatedness or unrelatedness. They also find it difficult to disregard how familiar or unfamiliar and how believable or unbelievable a conclusion is when judging whether or not it follows from a given set of premises. The accuracy of human judgments concerning consequence and non-consequence can thus be affected by such things as the familiarity and believability of the contents of the propositions involved.

This seems to raise questions concerning model construction as a means of proving consistency. More specifically, to the extent that a model con-

---

[an]See [22] for a useful summary of some of this research.

struction depends on testing of consequences, its effectiveness and reliability as a means of establishing consistency may depend on such characteristics as the abstractness or the familiarity of the interpretation on which it is based.

For example, judging a sentence to be a consequence when it isn't might cause a model constructor to overestimate the evidentness of a set of axioms (e.g. were a sentence wrongly judged to be a consequence also judged to be true and hence to confirm the axioms). On the other hand, it might cause her to underestimate it (e.g. were a sentence mistakenly judged to be a consequence also judged to be false and, so, to refute the axioms).[ao]

It would be wrong of course to assume that experts in model construction are as susceptible to such misestimations as their less expert counterparts. They are likely exceptional. Exceptional or not, though, they are human and their thinking cannot therefore be assumed to be immune to such influences as may be pervasive in human reasoning.

Model construction capable of serving as a means of establishing consistency might therefore have to be controlled not only for manifest satisfaction of the axioms, but for various more delicate characteristics such as the abstractness of the interpretations involved, or their familiarity.

For example, it might be that familiarity is positively correlated with axiomatic evidentness. But an interpretation interprets non-axioms and axioms alike. If familiarity of axioms were also to be correlated with familiarity among (some) non-axioms, it might lead to overestimation of relations of consequence.[ap]

There thus seems to be more to effective model construction (i.e. model construction that is effective in establishing the consistency of a set of introducing axioms) than simple identification of an interpretation under which the axioms have a high degree of intrinsic evidentness. To put it another way, there are indications of epistemological and psychological complexities in effective model construction that go beyond what is suggested by a

---

[ao]Similarly for misjudgments of non-consequence. A sentence judged to be true but incorrectly judged not to follow from a set of axioms might lead to their underestimation. Likewise, a sentence judged to be false and incorrectly judged not to follow from a set of axioms might lead to their overestimation.

[ap]It should perhaps also be noted that for theories whose received interpretations are of relatively high familiarity and/or evidentness, it might be more difficult (or unlikely) to produce a constructed model of comparable familiarity and/or evidentness. Determination of what stands to be gained by constructing another model for a given theory might thus overall be a more subtle and complex matter than it is currently generally taken to be.

simple reading of premise 3.

This complexity does not, however, affect the matter of central concern regarding premise 3, namely, whether model construction that is an effective means of establishing the consistency of a set of introducing axioms can also be expected to satisfy the CIC.

The answer to this question, I believe, is "Yes."

Satisfaction of the CIC is supposed ultimately to ensure that the truth of a set of introducing axioms is evident under the interpretation presumed to be provided by the CIC. Model construction is supposed to provide essentially the same type of thing. That is, it is supposed to ensure that the truth of a set of introducing axioms is evident under the interpretation provided by a given model construction.

In both cases, the evidentness mentioned appears to exceed what is properly regarded as being within the modeler's control. A modeler may be able to make different models of a theory the evidentness of whose modeling thereof is also different. She may even know how to produce a model of a theory whose modeling of it will be more (resp. less) evident than that of some other putative model of it.

From this, though, it does not follow that, given a putative model $\mathcal{M}$ for a theory T, a model constructor will be in a position to determine the extent to which $\mathcal{M}$ will evidently model T. This, I believe, is the salient similarity between modeling via satisfaction of the CIC and modeling via "model construction." What ultimately matters, in both cases, is whether the model provided evidently models the theory it is intended to model, and the extent to which this is so is not something the provider is free to determine once the putative model has been provided. To the extent that this is correct, premise 3 seems plausible, and also premise 4, for that matter.

6. Conclusion

My principal focus has been the Futility Argument. It presents what I regard as a credible if also largely forgotten challenge to claims urging the conceptual freedom of mathematics, a freedom that was defended by many of the leading foundational thinkers of the nineteenth and early twentieth centuries.

If I am not mistaken, it was an important influence shaping the development of Hilbert's so-called *direct* approach to consistency problems. It assumes renewed importance in light of the apparent practical failure of that approach. It also highlights deficiencies in our understanding of such basic things as the use of models to establish consistency and the knowledge that such use requires.

References

1. Ackermann, W. "Begründung des "tertium non datur" mittels der Hilbertschen Theorie der Widerspruchsfreiheit", *Mathematische Annalen* 93 (1924): 1–36.
2. Borga, M. and D. Palladino, "Logic and foundations of mathematics in Peano's school", *Modern Logic* 3 (1992): 18–44.
3. Brouwer, L. E. J. "Über die Bedeutung des Satzes vom ausgeschlossenen Dritten in der Mathematik, insbesondere in der Funktionentheorie", *Journal für die reine und angewandte Mathematik* 154 (1923): 1–7.
4. Brouwer, L. E. J. "On the significance of the principle of excluded middle in mathematics, especially in function theory", English translation of [3] in [63].
5. Brouwer, L. E. J. "Intuitionstische Betrachtungen über den Formalismus", *Koninklijke Akademie van wetenschappen de Amsterdam, Proceedings of the section of sciences* 31 (1928), 324–329. English translation with introduction in [63], 490–492. Page references are to this translation.
6. Brouwer, L. E. J. "Historical Background, Principles and Methods of Intuitionism", *South African Journal of Science* 49 (1952): 139–146
7. Brouwer, L. E. J. *Brouwer's Cambridge Lectures on Intuitionism*, D. van Dalen (ed.), Cambridge University Press, Cambridge, 1981.
8. Brown, H. C. Review of [60], *Journal of Philosophy, Psychology and Scientific Method* 3 (1906): 530–531.
9. Brown, H. C. "Infinity and the Generalization of the Concept of Number", *Journal of Philosophy, Psychology and Scientific Methods* 5 (1908): 628–634.
10. Burkhardt, H. *Einführung in die Theorie der analytischen Functionen einer complexen Veränderlichen*, Verlag Veit & Comp., Leipzig, 1897.
11. Cantor, G. "Ueber unendliche, lineare Punktmannichfaltigkeiten. 5. Fortsetzung", *Mathematische Annalen* 21 (1883): 545–591. Page references are to the reprinting in [12].
12. Cantor, G. *Grundlagen einer allgemeinen Mannigfalltigkeitslehre: ein mathematisch-philosophischer Versuch in der Lehre des Unendlichen*, Teubner, Leipzig, 1883.
13. Coolidge, J., *The Elements of Non-Euclidean Geometry*, Clarendon Press, Oxford, 1909.
14. Courant, R. & H. Robbins, *What is Mathematics?*, Oxford University Press, Oxford, 1981.
15. Dedekind, R. *Was sind und was sollen die Zahlen?*, Braunschweig: Vieweg, 1888. Reprinted in [16]. Page references are to this reprinting.
16. Dedekind, R. *Gesammelte mathematische Werke* vol. III, R. Fricke, E. Noether and O. Ore (eds.), F. Vieweg & Sohn, Braunschweig, 1932.
17. Detlefsen, M. "Formalism", in *Oxford Handbook of Philosophy of Mathematics and Logic*, Stewart Shapiro (ed.), ch. 8, 236–317, Oxford University Press, Oxford, 2005.

18. Durège, H. *Elements of the theory of functions of a complex variable, with especial reference to the methods of Riemann.* Authorized English trans. from the fourth German edition. G. Fisher & I. Schwatt, Philadelphia, 1896.

19. Eaton, R. *General Logic: An Introductory Survey*, C. Scribner's sons, New York, 1931.

20. Emch, A. "Consistency and Independence in Postulational Technique", *Philosophy of Science* 3 (1936): 185–196.

21. Evans, J. St. B. "On the conflict between logic and belief in syllogistic reasoning", *Memory and Cognition* 11 (1983): 295–306.

22. Evans, J. St. B. "In two minds: dual-process accounts of reasoning", *Trends in Cognitive Science* 7 (2003): 454–459.

23. Forgas, J. K. Williams and W. von Hippel (eds.), *Social judgments: Implicit and Explicit Processes*, Cambridge University Press, New York, 2003.

24. Frege, G. *Die Grundlagen der Arithmetik. Eine logisch-mathematische Untersuchung über den Begriff der Zahl*, W. Koebner, Breslau, 1884.

25. Frege, G. *Grundgesetze der Arithmetik, begriffsschriftlich abgeleitet* II, H. Pohle, Jena, 1903.

26. Frege, G. "Über die Grundlagen der Geometrie," *Jahresbericht der Deutschen Mathematiker-Vereinigung* 15 (1906): 293–309, 377–403, 423–430. English translation in [27].

27. Frege, G. *Gottlob Frege: On the Foundations of Geometry and Formal Theories of Arithmetic*, trans. and ed., with an introduction, by Eike-Henner W. Kluge, Yale University Press, New Haven, Conn., 1971.

28. Frege, G. *Gottlob Frege: Philosophical and Mathematical Correspondence*, G. Gabriel et al. eds., University of Chicago Press, Chicago, 1980.

29. Gauss, C. *Theoria residuorum biquadraticorum. Commentatio secunda, 1828–1831*, Dietrich, Göttingen, 1832. Reprinted with a German summary in [30], vol. II. Page references are to this reprinting.

30. Gauss, C. *Werke*, eleven volumes, Teubner, Leipzig, 1863–1903.

31. Gentzen, G. "Die Widerspruchsfreiheit der reinen Zahlentheorie", *Mathematische Annalen* 112 (1936): 493–565.

32. Gentzen, G. "Der Unendlichkeitsbegriff in der Mathematik", in *Semesterberichte Müunster*, WS 1936/37: 65–85.

33. Gentzen, G. "Unendlichkeitsbegriff und Widerspruchsfreiheit der Mathematik", in the *Travaux du IXeme Congrès internationale de philosophie*, vol. VI, Hermann, Paris, 1937.

34. Hankel, H., *Die Entwicklung der Mathematik in den letzten Jahrhunderten*, L. Fr. Fues'sche Sortiments-Buchhandlung, Tübingen, 1869.

35. Herschel, J. Review of Whewell's *History of the inductive sciences and Philosophy of the inductive sciences*, *Quarterly Review* 68 (1841): 177–238.

36. Hilbert, D. "Mathematische Probleme", *Göttinger Nachrichten* (1900): 253–297.

37. Hilbert, D. "Über die Grundlagen der Logik und der Arithmetik", reprinted in [38] as Anhang VII. Page references are to this reprinting.

38. Hilbert, D. *Grundlagen der Geometrie*, 4th ed., Teubner, Leipzig & Berlin, 1913.

39. Hilbert, D. *Probleme der mathematischen Logik*. Lecture notes (as recorded by Schönfinkel and Bernays), summer semester 1920.

40. Hilbert, D. "Über das Unendliche", *Mathematische Annalen* 95 (1926): 161–190.

41. Hilbert, D. "*Die Grundlagen der Mathematik*, mit Zusätzen von Hermann Weyl und Paul Bernays", *Hamburger Mathematische Einzelschriften* 5, Teubner, Leipzig, 1928.

42. Hilbert, D. "Naturerkennen und Logik", *Die Naturwissenschaften* 18 (1930): 959–963. Page references as in the reprinting in [44].

43. Hilbert, D. & P. Bernays, *Grundlagen der Mathematik*, vol. I, J. Springer, Berlin, 1934.

44. Hilbert, D. *Gesammelte Abhandlungen* vol. III, Springer, Berlin, 1935.

45. Hilbert, D. & W. Ackermann, *Grundzüge der theoretischen Logik*, 2nd. ed. of 1928 first edition, Berlin, J. Springer, 1938.

46. Hilbert, D. & W. Ackermann, *The Principles of Mathematical Logic*, authorized Engl. trans. of [45], Chelsea Pub. Co., New York, 1950.

47. Kant, I. *Kritik der reinen Vernunft*, J.F. Hartknoch, Riga, 1781 (2nd ed. 1787). English trans. in [48].

48. Kant, I. *Critique of Pure Reason*, P. Guyer and A. Wood (trans. and eds.), Cambridge University Press, 1998.

49. Keyser, C. "The human significance of mathematics", *Science* (new series) 42 (1089)(1915): 663–680.

50. König, J. *Neue Grundlagen der Logik, Arithmetik und Mengenlehre*, Veit, Leipzig, 1914.

51. Lieberman, M. "Reflective and reflexive judgment processes: a social cognitive neuroscience approach", in [23], 44–67.

52. Menger, K. "Die neue Logik", in *Krise und Neuaufbau in den Exakten Wissenschaften. Fünf Wiener Vorträge*, H. F. Mark *et al.* eds., F. Dueticke, Leipzig, 1933.

53. Menger, K. "The New Logic", *Philosophy of Science* 4 (1937): 299–336. English trans. of [52].

54. Moore, E. H. "On the foundations of mathematics", *Bulletin of the American Mathematical Society* 9 (1903): 402–424.

55. Nagel, E. "Intuition, Consistency, and the Excluded Middle", *Journal of Philosophy* 26 (1929): 477–489.

56. O'Hara, C. & D. Ward, *Introduction to Projective Geometry*, Oxford University Press, Oxford, 1937.

57. Peacock, G. *A treatise of algebra*, J. & J. J. Deighton, Cambridge, 1830.

58. Peano, G. "Super theorema de Cantor-Bernstein", reprinted in [59], vol. 1.

59. Peano, G., *Opere scelte 1, Analisi matematica–Calcolo numerico*. Ed. Cremonese, Rome, 1957.

60. Pieri, M. "Sur la compatibilité des axioms de l'arithmetique" *Revue de Métaphysique et de Morale* 14 (1906): 196–207.

61. Schopenhauer, A. *Die Welt als Wille und Vorstellung*, 2nd ed. Brockhaus, Leipzig, 1844. Reprinted in [62], vol. 2, 76.

62. Schopenhauer, A. *Arthur Schopenhauers Sämtliche Werke*, A. Schopenhauer

and J. Frauenstädt, eds.. Brockhaus, Leipzig, 1877.

63. Heijenoort, J. van *From Frege to Gödel: A sourcebook in mathematical logic 1879–1931*, Harvard University Press, Cambridge, MA 1967.

64. Neumann, J. von "Zur Hilbertschen Beweistheorie", *Mathematische Zeitschrift* 26 (1927): 1–46.

65. Veblen, O. & J. W. Young, *Projective Geometry* vol. I, Ginn & Co., Boston, 1910.

66. Weiss, P. "The Nature of Systems. II", *The Monist* 39(3) (1929): 440–472.

67. Weyl, H. "Diskussionsbemerkungen zu dem zweiten Hilbertschen Vortrag über die Grundlagen der Mathematik", *Abhandlungen aus dem mathematischen Seminar der Hamburgischen Universität* 6 (1928): 86–88. Reprinted in *Hamburger Einzelschriften* 5, pp. 22–24, Teubner, Leipzig, 1928. English translation in [63].

68. Weyl, H. "David Hilbert and his mathematical work", *Bulletin of the American Mathematical Society* 50 (1944): 612–654.

69. Young, J. W., W. W. Denton and U. G. Mitchell, *Lectures on Fundamental Concepts of Algebra and Geometry* Macmillan Co., New York, 1911.

A van Lambalgen theorem for Demuth randomness

David Diamondstone

Department of Mathematics, Victoria University of Wellington
Wellington, New Zealand
ddiamondstone@gmail.com

Noam Greenberg

Department of Mathematics, Victoria University of Wellington
Wellington, New Zealand
Noam.Greenberg@msor.vuw.az.nz

Dan Turetsky

Department of Mathematics, Victoria University of Wellington
Wellington, New Zealand
dturets@gmail.com

If $\mathcal{R}$ is a relativizable notion of randomness, we say that van Lambalgen's theorem holds for $\mathcal{R}$ if for all $A, B \in 2^\omega$, we have $A \oplus B \in \mathcal{R}$ if and only if $A \in \mathcal{R} \wedge B \in \mathcal{R}^A$. Van Lambalgen proved that this holds for Martin-Löf randomness. We show that van Lambalgen's theorem fails for Demuth randomness, but holds for the partial relativization Demuth$_{\mathrm{BLR}}$.

1. Introduction

1.1. *Partial relativization vs. full relativization*

Studies in algorithmic randomness have identified a hierarchy of effective randomness notions, of which the best known is Martin-Löf's. A notion of randomness is determined by a collection of statistical tests; these formalize the notion that a sequence is random if it lacks patterns which can be discerned in some sufficiently effective way. Formally, tests are null sets of reals,

The authors would like to thank the Marsden fund and André Nies for organizing the conference during which this research was carried out. We would also like to thank Kenshi Miyabe, who conjectured the result of this research. The second author was partially supported by the Marsden Fund and by a Rutherford Discovery Fellowship. The first and third authors were supported by the Marsden Fund, via postdoctoral positions.

and a randomness notion is defined by specifying a countable collection of tests, whose union is the resulting collection of non-random reals (which we here identify, via binary expansion, with infinite binary sequences). For example, a Martin-Löf test is the intersection of a nested sequence of sets $\langle U_n \rangle$ which are uniformly effectively open and which satisfy $\mu(U_n) \leqslant \epsilon(n)$, where μ is the usual Lebesgue measure and $\epsilon(n)$ is a computable sequence of rational numbers tending to 0. In general, the less effectivity we require, the larger the collections of tests and the stronger the resulting notion of randomness.

A particular way of expanding the collection of tests is by appealing to an oracle, a "black box" containing non-computable information. An oracle (such as the halting problem) may be sufficiently powerful to detect patterns in sequences, which cannot be found effectively. This process gives rise to the notion of *relative randomness*. Full relativization is the process of replacing all effective aspects of the definition of a statistical test by concepts which appeal to an oracle. To give an example, given an oracle A, an A-Martin-Löf test is a nested sequence $\langle U_n \rangle$ of sets which are A-effectively open (their basic open subsets can be enumerated with oracle A), such that $\mu(U_n) \leqslant \epsilon(n)$, where now ϵ is an A-computable sequence of rational numbers tending to 0.

Nies has pointed out that sometimes full relativization is not desirable. While trying to convert lowness notions to weak reducibilities, transitivity is usually obtained by letting only some of the computable processes allowed by the definition to appeal to an oracle, while barring others from doing so. The main example is the generalization of lowness for Martin-Löf randomness to obtain LR-reducibility: $A \leqslant_{\mathrm{LR}} B$ if every B-random sequence is also A-random, whereas full relativization to B would require every B-random sequence to be $A \oplus B$-random. Here only partial relativization results in a transitive relation.

When applied to randomness notions, often partial relativization does not give new notions. For example, in relativizing Martin-Löf randomness, one can require the measure-bounding function ϵ to be computable rather than A-computable, without changing the resulting notion of A-randomness. This is not always so.

The partial relativization of a randomness notion was first utilized by Franklin and Stephan[4] when studying lowness for Schnorr randomness, although this is not the context in which it interests us. An important basic tool in the study of relative randomness is van Lambalgen's theorem, which states that for every pair A, B of sets of natural numbers, the join $A \oplus B$ is

Martin-Löf random if and only if A is Martin-Löf random, and B is Martin-Löf random relative to A. Liang Yu[12] proved that this theorem fails when Martin-Löf randomness is replaced by Schnorr randomness. Franklin and Stephan defined an alternate relativization of Schnorr randomness (*truth-table Schnorr randomness*), and Miyabe[8] (later corrected[9]) showed that van Lambalgen's theorem holds for this notion of relative Schnorr randomness.

This prompts a general question: given a notion of randomness, is there one best way of defining its relativization to an oracle (which may not be necessarily the full relativization)? If so, what are the criteria for relativizations which are better than others? Alternately, should one think of different relativizations of a randomness notion as distinct notions, which happen to coincide when no oracles are present? In this context, Miyabe suggested that van Lambalgen's theorem should be a criterion for the "proper" relativization of a randomness notion.

Another example where a partially relativized randomness notion is better behaved than the full randomness notion comes from a paper of Bienvenu, Downey, Greenberg, Nies, and Turetsky.[2] They were interested in characterizing lowness for Demuth randomness—identifying when being Demuth random relative to a set A is equivalent to simply being Demuth random. In order to find their characterization, they were motivated to identify a partial relativization of Demuth randomness, Demuth$_{\mathrm{BLR}}$, which was easier to work with than the full relativization. Demuth randomness is defined using *Demuth tests*, which are Martin-Löf tests where the index of the nth member of the test can change a computably bounded number of times.

Definition 1.1. Given a set $W \subseteq 2^{<\omega}$, $[W]^{\prec}$ denotes the set of reals

$$\{Z \in 2^\omega : \exists \sigma \in W \, [\sigma \prec Z]\}.$$

A *Demuth test* is a sequence of c.e. open sets $\langle V_n \rangle$ such that $\mu(V_n) \leqslant 2^{-n}$ for all n, and there is an ω-c.e. function f such that $V_n = [W_{f(n)}]^{\prec}$.

A real Z *passes* the test $\langle V_n \rangle$ if Z is contained in only finitely many of the test elements. A real Z is *Demuth random* if it passes every Demuth test.

The full relativization of this definition to an oracle A replaces the c.e. open sets with open sets which are c.e. in A, and the ω-c.e. function by a function which is ω-c.e. relative to A; that is, both the approximation and the bound on the number of changes are A-computable. In contrast, a Demuth$_{\mathrm{BLR}}(A)$ test is a Demuth test relative to A where the function f is ω-*c.e. by A*—the approximation is A computable, but the bound on the number of changes is actually *computable*. The passing notion is the same for Demuth$_{\text{BLR}}$ tests and for Demuth tests: a real Z passes the test $\langle V_n \rangle$ if it is contained in only finitely many of the test elements.

By applying the partial relativization Demuth$_{\text{BLR}}$, Bienvenu et al. were able to characterize lowness for Demuth randomness. This is an example where understanding the partial relativization aids in understanding the full relativization. Bienvenu et al. write that the characterization of lowness for Demuth$_{\text{BLR}}$ is "the fundamental one", and the characterization of lowness for Demuth randomness is simply a corollary of that result and a related result of Downey and Ng.[3]

In the present work, we show that van Lambalgen's theorem holds for Demuth$_{\text{BLR}}$, but not for Demuth randomness. If one accepts Miyabe's thesis, this would imply that Demuth$_{\text{BLR}}$ is the correct relativization of Demuth randomness, rather than the full relativization. That the characterization of lowness for Demuth randomness[2] goes through Demuth$_{\text{BLR}}$, which is described as the more fundamental result, gives further evidence that Miyabe's thesis is correct, at least in the case of Demuth randomness.

1.2. Survey of van Lambalgen's theorem for various randomness notions

If $\mathcal{R}$ is a relativizable notion of randomness, we say that van Lambalgen's theorem holds for $\mathcal{R}$ if for every pair A, B, we have

$$A \oplus B \in \mathcal{R} \iff A \in \mathcal{R} \land B \in \mathcal{R}^A.$$

Van Lambalgen's theorem has been investigated for notions of randomness including Schnorr and computable randomness, n-randomness, and weak 1-randomness. Perhaps surprisingly, the right-to-left or "hard" direction of this equivalence, though harder to prove in the case of Martin-Löf randomness, holds for nearly all of the most-studied randomness notions. On the other hand, the easier to prove (in the case of Martin-Löf randomness) left-to-right direction is the one that is known to fail for several important randomness notions. We summarize the situation in Table 1.

We remark that in many of the cases where the "hard" direction is known to hold, the proof is a straightforward modification of the proof for Martin-Löf randomness. For Kolmogorov-Loveland randomness, it is not known whether the hard direction of van Lambalgen's theorem holds, but Merkle, Miller, Nies, Reimann, and Stephan[7] proved the weaker statement

$$A \oplus B \in \mathcal{KLR} \iff A \in \mathcal{KLR}^B \land B \in \mathcal{KLR}^A.$$

Randomness notion	"easy" direction	"hard" direction
weak 1-randomness	false	true
Schnorr	false[7][12]	true[59]
computable	false[7][12]	?
Kolmogorov–Loveland	true[7]	?
Martin-Löf	true[11]	true[11]
weak 2-randomness	false[1]	true[1]
n-randomness	true[11]	true[11]
Π_1^1	true[6]	true[6]

1.3. Notation

Notation is generally standard and follows Nies.[10] We use μ to denote the Lebesgue measure on 2^ω, and write $\mu(U|V)$ to denote the relative measure $\frac{\mu(U \cap V)}{\mu(V)}$ of U inside V. For $W \subseteq 2^{<\omega}$, we use $\mu(W)$ as an abbreviation for $\mu([W]^{\prec})$.

2. A van Lambalgen theorem for Demuth randomness

Theorem 2.1. $A \oplus B$ *is Demuth random if and only if* A *is Demuth random and* B *is* $\text{Demuth}_{\text{BLR}}(A)$ *random.*

We will prove the two directions separately.

Lemma 2.1. *If* $A \oplus B$ *is Demuth random, then* A *is Demuth random and* B *is* $\text{Demuth}_{\text{BLR}}(A)$ *random.*

Proof. By contraposition. If A is not Demuth random, it is clear that $A \oplus B$ is not Demuth random. Suppose $\langle [W_{g(n)}]^{\prec} \rangle$ is a $\text{Demuth}_{\text{BLR}}(A)$ test that B fails, where g has an A-computable approximation $\Phi^A(n,s)$ with mind-changes bounded by a computable function f. We can assume Φ is total and has mind-changes bounded by f on all oracles X, as we are only interested in the limit, and we can also assume that for all n, s, X, the measure $\mu(W_{\Phi^X(n,s)})$ is bounded by 2^{-n}.

What we want is to say that $A \oplus B$ fails the test

$$\left\langle \{X \oplus Y : Y \in [W_{\lim_s \Phi^X(n,s)}]^{\prec}\} \right\rangle_n.$$

This object is well-defined; the limit $\lim_s \Phi^X(n,s)$ always exists because Φ is total and has mind-changes bounded by f on all oracles. Furthermore, it must have measure at most 2^{-n} by Fubini's theorem, and captures $A \oplus B$ since $B \in [W_{\lim_s \Phi^A(n,s)}]^{\prec}$ for all n. However, it is not a Demuth test, because the natural approximation to it changes whenever $\Phi^X(n,s)$ changes, for *any* X, so potentially infinitely often, and certainly not bounded by f. To fix this, the idea is to enlarge each test element slightly by absorbing changes (increasing the measure by at most some set amount), until $\Phi^X(n,s)$ has changed at least once for some large measure of oracles, and only then changing our approximation to the nth test element.

To be more precise, for each n we define (uniformly in n) a (hopefully finite) sequence of *n-stages* $s_0 < s_1 < \ldots$ by $s_0 = 0$, and s_i is the least stage t after s_{i-1} such that

$$\mu\left(\bigcup_{s=s_{i-1}}^{t} \{X \oplus Y : Y \in [W_{\Phi^X(n,s)}]^{\prec}\} \right) > 2 \cdot 2^{-n}.$$

For all n, t, let $p_n(t)$ be the greatest *n-stage* $t' \leq t$, and let $s_n(t)$ be the least *n-stage* $t' > t$ (or $s_n(t) = \infty$ if there is no *n-stage* after t). Then we can define

$$V_n[t] = \bigcup_{p_n(t) \leq s < s_n(t)} \{X \oplus Y : Y \in [W_{\Phi^X(n,s)}]^{\prec}\},$$

and will have $\mu(V_n[t]) < 2 \cdot 2^{-n}$ for all t. As $V_n[t]$ is a c.e. open set (uniformly in n and t) and changes only at *n-stages*, $\langle \lim_t V_n[t] \rangle_n$ will be a Demuth test, provided we can exhibit a computable bound on the number of *n-stages*. If this is the case, $\lim_t V_n[t]$ will contain $\{X \oplus Y : Y \in [W_{\lim_s \Phi^X(n,s)}]^{\prec}\}$, and hence will contain $A \oplus B$.

We will show the number of *n-stages* is bounded by $2^n f(n)$. By the definition of s_i, we have

$$\mu\left(\bigcup_{s=s_{i-1}}^{s_i} \{X \oplus Y : Y \in [W_{\Phi^X(n,s)}]^{\prec}\} \right) > 2 \cdot 2^{-n}.$$

Furthermore,

$$\bigcup_{s=s_{i-1}}^{s_i} \{X \oplus Y : Y \in [W_{\Phi^X(n,s)}]^{\prec}\}$$

$$\subseteq \{X \oplus Y : Y \in [W_{\Phi^X(n,s_{i-1})}]^{\prec}\} \cup \{X \oplus Y : \Phi^X(n, \cdot)|_{[s_{i-1}, s_i]} \text{ non-constant}\}.$$

Thus

$$\mu\{X : \Phi^X(n, \cdot)|_{[s_{i-1}, s_i]} \text{ non-constant}\} > 2^{-n}.$$

Suppose there are more than $2^n f(n)$ n-stages, and for $1 \leqslant i \leqslant 2^n f(n)$, let

$$S_i = \{X : \Phi^X(n, \cdot)|_{[s_{i-1}, s_i]} \text{ non-constant}\}.$$

Since $\mu(S_i) > 2^{-n}$ for each i, by the pigeonhole principle there must be some X such that $X \in S_i$ for more than $f(n)$ distinct i, which implies that $\Phi^X(n, \cdot)$ changes more than $f(n)$ times. This contradicts the definition of Φ. □

The other direction is very similar to the proof of van Lambalgen's theorem for Martin-Löf randomness. In that proof, given a Martin-Löf test $\langle V_n \rangle$, two tests $\langle \hat{V}_n \rangle$ and $\langle U_n^X \rangle$ are constructed, the second being an oracle test, such that if $A \oplus B$ fails $\langle V_n \rangle$, then either A fails $\langle \hat{V}_n \rangle$ or B fails $\langle U_n^A \rangle$. Here we follow the same strategy, and observe that if the original test $\langle V_n \rangle$ was a Demuth test, then $\langle \hat{V}_n \rangle$ is also a Demuth test, and $\langle U_n^X \rangle$ is a Demuth$_{\mathrm{BLR}}(X)$-test.

Lemma 2.2. *If A is Demuth random and $B \in$ Demuth$_{\mathrm{BLR}}(A)$, then $A \oplus B$ is Demuth random.*

Proof. Again by contraposition. Suppose $A \oplus B$ is not Demuth random, and let $\langle V_n \rangle = \langle \lim_s V_n[s] \rangle$ be a test that it fails. Without loss of generality, assume there are infinitely many n such that $A \oplus B \in V_{2n}$ — if not, replace $2n$ with $2n + 1$. For strings σ, τ, let $[\sigma \oplus \tau]^{\prec}$ denote the clopen set $\{X \oplus Y : \sigma \prec X, \tau \prec Y\}$. Let

$$\hat{V}_n = \{\sigma : \mu(V_{2n} \mid [\sigma \oplus \varnothing]^{\prec}) > 2^{-n}\}.$$

Then $\mu(\hat{V}_n) \leqslant 2^{-n}$, or else V_{2n} violates the measure condition. Furthermore, the approximation $V_{2n}[s]$ to V_{2n} induces an approximation $\hat{V}_n[s]$ to $\hat{V}_n$ which is c.e. (uniformly in n and s) and changes only when $V_{2n}[s]$ changes, and so witnesses that $\langle \hat{V}_n \rangle$ is a Demuth test.

If A fails this Demuth test, then it is not Demuth random and we are finished. Otherwise, there is a finite set F such that for all n, m with $n \notin F$,

$$\mu(V_{2n} \mid [A{\upharpoonright}m \oplus \varnothing]) \leqslant 2^{-n}.$$

Fix $n \notin F$, and let

$$U_n = \{\tau : [A{\upharpoonright}|\tau| \oplus \tau]^{\prec} \subseteq V_{2n}\}.$$

Let U_n^l be the set of strings of length l in U_n. Then

$$\mu(U_n^l) \leqslant \mu(V_{2n} \mid [A{\upharpoonright}l \oplus \varnothing]^{\prec}) \leqslant 2^{-n}.$$

Moreover, U_n is closed upwards under $\prec$, and hence $\mu(U_n) = \lim_l \mu(U_n^l) \leqslant 2^{-n}$. Using the approximation $U_n[t]$ to U_n induced by the approximation $V_{2n}[t]$ to V_{2n}, we see that U_n is the limit of a sequence of open sets which are c.e. in A, and change only at stages when the approximation to V_{2n} changes (so at most $f(2n)$ times). Hence $\langle U_n \rangle$ is a $\text{Demuth}_{\text{BLR}}(A)$ test, which B fails. $\qquad\square$

Remark 2.1. As Demuth randomness is invariant under computable permutations, the results in this paper hold if the usual join is replaced by the Z-*join*

$$A \oplus_Z B \overset{\text{def}}{=} f_Z(A) \cup f_{\bar{Z}}(B),$$

where Z is some infinite, coinfinite computable set, and f_X is the function that enumerates X in increasing order.

3. Does a stronger version of van Lambalgen's theorem hold for Demuth randomness?

In the previous section, we showed that a version of van Lambalgen's theorem holds for the partial relativization of Demuth randomness. But what about the full relativization? Is it true that $A \oplus B$ is Demuth random if and only if A is Demuth random, and B is A-Demuth-random? Note that the "hard" direction is simply a weakening of Theorem 2.1, so the only question is the "easy" direction. This fails, because of the existence of a real in $\text{Demuth}_{\text{BLR}}(A)$ which is not Demuth random relative to A. Rod Downey and Keng Meng Ng[3] proved that lowness for Demuth randomness implies being computably dominated, by directly constructing, for each non-dominated set A, a Demuth random set B which is not Demuth random relative to B. Their construction can be partially relativized, to give the following theorem.

Theorem 3.1. *If A is not computably dominated, there is some $B \in \text{Demuth}_{\text{BLR}}(A)$ which is not Demuth random relative to A.*

Proof. The proof is a straightforward partial relativization of the Downey–Ng proof, so rather than rewriting their proof, we describe only the needed modifications to their proof.

Their proof is a construction, relative to a non-computably-dominated set A, of a real Z which is Demuth random, but not Demuth random relative to A. To make $Z \in \text{Demuth}_{\text{BLR}}(A)$, all that is needed is to replace the eth

Demuth test $\mathcal{U}^e = \langle U_x^e \rangle$ with the eth Demuth$_\mathrm{BLR}(A)$ test $\mathcal{U}^e(A) = \langle U_x^e(A) \rangle$. The rest of the construction is identical.

As this is an A-oracle construction, these sets can be uniformly enumerated just as easily as the sets U_x^e, so this does not affect the complexity of the construction. We still get the same property that $U_x^e(A)$ does not have a change of index until the eth partial order-function h_e converges on input x, and the number of changes of index is bounded by h_e. As these are Demuth$_\mathrm{BLR}$ tests, the functions h_e are computable, not merely A-computable. The verification now proceeds exactly as in Downey–Ng, and shows that there is a $\Delta_2^0(A)$ set Z which is not Demuth random relative to A, but which is not captured by any of the tests $\mathcal{U}^e(A)$, and is therefore Demuth$_\mathrm{BLR}(A)$ random. $\qquad\square$

We now get the following corollary.

Corollary 3.1. *There is a Demuth random real $A \oplus B$ such that B is not Demuth random relative to A. Moreover, A can be chosen to be an arbitrary Demuth random.*

Proof. Let A be Demuth random. By a result of Miller and Nies,[10] A is not computably dominated. By Theorem 3.1, there is some $B \in$ Demuth$_\mathrm{BLR}(A)\backslash$Demuth$(A)$. By Theorem 2.1, $A \oplus B$ is Demuth random. $\qquad\square$

References

1. G. Barmpalias, R. Downey, and K.M. Ng. Jump inversions inside effectively closed sets and applications to randomness. *J. Symbolic Logic*, 76(2):491–518, 2011.
2. L. Bienvenu, R. Downey, N. Greenberg, A. Nies, and D. Turetsky. Characterizing lowness for Demuth randomness. Unpublished, 20xx.
3. R. Downey and K.M. Ng. Lowness for Demuth randomness. In *Mathematical Theory and Computational Practice*, volume 5635 of *Lecture Notes in Computer Science*, pages 154–166. Springer Berlin / Heidelberg, 2009. Fifth Conference on Computability in Europe, CiE 2009, Heidelberg, Germany, July 19-July 24.
4. Johanna N. Y. Franklin and Frank Stephan. Schnorr trivial sets and truth-table reducibility. *J. Symbolic Logic*, 75(2):501–521, 2010.
5. Johanna N. Y. Franklin and Frank Stephan. van Lambalgen's theorem and high degrees. *Notre Dame J. Form. Log.*, 52(2):173–185, 2011.
6. G Hjorth and A Nies. Randomness in effective descriptive set theory. *J. London. Math. Soc.*, 75(2):495–508, 2007.

7. W. Merkle, J. Miller, A. Nies, J. Reimann, and F. Stephan. Kolmogorov–Loveland randomness and stochasticity. *Ann. Pure Appl. Logic*, 138:183–210, 2006.

8. K. Miyabe. Truth-table Schnorr randomness and truth-table reducible randomness. *Math. Log. Quart.*, 57(3):323–338, 2011.

9. Kenshi Miyabe and Jason Rute. van Lambalgen's theorem for uniformly relative Schnorr and computable randomness. Unpublished, 20xx.

10. A. Nies. *Computability and randomness*, volume 51 of *Oxford Logic Guides*. Oxford University Press, Oxford, 2009.

11. M. van Lambalgen. *Random Sequences*. University of Amsterdam, 1987.

12. Liang Yu. When van Lambalgen's theorem fails. *Proc. Amer. Math. Soc.*, 135:861–864, 2007.

Faithful Representations of Polishable Ideals

Dedicated to the memory of Greg Hjorth (1963–2011)

SU GAO

Department of Mathematics, University of North Texas
1155 Union Circle #311430
Denton, TX 76203, U.S.A.
* E-mail: sgao@unt.edu

We investigate faithful unitary representations for Polishable ideals, or more generally abelian Polish groups. We give various characterizations for the existence of such representations. The main technical result is to show that the density ideal does not admit any faithful unitary representations.

Keywords: faithful representation, unitary representation, Polishable ideal, summable ideal, density ideal, roundness

1. Introduction

Let $\mathcal{H}$ be an infinite-dimensional complex separable Hilbert space, henceforth called the Hilbert space. Let $U(\mathcal{H})$ or U_∞ (following Hjorth [11]) denote the group of all unitary transformations of $\mathcal{H}$ endowed with the strong (equivalently the weak) operator topology.

It was shown in [9] that any abelian Polish group is isomorphic to a topological quotient group of a closed subgroup of U_∞. This implies that any orbit equivalence relation of an action of an abelian Polish group is Borel reducible to an orbit equivalence relation of an action of U_∞ (c.f. [9] Theorem 4.4). However, it is still desirable to know which abelian Polish groups admit topological isomorphic embeddings into U_∞. Such embeddings are also known as continuous faithful unitary representations.

In [8] and [9] we already gave some characterizations for the existence of faithful unitary representations for general Polish groups and in particular for abelian Polish groups. In the current paper we will give a summary of all known characterizations, and use them to investigate the existence of faithful representations for Polishable ideals.

We will focus on two Polishable ideals: the summable ideal and the density ideal. We show that there are faithful representations for the summable ideal, and there are no faithful representations for the density ideal. For many other Polishable ideals we do not know if faithful representations exist for them.

2. Faithful representations for abelian Polish groups

In this section we summarize all known characterizations for abelian Polish groups to admit continuous faithful unitary representations. To begin, we first recall a characterization for general Polish groups in terms of positive definite functions. This has been a part of the folklore (see [16] §30); the statement and a complete proof can be found in [8].

Recall the definition of positive definite functions on a group.

Definition 2.1. *Let G be a group. A complex-valued function f on G is positive definite if for any $n \geq 1$ and arbitrary $g_1, \ldots, g_n \in G$, $c_1, \ldots, c_n \in \mathbb{C}$,*

$$\sum_{i,j=1}^{n} f(g_j^{-1} g_i) c_i \overline{c_j} \geq 0.$$

Theorem 2.1. *Let G be a Polish group and 1_G be its identity element. Let $\mathcal{F}$ be the collection of all continuous positive definite functions on G. Then the following are equivalent:*

(1) G admits a continuous faithful unitary representation.

(2) The topology of G is the weakest topology that makes functions in $\mathcal{F}$ continuous.

(3) For any open set $V \subseteq G$ with $1_G \in V$, there are functions $f_1, \ldots, f_n \in \mathcal{F}$ and open subsets $O_1, \ldots, O_n$ in $\mathbb{C}$ such that

$$1_G \in \bigcap_{i=1}^{n} f_i^{-1}(O_i) \subseteq V.$$

(4) For any closed set $F \subseteq G$ with $1_G \notin F$, there is an $f \in \mathcal{F}$ such that

$$\sup_{g \in F} |f(g)| < f(1_G).$$

(5) There is an $f \in \mathcal{F}$ such that for any closed set $F \subseteq G$ with $1_G \notin F$,

$$\sup_{g \in F} |f(g)| < f(1_G).$$

Since we are going to make use of a part of the proof in the rest of the paper, we reproduce the proof of $(1)\Rightarrow(2)$ and that of $(3)\Rightarrow(1)$ for the convenience of the reader. These proofs appeared in [8]. Note that $(2)\Rightarrow(3)$ is obvious. In the following we will use the fact that for any positive definite function f, $|f(g)| \leq 1_G$ for all $g \in G$.

$(1)\Rightarrow(2)$: For any $v \in \mathcal{H}$, let

$$f_v(g) = \langle g(v), v \rangle, \text{ for all } g \in U_\infty.$$

Then f_v is a continuous positive definite function on U_∞. The continuity is immediate from the definition of the weak operator topology on U_∞, and the positive definiteness is established by the standard computation:

$$\sum_{i,j=1}^{n} f_v(g_j^{-1}g_i)c_i\overline{c_j} = \sum_{i,j=1}^{n} \langle g_j^{-1}g_i(v), v \rangle c_i\overline{c_j} = \sum_{i,j=1}^{n} \langle g_i(v), g_j(v) \rangle c_i\overline{c_j}$$

$$= \sum_{i,j=1}^{n} \langle c_i g_i(v), c_j g_j(v) \rangle = \left\langle \sum_{i=1}^{n} c_i g_i(v), \sum_{j=1}^{n} c_j g_j(v) \right\rangle = \left\| \sum_{i=1}^{n} c_i g_i(v) \right\|^2 \geq 0.$$

Now by the polar identity

$$4\langle g(x), y \rangle = \langle g(x+y), x+y \rangle - \langle g(x-y), x-y \rangle$$
$$+ i\langle g(x+iy), x+iy \rangle - i\langle g(x-iy), x-iy \rangle,$$

any topology of U_∞ that makes the demonstrated positive definite functions continuous must make the functionals $f_{x,y}(g) = \langle g(x), y \rangle$ continuous as well. Thus (2) is proved.

$(3)\Rightarrow(1)$: Fix a countable family $\{f_n\}_{n\in\mathbb{N}}$ of continuous positive definite functions on G so that they generate a neighborhood basis of 1_G in the sense of (3). Without loss of generality we can assume $f_n(1_G) \leq 2^{-n}$ for all $n \in \mathbb{N}$.

Consider the space X of all complexed-valued functions on G with finite support, i.e., functions $x : G \to \mathbb{C}$ such that for all but finitely many $g \in G$, $x(g) = 0$. X is a linear space under addition and scalar multiplication. For $x, y \in X$, let

$$\langle x, y \rangle = \sum_{g,h\in G} \sum_{n\in\mathbb{N}} f_n(h^{-1}g)x(g)\overline{y(h)}.$$

This sesquilinear form is well defined since for any $g, h \in G$,

$$\left| \sum_n f_n(h^{-1}g) \right| \leq \sum_n |f_n(h^{-1}g)| \leq \sum_n f_n(1_G) < \infty.$$

Let $N = \{x \in X \mid \langle x, x \rangle = 0\}$. Then N is a linear subspace of X and $\langle \cdot, \cdot \rangle$ is defined on the quotient X/N, making it a pre-Hilbert space. Let H be the completion of X/N under the induced $\langle \cdot, \cdot \rangle$. Then H is a separable complex Hilbert space.

We consider the representation of G in $U(H)$ given by the definition that, for each $g \in G$ and $x \in X$,

$$T_g x(h) = x(g^{-1}h).$$

For any $g_0 \in G$ and $x, y \in X$, we have that

$$\langle T_{g_0} x, T_{g_0} y \rangle = \sum_{g,h,n} f_n(h^{-1}g) x(g_0^{-1}g) \overline{y(g_0^{-1}h)}$$
$$= \sum_{g,h,n} f_n(h^{-1}g) x(g) \overline{y(h)} = \langle x, y \rangle.$$

The map $g \mapsto T_g$ is obviously a group homomorphism. For any $g_0 \in G$ and $x \in X$, we have that

$$\langle T_{g_0} x, y \rangle = \sum_{g,h,n} f_n(h^{-1}g) x(g_0^{-1}g) \overline{y(h)} = \sum_{g,h,n} f_n(h^{-1}g_0 g) x(g) \overline{y(h)}.$$

It follows immediately from this and the continuity of f_n that $g \mapsto T_g$ is continuous.

We next check that $g \mapsto T_g$ is injective. For this it suffices to show that for $g_0 \neq 1_G$, $T_{g_0} \neq I$. Suppose $g_0 \neq 1_G$. There exists some $n \in \mathbb{N}$ such that $f_n(g_0) \neq f_n(1_G)$. Consider $x_0 \in X$ defined by

$$x_0(g) = \begin{cases} 1, & \text{if } g = g_0, \\ 0, & \text{otherwise.} \end{cases}$$

Then the n-th term of $\langle T_{g_0} x_0 - x_0, T_{g_0} x_0 - x_0 \rangle$ is

$$\sum_{g,h} f_n(h^{-1}g)(x_0(g_0^{-1}g) - x_0(g))\overline{(x_0(g_0^{-1}h) - x_0(h))}$$
$$= 2(f_n(1_G) - \operatorname{Re} f_n(g_0)) > 0.$$

It follows that $\langle T_{g_0} x_0 - x_0, T_{g_0} x_0 - x_0 \rangle > 0$, and therefore $T_{g_0} \neq I$.

Now to establish that $g \mapsto T_g$ is a topological group isomorphic embedding, the only thing that remains to be checked is that the inverse of $g \mapsto T_g$ is continuous. For this suppose $T_{g_m} \to T_{g_\infty}$, as $m \to \infty$, for group elements $g_m, g_\infty \in G$. We are to show that $g_m \to g_\infty$, as $m \to \infty$. Consider

$$x_0(g) = \begin{cases} 1, & \text{if } g = 1_G, \\ 0, & \text{otherwise,} \end{cases} \quad \text{and} \quad y_0(h) = \begin{cases} 1, & \text{if } g = g_\infty, \\ 0, & \text{otherwise.} \end{cases}$$

Then $\langle T_{g_m} x_0, y_0 \rangle \to \langle T_{g_\infty} x_0, y_0 \rangle$ as $m \to \infty$. But a straightforward computation shows that

$$\langle T_{g_m} x_0, y_0 \rangle = \sum_n f_n(g_\infty^{-1} g_m),$$

and

$$\langle T_{g_\infty} x_0, y_0 \rangle = \sum_n f_n(1_G).$$

These imply that for all $n \in \mathbb{N}$, $f_n(g_\infty^{-1} g_m) \to f_n(1_G)$ as $m \to \infty$. From our assumption on $\{f_n\}$ it follows that $g_\infty^{-1} g_m \to 1_G$, or $g_m \to g_\infty$, as $m \to \infty$. Thus (3)$\Rightarrow$(1) is proved.

Before continuing our discussion, we recall the following definition of Herer and Christensen ([10]).

Definition 2.2. *A topological group is exotic if it has no nontrivial continuous unitary representations.*

Thus being exotic is the opposite of having faithful representations, and yet we will see below that it can be characterized in terms of positive definition functions. The following is a corollary of the proof of Theorem 2.1.

Corollary 2.3. *A Polish group G is exotic iff there are no nontrivial continuous positive definite functions on G.*

Proof. If G is not exotic, then the proof of (1)$\Rightarrow$(2) shows that G has at least one nontrivial (i.e., non-constant) continuous positive definite function. On the other hand, a single nontrivial continuous positive definite function on G would induce a nontrivial strongly continuous unitary representation of G in some separable complex Hilbert space, by the proof of (3)$\Rightarrow$(1) of Theorem 2.1. □

Herer and Christensen constructed the first example of an exotic group ([10]). Later Banaszczyk found a class of exotic groups of the form E/K where E is an infinite dimensional normed space and K is a discrete subgroup of E ([3]). More recently Megrelishvili proved that $\text{Hom}_+([0,1])$, the group of all order-preserving homeomorphisms of $[0,1]$, is exotic ([14]). All these examples (when E is taken to be a separable Banach space in Banszczyk's examples) are Polish groups. Exotic Polish groups fail to have any nontrivial unitary representations, let alone faithful representations.

Turning back to faithful representations, we consider the Banach spaces $L_p([0,1])$ and l_p for $1 \leq p \leq 2$. The additive groups of these spaces are abelian Polish groups. The following result of Schoenberg [18] is well known.

Theorem 2.2. *For $1 \leq p \leq 2$, the function $e^{-|x|^p}$ is positive definite on $\mathbb{R}$, l_p or $L_p([0,1])$. Consequently, the additive groups of l_p and $L_p([0,1])$ for $1 \leq p \leq 2$ all admit continuous faithful unitary representations.*

It turns out that there is a universal abelian Polish group admitting continuous faithful unitary representations.

Notation 2.4. *Let $L_0([0,1], \mathbb{T})$ denote the multiplicative group of all Lebesgue measurable functions from $[0,1]$ into the unit circle $\mathbb{T}$ in $\mathbb{C}$ up to λ-a.e. equality. The topology on $L_0([0,1], \mathbb{T})$ is given by convergence in measure.*

$L_0([0,1], \mathbb{T})$ is obviously an abelian group. Moreover, $L_0([0,1], \mathbb{T})$ admits a continuous faithful unitary representation. To see this, we associate for each $f \in L_0([0,1], \mathbb{T})$ a transformation

$$T_f : L_2([0,1]) \longrightarrow L_2([0,1])$$
$$g \longmapsto fg$$

Then the map $f \mapsto T_f$ is a topological group isomorphic embedding from $L_0([0,1], \mathbb{T})$ into U_∞. Alternatively, noting that the constant one function $\mathbf{1}$ is the identity of $L_0([0,1], \mathbb{T})$, one can check that the metric

$$d(f,g) = \int_0^1 1 - e^{-|f(x)-g(x)|} dx$$

is an invariant (complete) compatible metric on $L_0([0,1], \mathbb{T})$ and the function

$$F(g) = e^{-d(g,\mathbf{1})}$$

is a continuous positive definite function on $L_0([0,1], \mathbb{T})$ separating $\mathbf{1}$ and closed sets of $L_0([0,1], \mathbb{T})$ not containing $\mathbf{1}$.

The following theorem is another folklore.

Theorem 2.3. *Let G be an abelian Polish group. Then G admits a faithful unitary representation iff*

(6) G is isomorphic to a subgroup of $L_0([0,1], \mathbb{T})$.

The numbering of the clause follows that in Theorem 2.1. The statement (in a stronger form) and a sketch of proof can be found in [9] (as Theorem 3.4 there). Since we are going to refer to the proof, we sketch the proof again for the convenience of the reader (the proof given below is more detailed than that in [9]). Basic facts about von Neumann algebras used in the proof can be found in [5].

Sketch of proof. We have seen the if direction above. For the only if direction, let G be an abelian closed subgroup of U_∞. Let $A \subset L(\mathcal{H})$ be the commutative sub-C^*-algebra generated by G. Let M be the abelian von Neumann algebra generated by A. Then M is countably generated (c.f., e.g., [21], §9, Corollary 25). Let D be a countable generating set in M.

Now let Q be a countable dense subset of G (in the weak operator topology) which contains the identity map. Let $B \subset L(\mathcal{H})$ be the sub-C^*-algebra generated by $D \cup Q$. Then by [5], Chapter 3, Corollary 1, M is the weak closure of B. Since B is a separable commutative C^*-algebra, it has a compact Polish spectrum K by the Gelfand-Naimark theorem. By Proposition 1 of Chapter 7 in [5], there is a positive Radon measure ν on K and an isometric isomorphism of normed *-algebras from M onto $L^\infty(K,\nu)$ which extends the Gelfand isomorphism from B onto $C(K,\mathbb{C})$, where $L^\infty(K,\nu)$ is endowed with the weak* topology as the dual of $L^1(K,\nu)$.

Every element of Q, being a unitary transformation, is under the Gelfand isomorphism correspondent to some element f of $C(K,\mathbb{C})$ such that $|f(x)| = 1$ for all $x \in K$ (since, by the proof of the Gelfand-Naimark theorem, for each $x \in K$, $f(x)$ is in the spectrum of the transformation correspondent with f). Since Q is dense in G (in the weak topology), it follows that every element of G is under the above isomorphism (from M onto $L^\infty(K,\nu)$) correspondent to some element f of $L^\infty(K,\nu)$ with $|f(x)| = 1$ for ν-a.e. $x \in K$. That is to say, there is an embedding from G into

$$\mathcal{G} = \{f : K \to \mathbb{C} \,|\, |f(x)| = 1 \text{ for } \nu\text{-a.e. } x \in K \text{ and } f \text{ is } \nu\text{-measurable}\},$$

where the group operation on $\mathcal{G}$ is multiplication and the topology is the weak* topology as the dual of $L^1(K,\nu)$. It is a part of a topological *-algebra embedding, hence it is in particular a topological group embedding.

It remains to isomorphically embed $\mathcal{G}$ into $L_0([0,1],\mathbb{T})$. Since ν is a Radon measure, we may assume that ν is a probability Borel measure. Let μ and δ be respectively the continuous and the discrete part of ν. Let $\alpha = \mu(K)$, $X = \{x_0, x_1, \dots\} \subseteq K$ be such that $\delta(x_i) > 0$ for all i and $\delta(K) = \sum_{x \in X} \delta(x)$. Let $Y = [0,\alpha] \subseteq [0,1], Y_0, Y_1, \dots$ be a partition of $[0,1]$ such that $\lambda(Y_i) = \delta(x_i)$ for all i. Let $\theta : K \to Y$ be a Borel isomorphism of

μ onto $\lambda \upharpoonright Y$. For each $f \in \mathcal{G}$, let $F_f \in L_0([0,1], \mathbb{T})$ be given by the obvious translation:

$$F_f(y) = \begin{cases} f(\theta^{-1}(y)), & \text{if } y \in Y, \\ f(x_i), & \text{if } y \in Y_i \text{ for some } i. \end{cases}$$

Then $f \mapsto F_f$ is a topological group isomorphic embedding from $\mathcal{G}$ into $L_0([0,1], \mathbb{T})$. $\qquad\square$

In another direction to strengthen Theorem 2.3, Aharoni, Maurey and Mitiyagin [2] investigated the case when G is a linear space. A more explicit characterization was obtained in this case. However, an implicit proof for this characterization had already appeared in an earlier paper [10] by Herer and Christensen.

Theorem 2.4. *Let G be a Polish linear space. Then the additive group G has a continuous faithful unitary representation iff G is linearly isomorphic to a subspace of $L_0([0,1])$, where $L_0([0,1])$ is the additive group of real-valued Lebesgue measurable functions on $[0,1]$ with the topology of convergence in measure.*

In particular, the theorem applies to all separable Banach spaces. It is well known that, for $p > 2$, no $L_p(\mu)$ space is isomorphic to a subspace of $L_0([0,1])$. (A proof can be found in [4], page 194.) Thus, in particular, no l_p for $p > 2$ admits continuous faithful unitary representations.

We now turn to another characterization for abelian Polish groups to admit continuous faithful unitary representations. To state the results we need to recall the notion of uniform homeomorphism between metric spaces.

Definition 2.5. *Let (X_1, d_1) and (X_2, d_2) be metric spaces. A uniform homeomorphism from (X_1, d_1) onto (X_2, d_2) is a homeomorphism j from X_1 onto X_2 such that both j and j^{-1} are uniformly continuous.*

If G is an abelian Polish group and d_1, d_2 are two invariant compatible metrics on G, then the identity map $i : G \to G$ is a uniform homeomorphism from (G, d_1) onto (G, d_2). Thus it is irrelevant which invariant metric is used when we talk about uniform homeomorphism from an abelian Polish group into some other metric space.

The following theorem is proved in [2] by Aharoni, Maurey and Mityagin. In the statement the space l_2 can be replaced by any infinite dimensional separable real or complex Hilbert space. We also recall the proof so as to draw some corollaries later.

Theorem 2.5. *Let G be an abelian Polish group. Then G admits continuous faithful unitary representations iff either of the following holds:*

(7) G is uniformly homeomorphic to a subset of l_2.
(7′) G is uniformly homeomorphic to a subset of the unit sphere of l_2.

Sketch of proof. We show the equivalence of (5) with (7) and (7′).

(5)⇒(7): Suppose f is a real-valued continuous positive definite function on G separating 1_G and closed sets not containing 1_G. Following the notation in the proof of (3)⇒(1) but using only a single function f rather than a sequence (f_n), let X, N, H and the inner product on H be defined. Define $j : G \to X$ by

$$j(g)(h) = \begin{cases} 1, & \text{if } g = h, \\ 0, & \text{otherwise.} \end{cases}$$

This induces an embedding from G into H, which we still denote by j. Then it is easy to see that, for any $g, h \in G$,

$$\langle j(g), j(h) \rangle = f(h^{-1}g),$$

and

$$\|j(g) - j(h)\|^2 = 2(f(1_G) - f(h^{-1}g)).$$

Thus j is uniformly continuous by the continuity of f, and j^{-1} is uniformly continuous by the separating property of f.

(7)⇔(7′): Note that in the proof of (5)⇒(7) if we normalize the function f so that $f(1_G) = 1$, then the image of j lies in the unit sphere of the Hilbert space. In particular l_2 itself is uniformly homeomorphic to a subset of its unit sphere. This gives (7)⇒(7′). The other direction is trivial.

(7′)⇒(5): Let j be a uniform homeomorphism from G onto a subset of the unit sphere of a real Hilbert space. By the amenability of G there exists a finitely additive invariant mean M on the bounded functions on G, i.e., for any real-valued bounded functions f_1, f_2 on G and $g \in G$, $M(f_1 + f_2) = Mf_1 + Mf_2$, $Mf_g = Mf$, where $f_g(x) = f(x + g)$ and $Mf_1 \geq 0$ for $f_1 \geq 0$. Then define

$$f(g) = M\langle j_g, j \rangle.$$

This f is a continous positive definite function on G separating 1_G and closed subsets. □

The proof of the direction (5)⇒(7) does not require that G is abelian. Thus we have established in effect that U_∞ is uniformly homeomorphic to a subset of l_2. This fact was explicitly stated in [15] and proved by a direct argument. Here the metric on U_∞ can be chosen to be any compatible one, as the proof is insensitive to other aspects of the metric. Characterization (7) can be used to prove that certain abelian Polish groups do not have continuous faithful unitary representations. Enflo [6] constructed a countable metric space not uniformly embeddable into l_2. On the other hand, a classical result of Banach-Mazur states that any separable metric space is isometric to a subset of $C([0, 1])$, the space of real-valued continuous functions on $[0, 1]$ (which is also an abelian Polish group under pointwise addition). Thus it follows that $C([0, 1])$ is not uniformly homeomorphic to any subset of l_2, and hence it does not have any continuous faithful unitary representations. Analogously, Aharoni [1] proved that any separable metric space is uniformly homeomorphic to a subset of the Banach space c_0. Hence the additive group of c_0 is not uniformly homeomorphic to any subset of l_2, and thus it does not have any continuous faithful unitary representations.

3. Faithful representations for Polishable ideals

Polishable ideals form an important class of abelian Polish groups. They are well studied by descriptive set theorists (c.f., e.g., Farah [7], Louveau and Velickovic [13], Oliver [17], Solecki [19] [20], Velickovic [22], etc.; also c.f. Kanovei [12]). An ideal $\mathcal{I}$ over $\mathbb{N}$ is an abelian group with the group operation $A \triangle B$, where $\triangle$ is the symmetric difference. As a group, $\mathcal{I}$ is of exponent 2, i.e., every element is an involution (i.e. of order 2). Recall that a Borel ideal $\mathcal{I}$ over $\mathbb{N}$ is *Polishable* if there is a Polish group topology on $\mathcal{I}$ giving rise to its Borel structure.

Our first result below is a characterization for Polish groups of exponent 2 to admit faithful unitary representations.

Notation 3.1. *Let $MALG_\triangle$ be the group of all Lebesgue measurable subsets of $[0, 1]$ up to λ-a.e. equality, where λ is the Lebesgue measure, with the group operation $A \triangle B$ being the symmetric difference and the topology given by the metric $\lambda(A \triangle B)$.*

The underlying space is the usual measure algebra of Lebesgue measurable sets. It does not make a difference if we replace $[0, 1]$ by other familiar spaces, such as the Cantor space $2^\mathbb{N}$. $MALG_\triangle$ is an also abelian Polish group of exponent 2. We have the following corollary from Theorem 2.3.

Corollary 3.2. *Let G be an abelian Polish group of exponent 2. Then*

G has a continuous faithful unitary representation iff G is isomorphic to a subgroup of $MALG_\triangle$. In particular, the statement applies to Polishable ideals.

Proof. For every Lebesgue measurable $A \subseteq [0,1]$, let χ_A be its characteristic function. Let $f_A = 2\chi_A - 1$. Then $A \mapsto f_A$ is a topological group isomorphic embedding from $MALG_\triangle$ into $L_0([0,1], \mathbb{T})$. On the other hand, every involution $g \in L_0([0,1], \mathbb{T})$ takes values in $\{-1, 1\}$ and hence can be identified with a Lebesgue measurable set by an inverse process. $\square$

Consider the summable ideal

$$I_0 = \left\{ A \subseteq \mathbb{N} : \sum_{n \in A} \frac{1}{n+1} < \infty \right\}$$

of subsets of $\mathbb{N}$ with the metric

$$d(A,B) = \sum_{n \in A \triangle B} \frac{1}{n+1}$$

for $A, B \in I_0$. d is a complete metric.

Proposition 3.3. *I_0 has continuous faithful unitary representations.*

Proof. There is an obvious isometric embedding from the space I_0 into l_1. Namely, for $A \in I_0$, define $g_A \in l_1$ by letting

$$g_A(n) = \begin{cases} \dfrac{1}{n+1}, & \text{if } n \in A, \\ 0, & \text{otherwise.} \end{cases}$$

Then $A \mapsto g_A$ is such an embedding. Now consider the function $f : I_0 \to \mathbb{R}$ defined by

$$f(A) = e^{-\|g_A\|}.$$

Then f is continuous on I_0 and moreover, f separates the identity of I_0 (the empty set) and closed sets of I_0 not containing its identity. To see that f is positive definite on I_0, it suffices to notice that, for $A, B \in I_0$, we have that $A^{-1} = A$ and

$$f(A \triangle B) = e^{-\|g_{A \triangle B}\|} = e^{-\|g_B - g_A\|}.$$

Thus the positive definiteness of f on I_0 follows from that of $e^{-\|x\|}$ on l_1. $\square$

In some sense, this proposition is not surprising since it is well known that the summable ideal is closely related to the Banach space l_1. Note, however, that I_0 is not isomorphic to a closed subgroup of l_1. In fact there is even no nontrivial homomorphism from I_0 into l_1. This is for a trivial algebraic reason: I_0 is torsion and l_1 is torsion free.

Next we consider another important example of Polishable ideal, the density ideal

$$I_d = \left\{ A \subseteq \mathbb{N} : \lim_n \frac{\operatorname{card}(A \cap n)}{n} = 0 \right\}$$

where $A \cap n = A \cap \{0, 1, \ldots, n-1\}$, endowed with the metric

$$d(A, B) = \sup_n \frac{\operatorname{card}((A \triangle B) \cap n)}{n}.$$

I_d is an abelian Polish group with the group operation $A \triangle B$. We demonstrate that I_d does not have any continuous faithful unitary representations.

Theorem 3.1. I_d *is not uniformly homeomorphic to any subset of l_2. Hence it does not have any continuous faithful unitary representations.*

The rest of this section is devoted to a proof of Theorem 3.1. A main ingredient of the proof is Enflo's notion of (generalized) roundness of metric spaces ([6]).

Definition 3.4. *Let (X, d) be a metric space. The roundness of (X, d) is the supremum of $p \geq 0$ such that for any $n > 1$ and arbitrary $x_1, \ldots, x_n, y_1, \ldots, y_n \in X$,*

$$\sum_{1 \leq i, j \leq n} d(x_i, y_j)^p \geq \sum_{1 \leq i < j \leq n} [d(x_i, x_j)^p + d(y_i, y_j)^p]. \qquad (3.1)$$

With the convention $0^0 = 1$ inequality (3.1) holds for any metric space for $p = 0$, since the left hand side is n^2 and the right hand side is $n(n-1)$. Thus the roundness for any metric space is ≥ 0. There are metric spaces whose roundness is ∞. However, if there are distinct $x, y, z \in X$ such that

$$d(x, z) = 2d(x, y) = 2d(y, z),$$

(which is called the *metric midpoint property* by Enflo,) then the roundness of X must be ≤ 2 (by choosing $x_1 = x$, $x_2 = z$ and $y_1 = y_2 = y$). It is easy to check that all Euclidean spaces $\mathbb{R}^n$ have roundness exactly 2. Similarly, l_2 also has roundness 2.

We next define a countable metric space N_∞ which will be uniformly homeomorphic to a subset of I_d but not of any space of positive roundness. N_∞ is constructed in stages, as follows.

For each $n > 1$, first define a metric space (P_n, d) as follows. The set P_n consists of 2^n many elements

$$p_0, p_1, \ldots, p_{2^n - 1}.$$

The metric is defined as

$$d(p_i, p_j) = \begin{cases} \dfrac{|i - j|}{2^n}, & \text{if } |i - j| \leq 2^{n-1}, \\ 1 - \dfrac{|i - j|}{2^n}, & \text{if } |i - j| > 2^{n-1}. \end{cases}$$

Next let $M_n = (P_n)^{n^n}$, i.e., M_n is the product of n^n copies of P_n, with the metric on M_n (also called d) defined by

$$d(\vec{x}, \vec{y}) = \sup\{d(x_k, y_k) \mid 1 \leq k \leq n^n\},$$

for $\vec{x} = (x_1, \ldots, x_{n^n})$ and $\vec{y} = (y_1, \ldots, y_{n^n})$. Finally, let $N_n = \bigsqcup_{1 < m \leq n} M_m$, the disjoint union of M_m for $1 < m \leq n$, with the metric (denoted d again) given by $d(x, y)$ whenever x, y belong to the same copy of M_m and 1 otherwise.

Eventually let $N_\infty = \bigsqcup_{m > 1} M_m$ with a similar metric as above. Then N_∞ is exactly the direct limit of all spaces N_n under the natural inclusions. This finishes the construction of N_∞. The definitions of P_n and M_n above are adapted from Enflo's proof in [6].

We check that N_∞ is uniformly homeomorphic to a subset of I_d. To see this we need to recall the notion of ε-isometry and the lemma following the definition.

Definition 3.5. *Let $\varepsilon \geq 0$ and $(X_1, d_1), (X_2, d_2)$ be metric spaces. An ε-isometry from (X_1, d_1) onto (X_2, d_2) is a homeomorphism $\varphi : X_1 \to X_2$ such that, for every $x, y \in X_1$,*

$$d_1(x, y) - \varepsilon \leq d_2(\varphi(x), \varphi(y)) \leq d_1(x, y) + \varepsilon.$$

Lemma 3.1. *For any $s \in 2^{<\mathbb{N}}, n > 1$ and $\varepsilon > 0$, there are $K > lh(s)$ and an ε-isometric embedding φ from M_n into I_d such that for any $A \subseteq \mathbb{N}$ in the image of φ, we have*

(i) $s \subset A$, i.e., for any $m < lh(s)$, $s(m) = 1$ iff $m \in A$; and
(ii) $A \subseteq [0, K)$, i.e., for any $m \geq K$, $m \notin A$.

Proof. We may assume $\varepsilon < 2^{-n-1}$ and $lh(s) > 0$. Let b_k, c_k, $1 \leq k \leq n^n$ be natural numbers such that

(a) $b_1 > \varepsilon^{-1} 2^{2n} \mathrm{lh}(s)$;

(b) for each $1 \leq k < n^n$, $b_{k+1} > \varepsilon^{-1} 2^{2n} c_k$;

(c) for each $1 \leq k \leq n^n$, $c_k > (1 + \varepsilon^{-1}) b_k$; and

(d) for each $1 \leq k \leq n^n$, $c_k - b_k$ is a multiple of 2^n.

The conditions (a)-(c) are simply largeness conditions for b_1 relative to $\mathrm{lh}(s)$, c_k relative to b_k, and b_{k+1} relative to c_k. By choosing these numbers consecutively and large enough these conditions can be easily arranged. The last condition is also easy to meet.

For a fixed natural number $l \geq 1$ and a $0,1$-sequence t of length l, we may think of t as a subset of $\{1, \ldots, l\}$ by identifying t with the set $\{1 \leq j \leq l \mid t(j) = 1\}$. If t_1 and t_2 are both $0,1$-sequences of length l, recall that the *Hamming distance* between t_1 and t_2 is given by

$$H(t_1, t_2) = \mathrm{card}(t_1 \triangle t_2).$$

For each $l \geq 1$ there exists a sequence $q_0, q_1, \ldots, q_{2l-1}$, where each q_j is a $0,1$-sequence of length l, such that

$$H(q_i, q_j) = \begin{cases} |i - j|, & \text{if } |i - j| \leq l, \\ 2l - |i - j|, & \text{if } |i - j| > l. \end{cases}$$

This can be seen easily by an induction on l. The base case $l = 1$ is trivial. In general, if $q_0, \ldots, q_{2l-1}$ is a sequence that works for l, then

$$q_0 {}^\frown 0, q_1 {}^\frown 0, \ldots, q_l {}^\frown 0, q_l {}^\frown 1, q_{l+1} {}^\frown 1, \ldots, q_{2l-1} {}^\frown 1, q_{2l-1} {}^\frown 0$$

is a sequence that works for $l + 1$.

Note that we may arbitrarily lengthen the sequences $q_0, \ldots, q_{2l-1}$ by appending 0s without altering the Hamming distance between any two elements in the sequence. For the rest of the proof we fix $l = 2^{n-1}$ and $q_0, \ldots, q_{2^n-1} = q_{2l-1}$, where each q_j has length 2^n, with the Hamming distances specified above.

Now for every $\vec{x} \in M_n$, we define $\varphi(\vec{x})$ to be a set $A \subseteq \mathbb{N}$ by the following requirements:

- if $m < \mathrm{lh}(s)$, then $m \in A$ iff $s(m) = 1$;
- if $\mathrm{lh}(s) \leq m < b_1$, then $m \notin A$;
- if $c_k \leq m < b_{k+1}$, then $m \notin A$;
- if $m \geq c_{n^n}$, then $m \notin A$;
- if $b_k \leq m < c_k$, and $x_k = p_i \in P_n$, then

$$m \in A \text{ iff } q_i(m - b_k \bmod 2^n) = 1.$$

Intuitively, the construction is to put down a large number of copies of q_i if the corresponding coordinate of the element of M_n is p_i.

It is easy to see that (i) and (ii) hold with $K = c_{n^n}$. To see that φ is an ε-isometry, it is straightforward to check, using conditions (a)-(d) and the construction above, that for any $\vec{x}, \vec{y} \in M_n$, letting $A = \varphi(\vec{x})$ and $B = \varphi(\vec{y})$, the density

$$\frac{\text{card}((A \triangle B) \cap m)}{m}$$

as a function of m approaches $d(x_k, y_k)$ from below with an error $< \varepsilon$ for each $1 \le k \le n^n$ when m approaches c_k, and its value is below ϵ when m approaches each b_k. Condition (c) guarantees that there are enough copies of q_i so that eventually the density is greater than $d(x_k, y_k) - \epsilon$. Thus

$$d(\vec{x}, \vec{y}) - \varepsilon \le d(A, B) \le d(\vec{x}, \vec{y}). \qquad \square$$

We are now ready to see that N_∞ is uniformly homeomorphic to a subset of I_d. For each $n > 1$ let $\varepsilon_n = 2^{-2n}$. By induction on n we define an embedding $\varphi_n : N_n \to I_d$. To begin with, let $\varphi_2 : N_2 = M_2 \to I_d$ be the ε_2-isometric embedding obtained by Lemma 3.1 with $s_2 = \langle 0 \rangle$. In general, assume that $\varphi_n : N_n \to I_d$ has been defined and moreover there is K_n such that for every $A \subseteq \mathbb{N}$ in the image of φ_n, $A \subseteq [0, K_n)$. Let $L_{n+1} > \varepsilon_n^{-1} K_n$ and s_{n+1} have length L_{n+1} and be defined by $s_{n+1}(m) = 1$ for all $m < L_{n+1}$. Let $\psi_{n+1} : M_{n+1} \to I_d$ be the ε_{n+1}-isometric embedding obtained by Lemma 3.1 with s_{n+1}. Letting $\varphi_{n+1} = \varphi_n \sqcup \psi_{n+1}$, this finishes the construction of the φ_n's. From our construction it makes sense to define φ as the increasing union of φ_n for $n > 1$.

We claim that φ is a uniformly homeomorphic embedding from N_∞ into I_d. In fact, for any $\varepsilon > 0$, let n_0 be such that $\varepsilon_{n_0} < \varepsilon$ and let

$$\delta = \min \left\{ \frac{1}{2^{n_0}}, \frac{\varepsilon - \varepsilon_{n_0}}{2} \right\}.$$

Then for $x, y \in N_\infty$ with $d(x, y) < \delta$, there is $n > n_0$ such that x, y are both in M_n. Thus

$$d(\varphi(x), \varphi(y)) < d(x, y) + \varepsilon_n < \delta + \varepsilon_{n_0} < \varepsilon.$$

The uniformity of φ^{-1} is similar.

To finish the proof of Theorem 3.1, it remains to show that N_∞ is not uniformly homeomorphic to any subset of a metric space of positive roundness. For this we use an argument similar to Enflo's proof in [6].

Consider a fixed M_n, where n is even. We need two more definitions.

Definition 3.6. *For $1 \leq m \leq n$, an m-segment in M_n is a pair $(\vec{x}, \vec{y})$ of elements such that*

$$card(\{1 \leq k \leq n^n \mid x_k \neq y_k\}) = n^m$$

and for all $1 \leq k \leq n^n$ with $x_k \neq y_k$, $d(x_k, y_k) = 2^{-m}$.

Definition 3.7. *For $1 \leq m \leq n$, a set S of elements of M_n is m-regular if $S = S_1 \cup S_2$ with $card(S_1) = card(S_2) = n$ such that, for all $\vec{x}, \vec{y} \in S$, the pair $(\vec{x}, \vec{y})$ is an m-segment in case $\vec{x}, \vec{y} \in S_1$ or $\vec{x}, \vec{y} \in S_2$, and an $m + 1$-segment otherwise.*

The following lemma is not an optimal result, but is sufficient for our purpose.

Lemma 3.2. *For n even and $1 \leq m \leq \dfrac{n}{2}$, there is an m-regular set in M_n.*

Proof. For $j = 1, \ldots, 2n$, let $I_j = \left[\dfrac{n^m}{2}(j-1), \dfrac{n^m}{2}j \right)$ and $J = [n^{m+1}, n^n)$. Then $I_1, \ldots, I_{2n}, J$ is a decomposition of $[0, n^n)$. Define $S_1 = \{\vec{x^1}, \ldots, \vec{x^n}\}$ and $S_2 = \{\vec{y^1}, \ldots, \vec{y^n}\}$ by letting, for $i = 1, \ldots, n$,

$$x_k^i = \begin{cases} p_{2^m}, & \text{if } k \in I_i, \\ p_0, & \text{if } k \in I_j \text{ for } 1 \leq j \leq n \text{ but } j \neq i, \\ p_{2^{m+1}}, & \text{if } k \in I_j \text{ for } n < j \leq 2n, \\ p_0, & \text{if } k \in J. \end{cases}$$

and

$$y_k^i = \begin{cases} p_{2^{m+1}}, & \text{if } k \in I_j \text{ for } 1 \leq j \leq n, \\ p_{2^m}, & \text{if } k \in I_{n+i}, \\ p_0, & \text{if } k \in I_j \text{ for } n < j \leq 2n \text{ but } j \neq n + i, \\ p_0, & \text{if } k \in J. \end{cases}$$

Then it is straightforward to check that $S = S_1 \cup S_2$ is m-regular. $\qquad\square$

Now suppose φ is an embedding of M_n into a metric space (X, d) of roundness $p > 0$. Fix $1 \leq m \leq \dfrac{n}{2}$. Then for an m-regular set

$$S = S_1 \cup S_2 = \{u_1, \ldots, u_n\} \cup \{v_1, \ldots, v_n\},$$

we have that

$$\sum_{i,j} d(\varphi(u_i), \varphi(v_j))^p \geq \sum_{i<j} [d(\varphi(u_i), \varphi(u_j))^p + d(\varphi(v_i), \varphi(v_j))^p] \qquad (3.2)$$

Let S_m be the set of all m-segments in M_n. Let $\theta_m = \text{card}(S_m)$. For any m-segment u, v in S_m, let $\beta_{u,v}$ be the number of m-regular sets in M_n that u and v both belong to. By the symmetry of the space M_n, if u, v and u', v' are both m-segments in M_n, then there is an isometry of M_n onto itself sending u to u' and v to v', and thus $\beta_{u,v} = \beta_{u',v'}$. Therefore we can define β_m to be this common number. Similarly, let η_m be the number of m-regular sets that any $m + 1$-segment belongs to. Finally, let ρ_m be the number of m-regular sets of M_n. Then we have the following equalities by simple counting:

$$\theta_m \beta_m = n(n-1)\rho_m \tag{3.3}$$

$$\theta_{m+1} \eta_m = n^2 \rho_m \tag{3.4}$$

Now apply inequality (3.2) for all m-regular sets in M_n and sum them up to get

$$\eta_m \sum_{(u,v)\in S_{m+1}} d(\varphi(u), \varphi(v))^p \geq \beta_m \sum_{(u,v)\in S_m} d(\varphi(u), \varphi(v))^p \tag{3.5}$$

Define

$$E_m = \left[\frac{\displaystyle\sum_{(u,v)\in S_m} d(\varphi(u), \varphi(v))^p}{\theta_m} \right]^{\frac{1}{p}}.$$

Then from (3.5) we get

$$\eta_m \theta_{m+1} E_{m+1}^p \geq \beta_m \theta_m E_m^p. \tag{3.6}$$

From (3.3) and (3.4) we get

$$\frac{\beta_m \theta_m}{\eta_m \theta_{m+1}} = \frac{n-1}{n}.$$

Thus (3.6) can be rewritten as

$$E_{m+1}^p \geq \frac{n-1}{n} E_m^p,$$

for all $1 \leq m \leq \dfrac{n}{2}$. By iteration we get

$$E_{\frac{n}{2}+1}^p \geq \left(\frac{n-1}{n}\right)^{\frac{n}{2}} E_1^p \geq \frac{1}{e} E_1^p,$$

or

$$E_{\frac{n}{2}+1} \geq e^{-\frac{1}{p}} E_1.$$

This implies that

$$\sup_{(u,v)\in S_{\frac{n}{2}+1}} d(\varphi(u),\varphi(v)) \geq e^{-\frac{1}{p}} \inf_{(u,v)\in S_1} d(\varphi(u),\varphi(v)). \qquad (3.7)$$

Now if M is uniformly homeomorphic via φ to a subset of a metric space (X,d) of roundness $p > 0$, then

$\delta = \inf\{d(\varphi(u),\varphi(v)) \mid (u,v) \text{ is a 1-segment in } M_n \text{ for some even } n \geq 4\} > 0.$

Therefore, by (3.7), for each even $n \geq 4$,

$\sup\{d(\varphi(u),\varphi(v)) \mid (u,v) \text{ is an } \frac{n}{2}+1\text{-segment in } M_n\} \geq e^{-\frac{1}{p}}\delta > 0.$

This is a contradiction to the uniformity of φ. Our proof of Theorem 3.1 is now complete.

Corollary 3.8. *The roundness of I_d is 0.*

We close this section with a few more remarks. It is well known that the density ideal is closely related to the Banach space c_0. However, there does not seem to be direct implications about embeddability of these spaces into U_∞. Enflo [6] indeed showed that any metric space which is universal with respect to uniformly homeomorphic embeddings must have roundness 0. Thus it follows from Aharoni's result [1] that c_0 has roundness 0. I_d does not share the universality property with c_0, since I_d is a zero-dimensional Polish space.

Acknowledgment

The author acknowledges the support of his research by the U.S. NSF grants DMS-0901853 and DMS-1201290.

Greg Hjorth had planned to attend the 12th Asian Logic Conference held in Wellington, New Zealand, in December 2011. But he never made it, because he had passed away unexpectedly in the January of the same year. Being his first PhD, I learned a lot from him and have a lot to thank him for. But mostly I would thank him for being a brave explorer in mathematics, traveling from one field of mathematics to another with ease and always leaving impressive results behind. He had always been interested in unitary group actions, and he had often checked with me on any progress about the group U_∞ (I tend to believe that he was the first to use this notation, as opposed to $U(\infty)$ before; his motivation was probably to create a nice contrast with the notation S_∞ for the infinite permutation group). It is his spirit of fearlessness that I tried to follow in writing this paper.

References

1. I. AHARONI, Every separable metric space is Lipschitz equivalent to a subset of c_0^+, *Israel J. Math.* 19 (1974), no. 3, 284–291.

2. I. AHARONI, B. MAUREY AND B. S. MITYAGIN, Uniform embeddings of metric spaces and of Banach spaces into Hilbert spaces, *Israel J. Math.* 52 (1985), no. 3, 251–265.

3. W. BANASZCZYK, On the existence of exotic Banach-Lie groups, *Math. Ann.* 264 (1983), 485–493.

4. Y. BENYAMINI AND J. LINDENSTRAUSS, *Geometric Nonlinear Functional Analysis*, Volume 1. American Mathematical Society Colloquium Publications 48, Providence, 2000.

5. J. DIXMIER, *Von Neumann Algebras*. North-Holland, 1981.

6. P. ENFLO, On a problem of Smirnov, *Ark. Mat.* 8 (1970), no. 2, 107–109.

7. I. FARAH, Ideals induced by Tsirelson submeasures, *Fund. Math.* 159 (1999), no. 3, 243–258.

8. S. GAO, Unitary group actions and Hilbertian Polish metric spaces, in *Logic and Its Applications*, 53–72. Contemporary Mathematics 380, American Mathematical Society, RI, 2005.

9. S. GAO AND V. PESTOV, On a universality property of some abelian Polish groups, *Fund. Math.* 179 (2003), no. 1, 1–15.

10. W. HERER AND J. P. R. CHRISTENSEN, On the existence of pathological submeasures and the construction of exotic topological groups, *Math. Ann.* 213 (1975), 203–210.

11. G. HJORTH, *Classification and Orbit Equivalence Relations*. Mathematical Surveys and Monographs, vol. 75, American Mathematical Society, Providence, 2000.

12. V. KANOVEI, *Borel Equivalence Relations. Structure and Classification*. University Lecture Series, 44. American Mathematical Society, Providence, RI, 2008.

13. A. LOUVEAU AND B. VELIČKOVIĆ, Analytic ideals and cofinal types, *Ann. Pure Appl. Logic* 99 (1999), no. 1-3, 171–195.

14. M. G. MEGRELISHVILI, Every semitopological semigroup compactification of $H_+[0,1]$ is trivial, *Semigroup Forum* 63 (2001), no. 3, 357–370.

15. M. G. MEGRELISHVILI, Reflexively but not unitarily representable topological groups, *Topology Proc.* 25 (2002), 615–625.

16. M. A. NAIMARK, *Normed Algebras*. Wolters-Noordhoff Publishing, 1972.

17. M.R. OLIVER, Borel cardinalities below c_0, *Proc. Amer. Math. Soc.* 134 (2006), no. 8, 2419–2425.

18. I. J. SCHOENBERG, Metric spaces and positive definite functions, *Trans. Amer. Math. Soc.* 44 (1938), no. 3, 522–536.

19. S. SOLECKI, Analytic ideals, *Bull. Symbolic Logic* 2 (1996), no. 3, 339–348.

20. S. SOLECKI, Analytic ideals and their applications, *Ann. Pure Appl. Logic* 99 (1999), no. 1-3, 51–72.

21. D. M. TOPPING, *Lectures on Von Neumann Algebras*. Van Nostrand Reinhold Company, London, 1971.

22. B. VELIČKOVIĆ, A note on Tsirelson type ideals, *Fund. Math.* 159 (1999), no. 3, 259–268.

144

FURTHER THOUGHTS ON DEFINABILITY IN THE URYSOHN SPHERE

ISAAC GOLDBRING

ABSTRACT. We discuss some basic geometry of sets definable in the Urysohn sphere using only finitely many parameters and briefly remark on the case of arbitrary definable sets. Then we discuss definable functions in the Urysohn sphere satisfying a special syntactic property.

1. INTRODUCTION

Understanding the sets and functions definable in a given structure is a common goal in model theory. While there are adequate notions of definability for *metric structures* using a continuous version of first-order logic (see [1]), there has been very little study of what sets and functions are definable in concrete metric structures. The author has undertaken the task of trying to understand definable functions in various metric structures, namely the *Urysohn sphere* (the unique universal and ultrahomogeneous metric space of diameter at most 1) [3] and Hilbert spaces (and some of their generic expansions) [4], [5].

Describing the sets definable in a particular metric structure appears to be a much harder task. In this paper, we give an adequate description of the sets definable in the Urysohn sphere defined using only *finitely* many parameters (definable sets in continuous logic are generally allowed to use countably many parameters in their definition). We then proceed to explain some of the difficulties involved in describing arbitrary definable subsets of the Urysohn sphere.

In [3], a reasonable description of the functions definable in the Urysohn sphere was given, although a complete characterization is still lacking. In the final section of this paper, we show how to completely characterize the definable functions in the Urysohn sphere that satisfy a certain syntactic requirement.

We assume familiarity with continuous logic; the unacquainted reader can consult [1]. However, since this is an article about definable sets, we repeat the definition of a definable set in a metric structure.

Definition 1.1. Suppose that $\mathcal{M}$ is an $\mathcal{L}$-structure and $A \subseteq M$.

(1) A continuous function $P : M^n \to [0,1]$ is an *A-definable predicate* if there are $\mathcal{L}(A)$-formulae $\varphi_n(x)$ such that the functions $\varphi_n^{\mathcal{M}}$:

Goldbring's work was partially supported by NSF grant DMS-1007144.

$M^n \to [0,1]$ converge uniformly to P. (Equivalently: there are $\mathcal{L}(A)$-formulae $\varphi_n(x)$ and a continuous function $u : [0,1]^{\mathbb{N}} \to [0,1]$ such that $P(x) = u((\varphi_n(x)))$ for all $x \in M^n$.)

(2) A closed set $X \subseteq M^n$ is A-*definable* if the function $d(x,X) : M^n \to [0,1]$ is A-definable.

We will use the following notation throughout this paper: $\mathfrak{U}$ denotes the Urysohn sphere, considered as a metric structure in the empty language consisting solely of the symbol d for the metric while $\mathbb{U}$ denotes an ω_1-saturated elementary extension of $\mathfrak{U}$. For $a \in \mathfrak{U}$ and $r \in [0,1]$, we set:

- $B(a;r) := \{x \in \mathfrak{U} \mid d(a,x) \leq r\}$, the *closed ball in* $\mathfrak{U}$ *centered at* a *with radius* r,
- $B^o(a;r) := \{x \in \mathfrak{U} \mid d(a,x) < r\}$, the *open ball in* $\mathfrak{U}$ *centered at* a *with radius* r, and
- $S(a;r) := \{x \in \mathfrak{U} \mid d(a,x) = r\}$, the *sphere in* $\mathfrak{U}$ *centered at* a *with radius* r.

At times, we will use the fact that, for $A \subseteq \mathfrak{U}$ (or $\mathbb{U}$), $\mathrm{dcl}(A) = \mathrm{acl}(A) = \overline{A}$, where $\overline{A}$ denotes the metric closure of A in $\mathfrak{U}$ (resp. $\mathbb{U}$); see [2, Fact 5.3] for a proof of this fact.

We would like to thank Julien Melleray for many useful discussions regarding this work.

2. FINITELY DEFINABLE SETS

In this section, we prove some properties about A-definable subsets of $\mathfrak{U}$, where $A \subseteq \mathfrak{U}$ is finite. The key observation is the following (we thank Ward Henson for a useful discussion concerning the last implication).

Lemma 2.1. *Suppose that T is an ω-categorical (continuous) theory and that $\mathcal{M} \models T$. For a closed set $X \subseteq M^n$, the following are equivalent:*

(1) X *is type-definable over $\emptyset$;*
(2) X *is definable over $\emptyset$;*
(3) X *is a zeroset over $\emptyset$;*
(4) X *is fixed setwise by* $\mathrm{Aut}(\mathcal{M})$.

Proof. (1)$\Rightarrow$(2) follows from ω-categoricity; see [1, Section 12]. (2)$\Rightarrow$(3) and (3)$\Rightarrow$(4) are always true. (4)$\Rightarrow$(1): Suppose that $c \in X$ and $\mathrm{tp}(c) = \mathrm{tp}(d)$. Since $\mathcal{M}$ is strongly ω-near-homogeneous ([1, Corollary 12.11]), there are, for $n \geq 1$, $\sigma_n \in \mathrm{Aut}(\mathcal{M})$ such that $\sigma_n(c) \to d$. Since each $\sigma_n(c) \in X$, we have $d \in \overline{X} = X$. It follows that $X = \bigcup_{p \in C} p(M^n)$ for some $C \subseteq S_n(T)$. (Here, $p(M^n)$ denotes the set of realizations of p in $\mathcal{M}$.) We claim that C is closed in the d-topology on $S_n(T)$. Indeed, suppose that p is in the d-closure of C and $a \models p$. Fix $\epsilon > 0$. Then there is $p' \in C$ such that $d(p,p') < \epsilon$. Thus, there are $a' \models p$ and $b \models p'$ such that $d(a',b) < \epsilon$. Since $b \in X$, we get that $d(a',X) \leq \epsilon$. However, $\mathrm{tp}(a) = \mathrm{tp}(a')$, so by invariance of X, we get that $d(a,X) \leq \epsilon$. Since X is closed, we get that $a \in X$, whence $p \in C$.

Since T is ω-categorical, C is closed in the *logic* topology on $S_n(T)$, whence there is a set $\Gamma(x_1, \ldots, x_n)$ of formulae such that $C = \{p \in S_n(T) \mid \Gamma \subseteq p\}$. It follows that X is type-defined by Γ. $\qquad\square$

Suppose that $A \subseteq \mathfrak{U}$ is relatively compact. Then, by compact homogeneity of $\mathfrak{U}$ (see [6, Section 4.5]), $\mathrm{Th}(\mathfrak{U}; (c_a)_{a \in A})$ is ω-categorical. Consequently, we have the following:

Corollary 2.2. *For relatively compact $A \subseteq \mathfrak{U}$ and closed $X \subseteq \mathfrak{U}^n$, the following are equivalent:*

(1) X *is type-definable over A;*

(2) X *is definable over A;*

(3) X *is a zeroset over A;*

(4) X *is fixed setwise by isometries of $\mathfrak{U}$ which fix A pointwise.*

In particular, the only $\emptyset$-definable sets in $\mathfrak{U}$ are $\emptyset$ and $\mathfrak{U}^n$.

The following lemma shows that certain topological and set-theoretic constructions preserve A-definability.

Lemma 2.3. *Suppose that $F, G \subseteq \mathfrak{U}$ are A-definable. Then:*

(1) ∂F *is A-definable;*

(2) $\overline{\mathrm{int}(F)}$ *is A-definable;*

(3) $\overline{\mathfrak{U} \setminus F}$ *is A-definable.*

(4) $F \cap G$ *is A-definable.*

(5) *Write $F = P \sqcup C$, where P is the perfect kernel of F and C is the scattered part of F. Then P and $\overline{C}$ are A-definable.*

Proof. These are all immediate from A-invariance. $\qquad\square$

For $\vec{a} = (a_1, \ldots, a_n) \in \mathfrak{U}^n$ and $\vec{r} = (r_1, \ldots, r_n) \in [0,1]^n$, set
$$S(\vec{a}; \vec{r}) := S(a_1; r_1) \cap \cdots \cap S(a_n; r_n).$$

Corollary 2.4. *Suppose that $F \subseteq \mathfrak{U}$ is closed. Set $A := \{a_1, \ldots, a_n\}$. Then F is A-definable if and only if there is a closed set $X \subseteq [0,1]^n$ such that $F = \bigcup_{\vec{r} \in X} S(\vec{a}; \vec{r})$.*

Proof. The "if" direction follows from the characterization of definability in terms of invariance under isometries fixing A. For the "only if" direction, let $X = \{\vec{r} \in [0,1]^n \mid \varphi(\vec{r}) = 0\}$, where $\varphi : [0,1]^n \to [0,1]$ is such that $d(x, F) = \varphi(d(x, a_1), \ldots, d(x, a_n))$. $\qquad\square$

Corollary 2.5. *$B(a; r)$ is A-definable if and only if $a \in A$.*

Proof. The "if" direction is clear. Now suppose that $B(a; r)$ is A-definable. Let ϕ be an isometry fixing A. Then ϕ is an isometry fixing $B(a; r)$, whence it fixes a. Thus $\{a\}$ is A-definable, implying that $a \in \mathrm{dcl}(A) = A$ since A is finite. $\qquad\square$

While the preceding corollary has a nice geometric proof, it is actually a special case of the following general result.

Lemma 2.6. *Suppose that F is A-definable. Then there is a finite $A_0 \subseteq A$ such that whenever F is B-definable, then $A_0 \subseteq \bar{B}$.*

Proof. This follows from the proof of the fact that $T_{\mathfrak{U}}$ has weak elimination of finitary imaginaries; see [2, Section 5]. □

Define a *(closed) annulus* in $\mathfrak{U}$ to be a set of the form

$$\bar{A}(a; r_1, r_2) := \{x \in \mathfrak{U} \mid r_1 \leq d(x, a) \leq r_2\},$$

where $0 \leq r_1 \leq r_2 \leq 1$. (An open annulus $A(a; r_1, r_2)$ is defined similarly, strengthening the inequality signs.) We call a the *center* of the annulus. Note that a closed ball centered at a of radius r is an annulus centered at a (take $r_1 = 0$, $r_2 = r$) and a sphere centered at a of radius r is an annulus centered at a (take $r_1 = r_2 = r$). Also $\{a\}$ is an annulus centered at a (take $r_1 = r_2 = 0$). By Corollary 2.4, every annulus is definable over its center. Conversely, we have:.

Corollary 2.7. *If F is a nonempty, connected $\{a\}$-definable subset of $\mathfrak{U}$, then F is a closed annulus centered at a.*

Proof. By Corollary 2.4, there is a nonempty closed $X \subseteq [0, 1]$ such that

$$F = \bigcup_{r \in X} S(a; r).$$

We must show that X is a closed subinterval of $[0, 1]$, that is we must show that X is convex. Suppose that $0 \leq r_1 < s < r_2 \leq 1$, where $r_1, r_2 \in X$. Suppose that $s \notin X$. Then $B^o(a; s) \cap F$ and $B(a; s)^c \cap F$ yield a disconnection of F, a contradiction. □

Fix an annulus $A := A(a; r, R)$. For $x \in A$, define the *local diameter of A at x* to be $\operatorname{diam}(x) := \sup\{d(x, y) \mid y \in A\}$. Define the *radius* of A to be the quantity $\inf\{\operatorname{diam}(x) \mid x \in A\}$.

Proposition 2.8. *For an annulus $A = A(a; r, R)$, the diamter of A is $2R$ and the radius of A is $r + R$.*

Proof. Since $d(a, x) = R$, $d(a, y) = R$, and $d(x, y) = 2R$ defines a metric space, it can be realized in $\mathfrak{U}$, whence $\operatorname{diam}(A) \geq 2R$. However, the triangle inequality yields $\operatorname{diam}(A) \leq 2R$, whence $\operatorname{diam}(A) = 2R$.

Next, suppose that $d(x, a) = r$. Then since we can realize $d(x, a) = r$, $d(y, a) = R$, $d(x, y) = r + R$ inside of $\mathfrak{U}$, we see that $\operatorname{diam}(x) \geq r + R$. However, for all $z \in \mathfrak{U}$, $d(x, z) \leq d(x, a) + d(a, z) \leq r + R$. Thus, $\operatorname{diam}(x) = r + R$, whence the radius of A is bounded above by $r + R$. By embedding an annulus in the euclidean plane of inner and outer radii r and R respectively in $\mathfrak{U}$, we see that $\operatorname{diam}(x) \geq r + R$ for each $x \in A$, whence the radius of A is at least $r + R$. □

Corollary 2.9. *Two annuli $A(a; r, R)$ and $A(a'; r', R')$ are isometric if and only if $r = r'$ and $R = R'$.*

Consequently, we see that the class of definable sets (even over one element) is quite exotic in the sense that there are continuum many non-isometric definable sets.

Define a *generalized annulus centered around* $(a_1, \ldots, a_n)$ to be a set of the form $\bigcup_{\vec{r} \in X} S(\vec{a}; \vec{r})$ where X is a nonempty sub*continuum* of $[0,1]^n$. The following is a generalization of Corollary 2.7.

Corollary 2.10. *If F is a nonempty connected A-definable subset of $\mathfrak{U}$, where $A = \{a_1, \ldots, a_n\}$, then F is a generalized annulus centered around $(a_1, \ldots, a_n)$.*

Proof. Write $F = \bigcup_{\vec{r} \in X} S(\vec{a}; \vec{r})$. Without loss of generality, we may suppose that $S(\vec{a}; \vec{r}) \neq \emptyset$ for each $\vec{r} \in X$. We will show that X is connected. Suppose, towards a contradiction, that there are disjoint open $O_1, O_2 \subseteq \mathbb{R}^n$ such that $X = (X \cap O_1) \cup (X \cap O_2)$. For $i = 1, 2$, set $F_i := \bigcup_{\vec{r} \in X \cap O_i} S(\vec{a}; \vec{r})$. Clearly each $F_i \neq \emptyset$ and $F = F_1 \sqcup F_2$. We must show that each F_i is open in F. Fix $i \in \{1, 2\}$ and write $O_i := \bigcup_\alpha \prod_{j=1}^n (b_j^\alpha, c_j^\alpha)$. Fix $y \in F_i$ and take α such that $(d(y, a_1), \ldots, d(y, a_n)) \in X \cap \prod_{j=1}^n (b_j^\alpha, c_j^\alpha)$. Fix $\epsilon > 0$ small enough such that $(d(y, a_j) - \epsilon, d(y, a_j) + \epsilon) \subseteq (b_j^\alpha, c_j^\alpha)$ for each $j = 1, \ldots, n$. It follows that if $z \in F$ is such that $d(y, z) < \epsilon$, then $z \in F_i$. $\square$

We now consider the decomposition of definable sets into their connected components. For the rest of this section, we assume that $A \subseteq \mathfrak{U}$ is finite.

Lemma 2.11. *Suppose that F is an A-definable set and that C is a connected component of F. Then C is A-definable.*

Proof. Fix $c \in C$. Then since every isometry of $\mathfrak{U}$ fixing Ac pointwise fixes C setwise, we have that C is Ac-definable. Suppose that there is $d \in C \setminus \{c\}$. Then C is Ad-definable. It follows that $d(x, C)$ is both Ac-definable and Ad-definable. Since $T_\mathfrak{U}$ admits weak elimination of finitary imaginaries, we have that $d(x, C)$ is A-definable.

It remains to show that every connected component of cardinality 1 is A-definable. Set

$$F' := \{c \in F \mid \{c\} \text{ is a connected component of } F\}.$$

Then F' is A-invariant, whence $\overline{F'}$ is A-definable. Thus, there is $X \subseteq [0, 1]^n$ such that

$$\overline{F'} = \bigcup_{\vec{r} \in X} (S(a_1; r_1) \cap \cdots \cap S(a_n; r_n)).$$

Fix $c \in F'$. Choose $\vec{r} \in X$ such that $c \in S := S(a_1; r_1) \cap \cdots \cap S(a_n; r_n)$. Since spheres in the Urysohn space are connected (see [6, Section 4.3]), we have $\{c\} = S$, whence $\{c\}$ is A-definable. $\square$

Corollary 2.12. *A Cantor set in $\mathfrak{U}$ cannot be A-definable.*

More generally:

Corollary 2.13. *If X is an A-definable compact set, then X is a finite subset of A.*

Proof. The connected components are A-definable, so generalized annuli. However, they are also compact and the only compact generalized annuli are one-element subsets of A (as compact subsets of $\mathfrak{U}$ have no interior). $\quad\square$

The next result says that, for sets defined over finitely many parameters, if there are infinitely many connected components, then the connected components cannot be a uniform distance from one another. For a subset C of $\mathfrak{U}$ and $\epsilon > 0$, we let $N(C, \epsilon)$ denote the open ϵ-neighborhood around C.

Corollary 2.14. *Suppose that F is A-definable. Then, given any $\epsilon > 0$, there are finitely many connected components $C_1, \ldots, C_n$ of F such that $F \subseteq \bigcup_{i=1}^n N(C_i; \epsilon)$.*

Proof. For $n \geq 1$, let $\varphi_n(x)$ be a formula with parameters from A such that $|d(x, F) - \varphi_n(x)| < \frac{1}{n}$ for all $x \in \mathfrak{U}$.

Fix $\epsilon > 0$ and let $(C_i \mid i < \alpha)$ enumerate the connected components of F. Since all of the predicates $d(x, C_i)$ are A-definable, we may find formulae $\psi_n^i(x)$ with parameters from A such that $|d(x, C_i) - \psi_n^i(x)| < \frac{1}{n}$. Fix $m \in \mathbb{N}$ such that $\frac{2}{m} < \epsilon$. Since the set of conditions

$$\{\varphi_n(x) \leq \frac{1}{n} \mid n \geq 1\} \cup \{\psi_n^i(x) \geq \frac{1}{n} \mid i < \alpha, n \geq m\}$$

is unsatisfiable, by ω-saturation, there are $i_1, \ldots, i_k < \alpha$ and $n_1, \ldots, n_k \geq m$ such that

$$\{\varphi_n(x) \leq \frac{1}{n} \mid n \geq 1\} \cup \{\psi_{n_j}^{i_j}(x) \geq \frac{1}{n_j} \mid j = 1, \ldots, k\}$$

is unsatisfiable, yielding the desired result. $\quad\square$

3. Arbitrary Definable Sets

Of course, being definable over a finite set of parameters is a very special thing in continuous logic. We would thus like to have some results concerning subsets of $\mathfrak{U}$ (or $\mathbb{U}$) defined over countably many parameters. The following example shows that our characterization of definability over finite parametersets fails for sets defined over a countably infinite set of parameters.

Example 3.1. Suppose that $A = \{a_n \mid n \in \mathbb{N}\}$ is a countable subset of $\mathbb{U}$ such that $d(a_i, a_j) = 1$ for all distinct $i, j \in \mathbb{N}$. Set

$$F := \{x \in \mathbb{U} \mid d(x, A) \geq \frac{1}{2}\},$$

a closed, A-invariant set. We claim that F is *not* A-definable. Indeed, if F were A-definable, then there would be a continuous function $\varphi : [0, 1]^{\mathbb{N}} \to [0, 1]$ such that $d(x, F) = \varphi((d(x, a_i)))$ for all $x \in \mathbb{U}$. Fix $\epsilon \in (0, \frac{1}{3})$. Choose $\delta > 0$ and $n \in \mathbb{N}^{>0}$ such that, for all $\vec{w}, \vec{z} \in [0, 1]^{\mathbb{N}}$, if $|w_i - z_i| < \delta$ for all $i < n$, then $|\varphi(\vec{w}) - \varphi(\vec{z})| < \epsilon$. Now take $x, y \in \mathbb{U}$ such that $d(x, a_i) = d(y, a_i) = 1$

for all $i \in \mathbb{N} \setminus \{n\}$ while $d(x, a_n) = \frac{1}{3}$ and $d(y, a_n) = \frac{1}{3} - \epsilon$. By the choice of n, we see that $|d(x, F) - d(y, F)| < \epsilon$. However, $d(x, F) = \frac{1}{2} - \frac{1}{3}$ and $d(y, F) = \frac{1}{2} - \frac{1}{3} + \epsilon$, a contradiction.

In all actuality, there is probably very little hope of classifying all definable subsets of $\mathfrak{U}$. Indeed, in any metric structure $\mathcal{M}$, any compact subset of M is definable (see [1, Proposition 9.19]). Since every compact metric space of diameter ≤ 1 embeds in $\mathfrak{U}$, it follows that *every compact metric space of diameter ≤ 1 is isometric to a definable subset of $\mathfrak{U}$.*

That being said, we would like to point out that understanding definable subsets of $\mathbb{U}$ will sometimes allow us to prove facts about definable subsets of $\mathfrak{U}$. The idea is to use the "canonical extension" of a definable subset of $\mathfrak{U}$ to a definable subset of $\mathbb{U}$. Since this canonical extension notion does not really appear in the literature, we now discuss it in more detail.

Suppose that $F \subseteq \mathfrak{U}$ is A-definable, where $A \subseteq \mathfrak{U}$ is countable. Then $P(x) := d(x, F)$ is an A-definable predicate in $\mathfrak{U}$. We know that there is a unique A-definable predicate Q in $\mathbb{U}$ extending P. Moreover, we have $(\mathfrak{U}, P) \preceq (\mathbb{U}, Q)$. (See [1, Theorem 9.8].) Now, since P is a distance predicate, it satisifes axioms (E1) and (E2) of [1, Section 9], whence by elementarity, Q satisfies (E1) and (E2). Let $\tilde{F} := \{x \in \mathbb{U} \mid Q(x) = 0\}$. By [1, Theorem 9.12], $Q(x) = d(x, \tilde{F})$. It follows that $F \subseteq \tilde{F}$, $\tilde{F}$ is A-definable and $\tilde{F} \cap \mathfrak{U} = A$.

Conversely, suppose that $E \subseteq \mathbb{U}$ is A-definable, where $A \subseteq \mathfrak{U}$ is countable. We claim that $E \cap \mathfrak{U}$ is an A-definable subset of $\mathfrak{U}$. Let $Q : \mathbb{U} \to [0, 1]$ be the A-definable predicate given by $Q(x) = d(x, E)$. Let $Q = P \upharpoonright \mathfrak{U}$, so $(\mathfrak{U}, P) \preceq (\mathbb{U}, Q)$. Then $Z(P)$ is an A-definable subset of $\mathfrak{U}$; but $Z(P) = Z(Q) \cap \mathfrak{U} = E \cap \mathfrak{U}$.

Corollary 3.2. *Suppose that $F, G \subseteq \mathfrak{U}$ are A-definable, where A is a countable subset of $\mathfrak{U}$. Suppose that $\tilde{F} \cap \tilde{G}$ is A-definable. Then $F \cap G$ is A-definable.*

Proof. $F \cap G = (\tilde{F} \cap \tilde{G}) \cap \mathfrak{U}$ is A-definable. $\square$

Along these same lines:

Corollary 3.3. *If $F \subseteq \mathfrak{U}$ is A-definable, where $A \subseteq \mathfrak{U}$ is countable, and the perfect kernel of $\tilde{F}$ is A-definable, then the perfect kernel of F is A-definable.*

Proof. The perfect kernel of $\tilde{X}$ is A-definable and the intersection of the perfect kernel of $\tilde{X}$ with $\mathfrak{U}$ is the perfect kernel of X. $\square$

Corollary 3.4. *Suppose $F \subseteq \mathfrak{U}$ is A-definable, where $A \subseteq \mathfrak{U}$ is countable. Then $\overline{\mathbb{U} \setminus \tilde{F}} \cap \mathfrak{U} = \overline{\mathfrak{U} \setminus F}$. Consequently, $\overline{\mathfrak{U} \setminus F}$ and ∂F are A-definable if $\overline{\mathbb{U} \setminus \tilde{F}}$ is A-definable.*

Proof. It is clear that $\overline{\mathfrak{U} \setminus F} \subseteq \overline{\mathbb{U} \setminus \tilde{F}} \cap \mathfrak{U}$. We now prove the other direction. Suppose that $x \in \overline{\mathbb{U} \setminus \tilde{F}} \cap \mathfrak{U}$. Let $x_n \in \mathbb{U} \setminus \tilde{F}$ be such that $d(x_n, x) \leq \frac{1}{n}$. Set

$\epsilon_n := d(x_n, \tilde{F}) > 0$. Then, for every n, we have

$$\mathbb{U} \models \inf_z \max(\epsilon_n \div Q(z), d(z, x) \div \frac{1}{n}) = 0.$$

By elementarity, the above condition is true in $\mathfrak{U}$, with $P(z)$ replacing $Q(z)$. Take $0 < \delta_n < \min(\epsilon_n, \frac{1}{n})$. Then there is $z_n \in \mathfrak{U}$ such that

$$\max(\epsilon_n \div P(z_n), d(z_n, x) \div \frac{1}{n}) \leq \delta_n.$$

Note then that $P(z_n) \geq \epsilon_n - \delta_n > 0$, whence $z_n \in \mathfrak{U} \setminus F$, and $d(z_n, x) \leq \frac{2}{n}$. It follows that $x \in \overline{\mathfrak{U} \setminus F}$. □

4. Special Definable Functions

In [3], the following theorem on definable functions in $\mathfrak{U}$ was proven:

Theorem 4.1. *If $f : \mathfrak{U}^n \to \mathfrak{U}$ is an A-definable function, where $A \subseteq \mathfrak{U}$ is countable, then either f is a projection function (namely, there is $i \in \{1, \ldots, n\}$ such that, for all $x = (x_1, \ldots, x_n) \in \mathfrak{U}^n$, $f(x) = x_i$) or else $f(\mathfrak{U}^n)$ is a relatively compact subset of $\overline{A}$.*

While this theorem can be used to draw many interesting corollaries, it still doesn't provide an exact characterization of the definable functions in $\mathfrak{U}$. In [3], the following conjecture appeared.

Conjecture 4.2. *If $f : \mathfrak{U}^n \to \mathfrak{U}$ is an A-definable function, where $A \subseteq \mathfrak{U}$ is countable, then either f is a projection function or else f is constantly equal to an element of $\overline{A}$.*

In [3], it is shown that if the conjecture is true for one-variable definable functions, then it is true for all definable functions.

While this conjecture remains open, it is the goal of this section to prove that the conjecture holds under a (strong) syntactic constraint on the definable functions. Indeed, the author's initial approach to studying definable functions in $\mathfrak{U}$ (which ultimately did not work) was to use the fact that the predicate $d(f(x), y)$ was approximable by formulae. More precisely, by quantifier elimination for $\text{Th}(\mathfrak{U})$, for every $n \geq 2$, there is a quantifier-free *restricted* formula $\varphi_n(x, y)$ with parameters from A such that

$$|d(f(x), y) - \varphi_n(x, y)| \leq 2^{-n}$$

for all $x, y \in \mathfrak{U}$. (See [1, Section 6] for the definition of a restricted formula.) Although in general we could not make this approach work, we can make it work for special kinds of φ_n:

Definition 4.3. We define what it means for a formula $\varphi(\vec{x})$ to be a **generalized atomic formula** by induction:

- If φ is atomic, then it is generalized atomic.
- If φ is generalized atomic, then so is $\frac{1}{2}\varphi$.
- Nothing else is a generalized atomic formula.

Definition 4.4. We define what it means for a formula $\varphi(\vec{x})$ to be a **special restricted formula** by induction:

- If φ is a generalized atomic formula, then it is a special restricted formula.
- If φ is a special restricted formula and ψ is a generalized atomic formula, then $\varphi \doteq \psi$ is a special restricted formula.
- Nothing else is a special restricted formula.

Remark 4.5. Note that if φ is a special restricted formula, then so is $\frac{1}{2}\varphi$ (not literally, but rather up to logical equivalence). This follow by induction and the fact that $\frac{1}{2}(a \doteq b) = \frac{1}{2}a \doteq \frac{1}{2}b$. Thus the only difference between a special restricted formula and a restricted formula is the way one is allowed to use the $\doteq$ connective.

Call a definable predicate P A-**special** if it can be approximated by special restricted formula with parameters from A plugged in. Likewise, call a definable function $f : \mathfrak{U}^n \to \mathfrak{U}$ A-**special** if the definable predicate $d(f(x), y)$ is A-special.

Proposition 4.6. *If $f : \mathfrak{U} \to \mathfrak{U}$ is an A-special definable function, then f is either the identity function or constantly equal to an element of $\bar{A}$.*

Proof. For $n \geq 2$, we let $\varphi_n(x, y)$ be a special restricted formula with parameters from A satisfying

$$|d(f(x), y) - \varphi_n(x, y)| \leq 2^{-n}$$

for all $x, y \in \mathfrak{U}$. Fix $n \geq 2$. Since φ_n is special, there are generalized atomic formulae $\psi_1(x, y), \ldots, \psi_m(x, y)$ with parameters from A such that

$$\varphi_n(x, y) = (\cdots ((\psi_1(x, y) \doteq \psi_2(x, y)) \doteq \psi_3(x, y)) \cdots) \doteq \psi_m(x, y).$$

From the identity $(a \doteq b) \doteq c = a \doteq (b + c)$, we see that we we have

$$\varphi_n(x, y) = \psi_1(x, y) \doteq (\psi_2(x, y) + \cdots + \psi_n(x, y))$$

for all $x, y \in \mathfrak{U}$.

First suppose that $\psi_1(x, y) = 2^{-k} d(x, a)$ for some $k \geq 0$ and some $a \in A$. Then $\varphi_n(a, y) = 0$ for all $y \in \mathfrak{U}$, implying that $d(f(a), y) \leq 2^{-n}$ for all $y \in \mathfrak{U}$, which is a contradiction. Thus this case is impossible.

Next suppose that $\psi_1(x, y) = 2^{-k} d(y, a)$ for some $k \geq 0$ and some $a \in A$. Then $\varphi_n(x, a) = 0$ for all $x \in \mathfrak{U}$, implying that $d(f(x), a) \leq 2^{-n}$ for all $x \in \mathfrak{U}$, i.e. that image$(f) \subseteq B(a; 2^{-n})$.

Next suppose that $\psi_1(x, y) = 2^{-k} d(x, y)$ for some $k \geq 0$. Then $\varphi_n(x, x) = 0$ for all $x \in \mathfrak{U}$, implying that $d(f(x), x) \leq 2^{-n}$ for all $x \in \mathfrak{U}$.

Finally suppose that $\psi_1(x, y)$ has no occurrences of x or y. Then ψ_1 is either the constant 0 or the constant 2^{-k} for some $k \geq 1$ or the constant $2^{-k} d(a, a')$ for some $a, a' \in A$ and some $k \geq 0$. This case requires some work.

Let us denote $\psi_1(x, y)$ by the constant c. We may rewrite $\psi_2 + \cdots + \psi_m$ as

$$\sum_{i=1}^{p} 2^{-l_i} d(x, y) + \sum_{j=1}^{q} 2^{-m_j} d(x, a_j) + \sum_{k=1}^{r} 2^{-n_k} d(y, b_k) + s,$$

where $l_i, m_j, n_k \geq 0$ and s is a constant which appears by summing together those ψ_i's which have no x or y in them. Choose $x_0 \in \mathfrak{U}$ such that $d(x_0, a_j) = 1$ for all $j = 1, \ldots, q$. Choose $y_0 \in \mathfrak{U}$ such that $d(x_0, y_0) = 1$, $d(f(x_0), y_0) = 1$, and $d(y_0, b_k) = 1$ for all $k = 1, \ldots, r$. Then

$$\varphi_n(x_0, y_0) = c \doteq (\sum_{i=1}^{p} 2^{-l_i} + \sum_{j=1}^{q} 2^{-m_j} + \sum_{k=1}^{r} 2^{-n_k} + s) =: c',$$

implying that $c' \geq 1 - 2^{-n}$. However, $\varphi_n(x, y) \geq c'$ for all $x, y \in \mathfrak{U}$, implying that $d(f(x), y) \geq 1 - 2^{-n+1}$ for all $x, y \in \mathfrak{U}$, which is a contradiction. Thus this case is impossible.

To summarize, by knowing that φ_n approximates $d(f(x), y)$ up to an error of 2^{-n}, we learn that either image$(f) \subseteq B(a; 2^{-n})$ for some $a \in A$ or that $\|f - \mathrm{id}_{\mathfrak{U}}\|_\infty \leq 2^{-n}$. Let us call these options $(\mathrm{I})_n$ and $(\mathrm{II})_n$. If option $(\mathrm{I})_n$ happens for infinitely many n, then we see that f must be constantly equal to an element of $\bar{A}$. If option $(\mathrm{II})_n$ happens for infinitely many n, then $f = \mathrm{id}_{\mathfrak{U}}$. $\square$

Corollary 4.7. *If* $f : \mathfrak{U}^n \to \mathfrak{U}$ *is an A-special definable function, then either* f *is a projection function or constantly equal to an element of* $\bar{A}$.

Proof. One proves this by induction on n, mimicking the proof of [3, Proposition 4.8], using Proposition 4.6 above to cover the base case. One will need to use the observation that if $f : \mathfrak{U}^n \to \mathfrak{U}$ is A-special, then the functions f_b and f^c are Ab-special and Ac-special respectively. $\square$

REFERENCES

[1] I. Ben Yaacov, A. Berenstein, C. W. Henson, A. Usvyatsov, *Model theory for metric structures*, Model theory with applications to algebra and analysis. Vol. 2, pgs. 315-427, London Math. Soc. Lecture Note Ser. (350), Cambridge Univ. Press, Cambridge, 2008.

[2] C. Ealy and I. Goldbring, *Thorn-forking in continuous logic*, Journal of Symbolic Logic, Volume 77 (2012), 63-93.

[3] I. Goldbring, *Definable functions in Urysohn's metric space*, to appear in the Illinois Journal of Mathematics.

[4] I. Goldbring, *Definable operators on Hilbert spaces*, Notre Dame Journal of Formal Logic, **53**, Number 2(2012), 193-201.

[5] I. Goldbring, *An approximate Herbrand's theorem and definable functions in metric structures*, Math. Logic Quarterly, **58**, Issue 3 (2012), 208-216.

[6] J. Melleray, *Some geometric and dynamical properties of the Urysohn space*, Topology and its Applications **155**(2008), 1531-1560.

154

DEPARTMENT OF MATHEMATICS, STATISTICS, AND COMPUTER SCIENCE, UNIVERSITY OF ILLINOIS AT CHICAGO, SCIENCE AND ENGINEERING OFFICES M/C 249, 851 S. MORGAN ST., CHICAGO, IL, 60607-7045

E-mail address: isaac@math.uic.edu

URL: http://www.math.uic.edu/~isaac

Simple completeness proofs
for some spatial logics of the real line

Ian Hodkinson

Department of Computing, Imperial College London,
London, SW7 2AZ, UK
E-mail: i.hodkinson@imperial.ac.uk
www. doc. ic. ac. uk/~imh

McKinsey–Tarski (1944), Shehtman (1999), and Lucero-Bryan (2011) proved completeness theorems for modal logics with modalities $\Box$, $\Box$ and $\forall$, and $[\partial]$ and $\forall$, respectively, with topological semantics over the real numbers. We give short proofs of these results using lexicographic sums of linear orders.

Keywords: modal logic; topological semantics; lexicographic sums of linear orders

1. Introduction

This paper contains no new results at all. Its sole aim is to present what I believe are new and simple completeness proofs of some modal logics of the real line $\mathbb{R}$. They are often regarded as *spatial logics* — see [1] for example. The paper is deliberately kept short, with little historical background. There are three main theorems:

(1) If $\Box$ is read as the interior operator in the standard topology on $\mathbb{R}$, the logic of $\mathbb{R}$ is S4 — proved by McKinsey–Tarski [13]. This result was the first in the field. Interest in it is undergoing a renaissance and several alternative proofs have recently appeared [1,2,8,10,14,15]. So yet another proof will do no harm and may be of interest.
(2) The logic of $\mathbb{R}$ with $\Box$ and the universal modality $\forall$ is S4UC — proved by Shehtman [19].
(3) If we replace $\Box$ by a different box $[\partial]$, to be read as the coderivative operator, then the logic of $\mathbb{R}$ with $[\partial]$ and $\forall$ is KD4G$_2$.UC — proved by Lucero-Bryan [12].

The logic of $\mathbb{R}$ with $[\partial]$ alone is KD4G$_2$: this was proved by Shehtman [21], and later by Lucero-Bryan [12]. We will not prove it here. It can be done by removing parts of the proof of (3), which the reader may wish to do.

One may wonder whether the proofs would go through with $\forall$ replaced by the stronger *difference operator* $[\neq]$. However, Kudinov [9] has shown that the logic of $\mathbb{R}$ with $\square$ and $[\neq]$ is not finitely axiomatisable, and his argument appears to work for $[\partial]$ and $[\neq]$ as well.

Completeness proofs for modal logics with topological semantics over $\mathbb{R}$ often start by applying methods from classical modal logic, of varying sophistication, and end by applying topological techniques. Our proof proceeds like this as well, but with two differences. First, our use of modal logic is relatively straightforward. All we need is the *finite model property* for the logics, so that we can argue by induction on the size of parts of the finite model. For some of the logics the finite model property is nontrivial to establish, but we have nothing new to contribute here so we omit proofs and simply cite the literature. (It is worth noting here that we presuppose some familiarity with basic modal logic.) Second, we use very little topology. Instead, we use *lexicographic sums of linear orders*. Although these are very well known in some circles, a substantial part of the paper is devoted to introducing them, in the hope that they become known a little more widely, and to make the paper more self-contained.

The layout of the paper is simple. We describe syntax and semantics in §2 and lexicographic sums in §3. The three completeness proofs are in §§4–6, and we conclude in §7 with a couple of open questions.

We use standard notation such as $\mathbb{Z}$, $\mathbb{Q}$, $\mathbb{R}$. We often identify (notationally) a structure with its domain. For a map $f : X \to Y$ and subsets $X' \subseteq X$, $Y' \subseteq Y$, we write $f(X') = \{f(x) : x \in X'\}$, $\mathrm{rng}(f) = f(X)$, and $f^{-1}(Y') = \{x \in X : f(x) \in Y'\}$. The cardinality of a set X is denoted by $|X|$.

2. Definitions

We will study the logic of $\mathbb{R}$ in three sublanguages of the following ambient language $\mathcal{L}$. We fix a countably infinite set PV of propositional variables (or 'atoms').

2.1. Syntax — $\mathcal{L}$-formulas

The formulas of $\mathcal{L}$ are as follows:

(1) $\top$ is an $\mathcal{L}$-formula.

(2) Any atom $p \in PV$ is an $\mathcal{L}$-formula.

(3) If φ, ψ are $\mathcal{L}$-formulas then so are $\neg\varphi$ and $(\varphi \wedge \psi)$.

(4) If φ is an $\mathcal{L}$-formula then $\Box\varphi$, $[\partial]\varphi$, and $\forall\varphi$ are also $\mathcal{L}$-formulas.

We will write $\mathcal{L}$ for the set of all $\mathcal{L}$-formulas, $\mathcal{L}_\Box$ for the set of $\mathcal{L}$-formulas not involving $[\partial]$ or $\forall$, $\mathcal{L}_{[\partial]}$ for the set of $\mathcal{L}$-formulas not involving $\Box$ or $\forall$, $\mathcal{L}_{\Box\forall}$ for the set of $\mathcal{L}$-formulas not involving $[\partial]$, and $\mathcal{L}_{[\partial]\forall}$ for the set of $\mathcal{L}$-formulas not involving $\Box$. We will use the standard abbreviations: $\bot = \neg\top$, $\varphi \vee \psi = \neg(\neg\varphi \wedge \neg\psi)$, $\varphi \to \psi = \neg(\varphi \wedge \neg\psi)$, $\Diamond\varphi = \neg\Box\neg\varphi$, $\langle\partial\rangle\varphi = \neg[\partial]\neg\varphi$, and $\exists\varphi = \neg\forall\neg\varphi$. We adopt the usual binding conventions for the connectives and omit parentheses where no ambiguity results.

2.2. Kripke semantics

Although Kripke semantics is not the main concern of the paper, our proofs will use Kripke semantics for $\mathcal{L}$-formulas. A *binary relation* on a set W is a subset $R \subseteq W \times W$. We will write any of $R(w, u)$, Rwu, and wRu to denote that $(w, u) \in R$. For $w \in W$, we write $R(w)$ for the set $\{u \in W : Rwu\}$. For $X \subseteq W$ we write $R \restriction X$ for the binary relation $R \cap (X \times X)$ on X.

A *Kripke frame* is a pair $\mathcal{F} = (W, R)$, where W is a nonempty set and R a binary relation on W. A Kripke frame (W', R') is said to be a *generated subframe* of $\mathcal{F}$ if $W' \subseteq W$, $R' = R \restriction W'$, and $R(w) \subseteq W'$ for every $w \in W'$. An *assignment into* $\mathcal{F}$ is a map $g : PV \to \wp(W)$, where $\wp$ denotes the power set operation, and a *Kripke model* is a triple (W, R, g), where $\mathcal{F} = (W, R)$ is a Kripke frame and g an assignment into $\mathcal{F}$.

For a Kripke model $\mathcal{M} = (W, R, g)$ an element $w \in W$, and a formula $\varphi \in \mathcal{L}$, we define $\mathcal{M}, w \models \varphi$ ('φ is true in $\mathcal{M}$ at w') by induction on φ as follows:

(1) $\mathcal{M}, w \models \top$

(2) $\mathcal{M}, w \models p$ iff $w \in g(p)$, for $p \in PV$

(3) $\mathcal{M}, w \models \neg\varphi$ iff $\mathcal{M}, w \not\models \varphi$

(4) $\mathcal{M}, w \models \varphi \wedge \psi$ iff $\mathcal{M}, w \models \varphi$ and $\mathcal{M}, w \models \psi$

(5) $\mathcal{M}, w \models \Box\varphi$ iff $\mathcal{M}, u \models \varphi$ for every $u \in R(w)$

(6) $\mathcal{M}, w \models [\partial]\varphi$ iff $\mathcal{M}, u \models \varphi$ for every $u \in R(w)$

(7) $\mathcal{M}, w \models \forall\varphi$ iff $\mathcal{M}, u \models \varphi$ for every $u \in W$

We make no distinction between $\Box$ and $[\partial]$ in Kripke semantics. We will always consider the two boxes separately, so this will not be a problem for us.

As usual, an $\mathcal{L}$-formula φ is said to be *satisfied* in a Kripke model $\mathcal{M} = (W, R, h)$ if $\mathcal{M}, w \models \varphi$ for some $w \in W$, and *valid* in a Kripke frame $\mathcal{F} = (W, R)$ if $(W, R, h), w \models \varphi$ for every assignment h into $\mathcal{F}$ and every $w \in W$.

2.3. Linear orders

A *linear order* is a structure $(I, <)$, where I is a nonempty set and $<$ a binary relation on I with the following properties:

(1) $\forall x \neg (x < x)$ irreflexivity
(2) $\forall xyz(x < y \wedge y < z \to x < z)$ transitivity
(3) $\forall xy(x < y \vee x = y \vee y < x)$ linearity

We let $x \leq y$ abbreviate $x < y \vee x = y$ as usual. In line with our general convention, we will often identify (notationally) a linear order $(I, <)$ with its domain I. For example, $(\mathbb{Z}, <)$ and $(\mathbb{R}, <)$ are linear orders, and we often write them simply as $\mathbb{Z}, \mathbb{R}$. A subset $D \subseteq I$ is said to be *dense* if for every $i, j \in I$ with $i < j$, there is $d \in D$ with $i < d < j$. The order I itself is *dense* if I is a dense subset of I. Linear orders $(I, <)$, $(I', <')$ are said to be *isomorphic* (in symbols, $(I, <) \cong (I', <')$) if there is a bijection $f : I \to I'$ such that $i < j$ iff $f(i) <' f(j)$ for all $i, j \in I$; we say that $f : I \to I'$ is an *isomorphism*.

2.4. Linear models

We give $\mathcal{L}$-formulas semantics over a linear order $(I, <)$ as follows. An *assignment (into I)* is a map $h : PV \to \wp(I)$. A *linear model (over I)* is a triple $M = (I, <, h)$, where $(I, <)$ is a linear order and h an assignment into I. We write $\mathrm{dom}(M)$ (the *domain* of M) for the set I, and $\mathrm{supp}(M)$ (the *support* of M) for the set $\{p \in PV : h(p) \neq \emptyset\}$. For a linear order $(I', <')$ and a linear model $M' = (I', <', h')$, we say that M is *isomorphic* to M', and write $M \cong M'$, if there is an isomorphism $f : (I, <) \to (I', <')$ with $h'(p) = f(h(p))$ for every $p \in PV$. We say that M' is a *submodel* of M, and write $M' \subseteq M$, if $I' \subseteq I$, $<' = < \upharpoonright I'$, and $h'(p) = h(p) \cap I'$ for every $p \in PV$. We say that M' is an *initial submodel* of M if $M' \subseteq M$ and whenever $i \in I$, $i' \in I'$, and $i < i'$, we have $i \in I'$. We say that M' is a *final submodel* of M if $M' \subseteq M$ and whenever $i \in I$, $i' \in I'$, and $i' < i$, we have $i \in I'$.

For a linear model $M = (I, <, h)$ and a point $x \in I$, we define $M, x \models \varphi$ by induction on φ, as follows:

(1) $M, x \models \top$

(2) $M, x \models p$ iff $x \in h(p)$, for $p \in PV$

(3) $M, x \models \neg\varphi$ iff $M, x \not\models \varphi$

(4) $M, x \models \varphi \wedge \psi$ iff $M, x \models \varphi$ and $M, x \models \psi$

(5) $M, x \models \Box\varphi$ iff there exist $y, z \in I$ with $y < x < z$ and such that $M, t \models \varphi$ for all $t \in I$ with $y < t < z$

(6) $M, x \models [\partial]\varphi$ iff there exist $y, z \in I$ with $y < x < z$ and such that $M, t \models \varphi$ for all $t \in I$ with $y < t < z$ and $t \neq x$

(7) $M, x \models \forall\varphi$ iff $M, y \models \varphi$ for all $y \in I$

An $\mathcal{L}$-formula φ is said to be *satisfiable over* $\mathbb{R}$ if there exist an assignment h into $\mathbb{R}$, and a point $x \in \mathbb{R}$, such that $(\mathbb{R}, h), x \models \varphi$. The formula φ is said to be *valid over* $\mathbb{R}$ if $(\mathbb{R}, h), x \models \varphi$ for every assignment h into $\mathbb{R}$ and every $x \in \mathbb{R}$. (There is a potential ambiguity here since $(\mathbb{R}, <)$ is also a Kripke frame, but in this paper we never consider Kripke semantics in $(\mathbb{R}, <)$.) Clearly, φ is valid over $\mathbb{R}$ iff $\neg\varphi$ is not satisfiable over $\mathbb{R}$. Let L denote the set of $\mathcal{L}$-formulas that are valid over $\mathbb{R}$ — the *logic of* $\mathbb{R}$. We define $\mathsf{L}_\Box = \mathsf{L} \cap \mathcal{L}_\Box$, $\mathsf{L}_{\Box\forall} = \mathsf{L} \cap \mathcal{L}_{\Box\forall}$, and $\mathsf{L}_{[\partial]\forall} = \mathsf{L} \cap \mathcal{L}_{[\partial]\forall}$ — the logics of $\mathbb{R}$ in each of the respective sublanguages. Our main aim is to give simple completeness proofs for these three logics. (There is little point in considering L itself, since $\Box$ can be expressed with $[\partial]$.)

3. Construction of linear models

We now recall some well known information about lexicographic sums of monadic expansions of linear orders. Sources include [3,11,18]. Taken further, this becomes an extremely powerful model-theoretic method and we cite [5,7,16,17,22] as further reading.

3.1. *Lexicographic sums*

Let $(J, <_J)$ be a linear order, and for each $j \in J$ let $M_j = (I_j, <_j, h_j)$ be a linear model. We write

$$M = \sum_{j \in J} M_j$$

for the linear model $(I, <, h)$, where $I = \{\langle i, j \rangle : j \in J, i \in I_j\}$, $<$ is defined lexicographically by $\langle i, j \rangle < \langle i', j' \rangle$ iff $j <_J j'$ or $(j = j'$ and $i <_j i')$, and $h(p) = \bigcup_{j \in J}(h_j(p) \times \{j\}) = \{\langle i, j \rangle : j \in J, i \in h_j(p)\}$ for each $p \in PV$. It can be verified that $(I, <)$ is a linear order. When $J = (\{0, \ldots, n - 1\}, <)$, we may write M as $\sum_{j < n} M_j$. When $J = (\{0, 1\}, <)$, we may write M as

$M_0 + M_1$. Up to isomorphism, $+$ is associative (though not commutative), so we may omit brackets in finite sums.

For $j \in J$, we let $M \upharpoonright j$ denote the submodel of M with domain $I_j \times \{j\}$. It is isomorphic to M_j (the isomorphism is $\langle i, j \rangle \mapsto i$). We will sometimes identify the two, and so regard M_j as a submodel of M via this isomorphism.

3.2. Intervals of $\mathbb{R}$

An *interval of* $\mathbb{R}$ is a nonempty convex subset $X \subseteq \mathbb{R}$, regarded implicitly as a linear order $(X, < \upharpoonright X)$. An interval is *open* if it has no least element and no greatest element. We will use standard notation for intervals: $[x, y] = \{z \in \mathbb{R} : x \leq z \leq y\}$, (x, y), $[x, y)$, etc. We will be interested in linear models whose domains are (isomorphic to) intervals of $\mathbb{R}$. The following is a trivial but useful case.

Definition 3.1. For $p \in PV$ we will let $\widehat{p}$ denote the one-point linear model $(\{0\}, \emptyset, h)$, where $h(p) = \{0\}$ and $h(q) = \emptyset$ for each $q \in PV \setminus \{p\}$.

Example 3.1. Let $p, q \in PV$. Let $M_j = \widehat{p}$ for each $j \in \mathbb{Q}$ and $M_j = \widehat{q}$ for each $j \in \mathbb{R} \setminus \mathbb{Q}$. Then $\sum_{j \in \mathbb{R}} M_j$ is isomorphic to the linear model $M = (\mathbb{R}, <, h)$, where $h(p) = \mathbb{Q}$, $h(q) = \mathbb{R} \setminus \mathbb{Q}$, and $h(r) = \emptyset$ for every $r \in PV \setminus \{p, q\}$.

In the example, the underlying order of M was isomorphic to $\mathbb{R}$. This is an instance of a more general phenomenon:

Proposition 3.1. *Let $(J, <)$ be a linear order, and for each $j \in J$ let M_j be a linear model over an interval of $\mathbb{R}$. Suppose that one of the following conditions holds:*

(1) $(J, <) = (\{0, 1, \ldots, n\}, <)$ for some integer $n \geq 0$, M_j has a greatest element and no least element for each $j \in \{0, 1, \ldots, n - 1\}$, and M_n has no least element and no greatest element.

(2) $(J, <) = (\mathbb{Z}, <)$, and for each $j \in \mathbb{Z}$, M_j has a greatest element and no least element.

(3) $(J, <) = (\mathbb{R}, <)$, each M_j has a least and a greatest element, and $\mathrm{dom}(M_j)$ is a singleton whenever $j \in \mathbb{R} \setminus \mathbb{Q}$.

Then the underlying order of $\sum_{j \in J} M_j$ is isomorphic to $\mathbb{R}$.

Proof. It is well known and easily proved that a linear order is isomorphic to $\mathbb{R}$ iff it has no least element, no greatest element, is separable (has a countable dense subset), (hence) is dense, and is Dedekind complete (any nonempty subset with an upper bound has a least upper bound). It is easily checked that $\sum_{j \in J} M_j$ has these properties in each case. □

3.3. Shuffles

An important and attractive type of lexicographic sum is the so-called *shuffle*. Shuffles give us an exceedingly simple way to define relatively complicated linear models.

Let $\mathcal{N}$ be a countable set of linear models, where each $N \in \mathcal{N}$ is based on an interval of $\mathbb{R}$ with a least element and a greatest element (such as $[0, 1]$ or $\{0\}$). Let N_0 be a linear model based on a singleton interval. A *shuffle choice map* is a map $s : \mathbb{R} \to \mathcal{N} \cup \{N_0\}$ such that:

(1) $s^{-1}(N)$ is a dense subset of $\mathbb{R}$ for each $N \in \mathcal{N}$.
(2) $s(x) = N_0$ for each irrational $x \in \mathbb{R}$.

Since $\mathbb{Q}$ can be partitioned into infinitely many dense subsets, it is not difficult to show that shuffle choice maps exist. Choose a shuffle choice map s, and define

$$M = \text{Shuffle}(\mathcal{N} \; ; \; N_0) = \sum_{j \in \mathbb{R}} s(j). \tag{1}$$

By proposition 3.1(3), M is a linear model whose underlying order is isomorphic to $\mathbb{R}$, and *we will regard it as actually having $\mathbb{R}$ as its underlying order*. Its formal form depends on s and the isomorphism to $\mathbb{R}$, but the specific choices are immaterial here, and in any case, an argument similar to the proof that any countable dense linear order is isomorphic to $(\mathbb{Q}, <)$ will show that, up to isomorphism, M is independent of these choices. Whenever we use the Shuffle notation as in equation (1), we will assume that they have been tacitly chosen.

Let M be the shuffle above. An element of M is said to be an *M-endpoint* if it is a least or greatest element of $M \upharpoonright j$ for some $j \in \mathbb{R}$ (see §3.1 for the definition of $M \upharpoonright j$).

Lemma 3.1. *Let $x \in M$ and $p \in \text{supp}(M)$. (See §2.4 for $\text{supp}(M)$.)*

(1) If x is an M-endpoint, then $M, x \models \langle \partial \rangle p$ and $M, x \models \Diamond p$.
(2) If x is not an M-endpoint, suppose that $x \in M \upharpoonright j$ for (unique) $j \in \mathbb{R}$. Then $M, x \models \Diamond p$ iff $M \upharpoonright j, x \models \Diamond p$, and $M, x \models \langle \partial \rangle p$ iff $M \upharpoonright j, x \models \langle \partial \rangle p$.
(3) There are $y, z \in \mathbb{R}$ with $y < x < z$, $M, y \models p$, and $M, z \models p$.

Proof sketch. For part 1, suppose that x is the greatest point of $M \upharpoonright j$, for some $j \in \mathbb{R}$. Let $y > x$ be given. Plainly, $y \in M \upharpoonright k$ for some $k \in \mathbb{R}$ with $k > j$. Pick $N \in \mathcal{N} \cup \{N_0\}$ and $t \in N$ with $N, t \models p$. As $s^{-1}(N)$ is dense in $\mathbb{R}$, we may find $l \in (j, k)$ with $s(l) = N$, so $M \upharpoonright l \cong N$. Let z be the element of $M \upharpoonright l$ corresponding to t under this isomorphism. Then $M, z \models p$ and $z \in (x, y)$. Since y was arbitrary, $M, x \models \langle \partial \rangle p \wedge \Diamond p$. The case where x is the least point of $M \upharpoonright j$ is similar. Part 2 holds simply because the part of $M \upharpoonright j$ excluding its endpoints is an open interval of $\mathbb{R}$ containing x. Finally, part 3 holds because for each $N \in \mathcal{N} \cup \{N_0\}$ there are arbitrarily large and small $j \in \mathbb{R}$ with $s(j) = N$. □

Example 3.2. Let $p, q \in PV$.

(1) Shuffle($\{\widehat{p}\}$; $\widehat{q}$) is, up to isomorphism, the model M of example 3.1.
(2) Let $N = \widehat{p} + \text{Shuffle}(\emptyset ; \widehat{p}) + \widehat{p}$. This is a linear model whose underlying order is isomorphic to $[0, 1]$, and all its points satisfy p and only p.
(3) $S = \text{Shuffle}(\{N\}$; $\widehat{q})$ is a linear model that can be described up to isomorphism as: each rational in $\mathbb{R}$ is replaced by a non-singleton closed interval of $\mathbb{R}$ whose points satisfy only p, and each irrational is left intact and made to satisfy only q. The S-endpoints are the endpoints of the closed intervals and the intact irrationals. The endpoints satisfy $\langle \partial \rangle p \wedge \langle \partial \rangle q$, while the other points satisfy $\Box(p \wedge \neg q)$. The underlying order of S is isomorphic to $\mathbb{R}$.
(4) Shuffle($\{N, \widehat{p}\}$; $\widehat{q}$) is rather different: again up to isomorphism, we split $\mathbb{Q}$ into two dense subsets, replace the points of the first set by copies of N, make the points of the second set satisfy only p, and make the irrationals satisfy only q as before. This model is not isomorphic to S above, because it has singleton subintervals (endpoints) satisfying p that are not part of any longer interval whose points satisfy p. However, perhaps surprisingly, an Ehrenfeucht–Fraïssé game will show that it is indistinguishable from S in $\mathcal{L}$.

Armed with these devices and standard modal methods, we will be able to prove completeness theorems for $\mathbb{R}$ really rather easily. There are two main steps. First, given an appropriate finite Kripke frame (W, R), by applying shuffle we obtain a linear model whose domain is isomorphic to $\mathbb{R}$, and which gives rise to a certain function $f : \mathbb{R} \to W$. Second, given an assignment, say g, into (W, R), we define a new assignment $h = f^{-1} \circ g$ into $\mathbb{R}$, yielding a linear model $M = (\mathbb{R}, <, h)$. From these two main steps we can prove the following 'satisfaction' lemma: $M, x \models \varphi$ iff $(W, R, g), f(x) \models$

φ, for each $x \in \mathbb{R}$ and $\mathcal{L}$-formula φ in the appropriate fragment. The finite model property for each of the logics under consideration yields a finite model (W, R, g) satisfying any given consistent formula φ, which is turned into a linear model satisfying φ as above.

4. The logic of $\mathbb{R}$ with $\square$

We start with the classical result of McKinsey and Tarski [13] that the logic of $\square$ over $\mathbb{R}$ is S4. So in this section we work with the language $\mathcal{L}_\square$ whose formulas involve only $\top$, atoms, the boolean operations, and $\square$. Recall that S4 is the smallest set of $\mathcal{L}_\square$-formulas that contains the axioms:

(1) all propositional tautologies
(2) $\square(p \to q) \to (\square p \to \square q)$ normality
(3) $\square p \to p$ reflexivity
(4) $\square p \to \square\square p$ transitivity

and is closed under the inference rules:

(1) modus ponens: $\dfrac{\varphi,\ \varphi \to \psi}{\psi}$

(2) generalisation (or necessitation) for $\square$: $\dfrac{\varphi}{\square\varphi}$

(3) substitution: $\dfrac{\varphi(p)}{\varphi(\psi/p)}$

Let (W, R) be a finite Kripke frame in which all the axioms of S4 are valid, and such that $W \subseteq PV$ (this will allow us to define models $\widehat{w}$ and write $\mathcal{L}$-formulas such as $\Diamond w$, for $w \in W$). It follows that R is reflexive and transitive. Define a binary relation $R^\bullet$ on W by $R^\bullet wu$ iff $Rwu \wedge \neg Ruw$. For each $w \in W$ define $C(w) = \{u \in W : Rwu \wedge Ruw\}$.

Definition 4.1. For each $w \in W$, we define a linear model N_w over $\mathbb{R}$ by complete induction on $|R(w)|$:

$$N_w = \text{Shuffle}\big(\{\widehat{w} + N_u + \widehat{w} : u \in R^\bullet(w)\} \cup \{\widehat{u} : u \in C(w)\}\ ;\ \widehat{w}\big).$$

We check that N_w is well defined. If $u \in R^\bullet(w)$ then $R(u) \subseteq R(w)$ by transitivity of R, and plainly, $w \in R(w) \setminus R(u)$. So $|R(u)| < |R(w)|$. Inductively, N_u is a well defined linear model over $\mathbb{R}$. So the underlying order of $\widehat{w} + N_u + \widehat{w}$ is isomorphic to an interval of $\mathbb{R}$ with a least and a greatest element. All models $\widehat{u}, \widehat{w}$ are based on singleton intervals, so N_w is a legal shuffle and a well defined linear model over $\mathbb{R}$.

Since N_w is 'made' wholly from linear models of the form $\widehat{u}$ for $u \in R(w)$, we have $\mathrm{supp}(N_w) \subseteq R(w)$, and for each $x \in \mathbb{R}$ there is a unique $u \in R(w)$ with $N_w, x \models u$. (This can be proved formally by a trivial induction on $|R(w)|$.) We write $f_w(x)$ for this u. So we have defined a map $f_w : \mathbb{R} \to W$. For all $x \in \mathbb{R}$ and $u \in W$,

$$N_w, x \models u \iff f_w(x) = u. \tag{2}$$

Lemma 4.1. *For every $w \in W$, the following hold:*

(1) We have $\mathrm{rng}(f_w) = \mathrm{supp}(N_w) = R(w)$.
(2) For each $v \in W$ and $x \in \mathbb{R}$ we have $N_w, x \models \Diamond v$ iff $R(f_w(x), v)$.

Proof. By induction on $|R(w)|$. Assume the lemma inductively for every $u \in W$ with $|R(u)| < |R(w)|$ — in particular, every $u \in R^\bullet(w)$. By definition, $\mathrm{supp}(N_w) = \{u \in PV : \exists x \in \mathbb{R}(N_w, x \models u)\}$. By the above, this is $\{u \in W : \exists x \in \mathbb{R}(u = f_w(x))\} = \mathrm{rng}(f_w)$. Noting that $\mathrm{supp}(\widehat{u}) = \{u\}$, it follows from the definition of N_w that $\mathrm{supp}(N_w) = \bigcup\{\{w\} \cup \mathrm{supp}(N_u) : u \in R^\bullet(w)\} \cup C(w)$. Inductively, this is $\bigcup\{\{w\} \cup R(u) : u \in R^\bullet(w)\} \cup C(w) = R(w)$. This proves part 1. For part 2, let v, x be given. There are two cases.

Case 1. If x is an N_w-endpoint, then inspection of the definition of N_w shows that $f_w(x) = w$ or $f_w(x) \in C(w)$. Either way, $f_w(x) \in C(w)$. So by transitivity of R we have $R(f_w(x)) = R(w)$, and hence $R(f_w(x), v)$ iff $v \in R(w)$. By part 1, $R(w) = \mathrm{supp}(N_w)$. Also, $v \in \mathrm{supp}(N_w)$ iff $N_w, x \models \Diamond v$ ($\Rightarrow$ is by lemma 3.1, and $\Leftarrow$ is trivial). Stringing all this together, we see that $R(f_w(x), v)$ iff $N_w, x \models \Diamond v$.

Case 2. If not, then $x \in N_w \upharpoonright j \cong \widehat{w} + N_u + \widehat{w}$ for some $j \in \mathbb{R}$ and $u \in R^\bullet(w)$. We identify N_u with the submodel of $N_w \upharpoonright j \subseteq N_w$ as usual. As x is not a N_w-endpoint, we have $x \in N_u$. By lemma 3.1(2), $N_w, x \models \Diamond v$ iff $N_w \upharpoonright j, x \models \Diamond v$. The least and greatest points of $N_w \upharpoonright j$ do not affect $\Diamond$, so this is iff $N_u, x \models \Diamond v$. Inductively, this is iff $R(f_u(x), v)$. But plainly, $f_u(x) = f_w(x)$. $\square$

Now fix $w \in W$, and write N_w and f_w simply as N and f, respectively. Let $g : PV \to \wp(W)$ be an assignment and let $\mathcal{M}$ be the Kripke model (W, R, g). Define an assignment $h : PV \to \wp(\mathbb{R})$ by $h(p) = f^{-1}(g(p)) = \{x \in \mathbb{R} : \mathcal{M}, f(x) \models p\}$, for each $p \in PV$. Let M be the linear model $(\mathbb{R}, <, h)$. Of course, M depends on $\mathcal{M}$ and w. We now have two linear models N, M over $\mathbb{R}$ and we will use them both below.

Lemma 4.2. *For every $\psi \in \mathcal{L}_\square$ and $x \in \mathbb{R}$, we have $M, x \models \psi$ iff $\mathcal{M}, f(x) \models \psi$.*

Proof. By induction on ψ. The atomic and boolean cases are easy and we omit them. Assume the lemma for ψ, and consider $\lozenge\psi$. First suppose that $\mathcal{M}, f(x) \models \lozenge\psi$, so there is $u \in W$ with $R(f(x), u)$ and $\mathcal{M}, u \models \psi$. By lemma 4.1, $N, x \models \lozenge u$. Inductively, every $y \in \mathbb{R}$ with $N, y \models u$ satisfies $M, y \models \psi$ (since $f(y) = u$). It follows that $M, x \models \lozenge\psi$.

Conversely, suppose that $M, x \models \lozenge\psi$. We claim that for some $u \in W$, every open interval of $\mathbb{R}$ containing x contains a point y with $M, y \models \psi$ and $f(y) = u$. For if not, for each u there is an open interval O_u containing x but containing no such point y. Let $O = \bigcap_{u \in W} O_u$. Since W is finite, O is again an open interval containing x. But $M, x \models \lozenge\psi$, so O contains a point y with $M, y \models \psi$. Let $f(y) = u$. Then $y \in O_u$, contradicting the definition of O_u. This proves the claim.

Let u be as in the claim. Plainly, $N, x \models \lozenge u$, so by lemma 4.1, $R(f(x), u)$. Also, inductively we have $\mathcal{M}, u \models \psi$. Hence, $\mathcal{M}, f(x) \models \lozenge\psi$ as required. $\square$

Theorem 4.1 (McKinsey–Tarski, 1944). *The $\mathcal{L}_\square$-logic $\mathsf{L}_\square$ of $\mathbb{R}$ is S4.*

Proof. It is easy to check that the S4 axioms are valid over $\mathbb{R}$, and that the inference rules preserve validity. So $\text{S4} \subseteq \mathsf{L}_\square$. For the converse, take $\varphi \in \mathcal{L}_\square$ with $\varphi \notin \text{S4}$. We will show that $\neg\varphi$ is satisfiable over $\mathbb{R}$, so that $\varphi \notin \mathsf{L}_\square$, completing the proof.

It is known that S4 has the *finite model property*. (This can be proved by filtration: see, e.g., [4, corollary 5.32].) So $\neg\varphi$ is satisfied in a finite Kripke model $\mathcal{M} = (W, R, g)$ such that all the S4-axioms are valid in the Kripke frame (W, R). It is immaterial what the elements of W are, so we may assume without loss of generality that $W \subseteq PV$. Choose $w \in W$ such that $\mathcal{M}, w \models \neg\varphi$.

We now suppose that W, R, $\mathcal{M}$, and w are as in the foregoing discussion. This can be done without loss of generality. Define f and M as above. By lemma 4.1(1), there is $x \in \mathbb{R}$ with $f(x) = w$. As $\mathcal{M}, w \models \neg\varphi$, lemma 4.2 yields $M, x \models \neg\varphi$. So $\neg\varphi$ is satisfiable over $\mathbb{R}$ and $\varphi \notin \mathsf{L}_\square$ as required. $\square$

The map f is an *interior map,* as in several other proofs of this result. It is worth noting that the proof transforms a finite Kripke model satisfying a formula effectively into an explicit and simple description of a model over $\mathbb{R}$ that satisfies the formula. It is easy to write down the description in practice, using shuffles.

Example 4.1. The formula $\varphi = p \wedge \Diamond(\neg p \wedge \Diamond p)$ is plainly true at 0 in the Kripke model $\mathcal{M} = (\{0, 1, 2\}, \leq, g)$ where $g(p) = \{0, 2\}$. We assume as above that $0, 1, 2 \in PV$, and construct three linear models:

$$N_2 = \mathrm{Shuffle}(\{\widehat{2}\} \, ; \, \widehat{2})$$
$$N_1 = \mathrm{Shuffle}(\{\widehat{1} + N_2 + \widehat{1}, \widehat{1}\} \, ; \, \widehat{1})$$
$$N_0 = \mathrm{Shuffle}(\{\widehat{0} + N_1 + \widehat{0}, \widehat{0} + N_2 + \widehat{0}, \widehat{0}\} \, ; \, \widehat{0})$$

The underlying order of N_0 is $\mathbb{R}$. We define $f = f_0 : \mathbb{R} \to \{0, 1, 2\}$ as above. So for each $i \in \{0, 1, 2\}$, $f^{-1}(i)$ is the set of points of N_0 lying in 'copies' of $\widehat{i}$. We define $h : PV \to \wp(\mathbb{R})$ by $h(q) = f^{-1}(g(q))$, for $q \in PV$. So $h(p) = f^{-1}(0) \cup f^{-1}(2)$. We define M to be the linear model $(\mathbb{R}, <, h)$. Then $M, x \models \varphi$ for any x in a copy of $\widehat{0}$. Indeed it is plain that any such x satisfies p and has arbitrarily close to it points y in copies of $\widehat{1}$. Such y satisfy $\neg p$, and have points z in copies of $\widehat{2}$ arbitrarily close to them; such z satisfy p.

5. The logic of $\mathbb{R}$ with $\Box$ and $\forall$

We now move on to the language $\mathcal{L}_{\Box\forall}$ containing formulas using both $\Box$ and $\forall$. In [20], Shehtman showed that the logic of $\mathbb{R}$ in this language is S4UC. The logic S4UC is the smallest set of $\mathcal{L}_{\Box\forall}$-formulas closed under the inference rules of modus ponens, generalisation for both $\Box$ and $\forall$, and substitution, and containing the following axioms:

(1) all propositional tautologies
(2) $\Box(p \to q) \to (\Box p \to \Box q)$ normality
(3) $\Box p \to p$ reflexivity
(4) $\Box p \to \Box\Box p$ transitivity
(5) $\forall(p \to q) \to (\forall p \to \forall q)$ normality
(6) $\forall p \to p$ reflexivity
(7) $\forall p \to \forall\forall p$ transitivity
(8) $\exists\forall p \to p$ symmetry
(9) $\forall p \to \Box p$ 'U'
(10) $\forall(\Box p \vee \Box\neg p) \to \forall p \vee \forall\neg p$ connectedness, 'C'

Let (W, R) be a finite Kripke frame in which all the axioms of S4UC are valid and such that $W \subseteq PV$. So R is reflexive and transitive. We will apply the same idea as in the preceding section, but since $\forall$ is in the language, we need to arrange that the map $f : \mathbb{R} \to W$ is surjective. To do this, we will use that (W, R) is connected.

Definition 5.1. Let $\mathcal{F} = (W, R)$ be a Kripke frame. A *connected compo-nent* of $\mathcal{F}$ is a minimal nonempty subset $D \subseteq W$ such that for all $w \in W$:

- if $w \in D$ then $R(w) \subseteq D$
- if $w \in W \setminus D$ then $R(w) \subseteq W \setminus D$

If k is an integer, $\mathcal{F}$ is said to be *k-connected* if it has at most k connected components, and *connected* if it is 1-connected. A Kripke model is said to be connected or k-connected if its frame has this property.

The slight differences of definition 5.1 from definitions in [12,20] will not matter, since we will not formally use any results involving connectedness from those papers.

Indeed, (W, R) is connected. This is easy to see (cf. [20, lemma 8]). For if D is a connected component of (W, R), let g be an assignment into (W, R) with $g(p) = D$, and let $w \in W$. Then $(W, R, g), w \models \forall(\Box p \vee \Box \neg p)$. Axiom C is valid in (W, R), so $(W, R, g), w \models \forall p \vee \forall \neg p$, and hence $D = W$ or $D = \emptyset$. Since D is nonempty, $D = W$.

As (W, R) is finite and connected, a little thought shows that there exist points $d_0, u_0, d_1, u_1, \ldots, d_{n-1}, u_{n-1}, d_n \in W$, for some finite n, such that:

- $Ru_j d_j$ and $Ru_j d_{j+1}$ for each $j < n$
- $W = \bigcup_{j<n} R(u_j)$

For each $w \in W$, let N_w be the linear model of definition 4.1, with under-lying order $\mathbb{R}$.

Lemma 5.1. *For each $x \in \mathbb{R}$ and $u \in W$, we have $u \in \mathrm{supp}(N_w)$ iff there are $y, z \in \mathbb{R}$ with $y < x < z$, $N_w, y \models u$, and $N_w, z \models u$.*

Proof. $\Rightarrow$ is immediate from lemma 3.1(3), and $\Leftarrow$ is trivial. $\qquad \square$

Now define

$$N = \Big(\sum_{j<n} (N_{d_j} + \widehat{u_j} + N_{u_j} + \widehat{u_j}) \Big) + N_{d_n}. \tag{3}$$

In effect, N is the finite sum

$$N_{d_0} + \widehat{u_0} + N_{u_0} + \widehat{u_0} + N_{d_1} + \widehat{u_1} + N_{u_1} + \widehat{u_1} + \cdots + N_{u_{n-1}} + \widehat{u_{n-1}} + N_{d_n}.$$

A couple of applications of proposition 3.1(1) show that N is a linear model whose underlying linear order is isomorphic to $\mathbb{R}$. As usual, we will assume that its underlying order is actually $\mathbb{R}$, and that each of the N_{d_j}, N_{u_j}, and two copies of $\widehat{u_j}$ are submodels of N.

As in §4, for each $x \in \mathbb{R}$ there is a unique $u \in W$ with $N, x \models u$, and we write $f(x)$ for this u. Thus, $f : \mathbb{R} \to W$. By lemma 4.1(1) and the choice of u_j, we have

$$\text{rng}(f) = \text{supp}(N) \supseteq \bigcup_{j<n} \text{supp}(N_{u_j}) = \bigcup_{j<n} R(u_j) = W.$$

So f is surjective.

Lemma 5.2. *Let* $x \in \mathbb{R}$ *and* $w \in W$. *Then* $N, x \models \Diamond w$ *iff* $R(f(x), w)$.

Proof. If $x \in N_{d_j}$ for some $j \leq n$ or $x \in N_{u_j}$ for some $j < n$, the result follows from lemma 4.1(2), since these submodels are based on open intervals of $\mathbb{R}$. Suppose for some $j < n$ that x is in the submodel $\widehat{u_j}$ that is preceded by N_{d_j} and followed by N_{u_j}. Clearly, $N, x \models \Diamond w$ iff (a) arbitrarily large elements of N_{d_j} satisfy w, or (b) $N, x \models w$, or (c) arbitrarily small elements of N_{u_j} satisfy w. Now by lemmas 5.1 and 4.1(1), (a) holds iff $w \in \text{supp}(N_{d_j}) = R(d_j)$, and (c) holds iff $w \in \text{supp}(N_{u_j}) = R(u_j)$. Plainly, (b) holds iff $w = u_j$. So $N, x \models \Diamond w$ iff $w \in R(d_j) \cup \{u_j\} \cup R(u_j)$. Because R is reflexive and transitive and $Ru_j d_j$, we have $R(d_j) \cup \{u_j\} \cup R(u_j) = R(u_j) = R(f(x))$. So $N, x \models \Diamond w$ iff $R(f(x), w)$, as required. The argument when x is in the submodel $\widehat{u_j}$ between N_{u_j} and $N_{d_{j+1}}$ is similar, using that $Ru_j d_{j+1}$. $\square$

Now, as before, let $g : PV \to \wp(W)$ be an assignment into W, and let $\mathcal{M}$ be the Kripke model (W, R, g). Define the linear model $M = (\mathbb{R}, <, h)$, where $h(p) = f^{-1}(g(p))$ for each $p \in PV$.

Lemma 5.3. *For every* $\psi \in \mathcal{L}_{\Box\forall}$ *and* $x \in \mathbb{R}$, *we have* $M, x \models \psi$ *iff* $\mathcal{M}, f(x) \models \psi$.

Proof. The proof is the same as for lemma 4.2, but there is an additional case: $\forall \psi$. So assume the lemma inductively for ψ, and let $x \in \mathbb{R}$ be given. If $\mathcal{M}, f(x) \models \forall \psi$, then $\mathcal{M}, w \models \psi$ for all $w \in W$. Inductively, $M, y \models \psi$ for all $y \in \mathbb{R}$, and we obtain $M, x \models \forall \psi$. Conversely, assume that $M, x \models \forall \psi$, and let $w \in W$ be given. As f is surjective, we can find $y \in \mathbb{R}$ with $f(y) = w$. By assumption, $M, y \models \psi$, and inductively, $\mathcal{M}, f(y) \models \psi$ as well. Since w was arbitrary, we obtain $\mathcal{M}, f(x) \models \forall \psi$ as required. $\square$

Theorem 5.1 (Shehtman, 1999). *The $\mathcal{L}_{\Box\forall}$-logic $\mathsf{L}_{\Box\forall}$ of $\mathbb{R}$ is* S4UC.

Proof. Again it is easy to check soundness: that S4UC $\subseteq \mathsf{L}_{\Box\forall}$. (Axiom C is valid over $\mathbb{R}$ because $\mathbb{R}$ is connected: it cannot be written as the union of two disjoint nonempty open sets.) For the converse, we take a formula $\varphi \notin$ S4UC and show that $\neg\varphi$ is satisfiable over $\mathbb{R}$, so that $\varphi \notin \mathsf{L}_{\Box\forall}$.

By [20, theorem 10] (proved by filtration), S4UC has the finite model property. So we may take a finite Kripke model $\mathcal{M} = (W, R, g)$ satisfying $\neg\varphi$ and in whose frame (W, R) all axioms of S4UC are valid. We may assume that W, R, g are the same as above. Define f, M as above. Take $w \in W$ with $\mathcal{M}, w \models \neg\varphi$. As f is surjective, we may find $x \in \mathbb{R}$ with $f(x) = w$. By lemma 5.3, $M, x \models \neg\varphi$. Thus, $\neg\varphi$ is satisfiable over $\mathbb{R}$, which completes the proof. □

6. The logic of $\mathbb{R}$ with $[\partial]$ and $\forall$

Finally we consider the language $\mathcal{L}_{[\partial]\forall}$ containing formulas using $[\partial]$ and $\forall$ but not $\Box$. Actually we will use $\Box$, but as an *abbreviation:* $\Box\varphi$ will abbreviate $\varphi \wedge [\partial]\varphi$. (It can be checked that the semantics of $\Box$ in linear models — though not in Kripke models — is as in earlier sections.)

The logic of $\mathbb{R}$ in the language $\mathcal{L}_{[\partial]}$ is KD4G$_2$ — this was conjectured by Shehtman [19] and proved by Shehtman [21] and Lucero-Bryan [12, theorem 4.5]. Lucero-Bryan goes on to show [12, corollary 5.27] that the logic of $\mathbb{R}$ in the language $\mathcal{L}_{[\partial]\forall}$ is KD4G$_2$.UC. The logic KD4G$_2$.UC is the smallest set of $\mathcal{L}_{[\partial]\forall}$-formulas closed under the standard inference rules (modus ponens, generalisation for $[\partial]$ and $\forall$, and substitution) and containing the following axioms:

(1) all propositional tautologies
(2) $[\partial](p \to q) \to ([\partial]p \to [\partial]q)$ normality
(3) $\langle\partial\rangle\top$ seriality
(4) $[\partial]p \to [\partial][\partial]p$ transitivity
(5) $\forall(p \to q) \to (\forall p \to \forall q)$ normality
(6) $\forall p \to p$ reflexivity
(7) $\forall p \to \forall\forall p$ transitivity
(8) $\exists\forall p \to p$ symmetry
(9) $\forall p \to [\partial]p$ 'U'
(10) $\forall(\Box p \vee \Box\neg p) \to \forall p \vee \forall\neg p$ connectedness, 'C'
(11) $[\partial](\bigvee_{0 \le i \le 2} \Box\varphi_i) \to \bigvee_{0 \le i \le 2}[\partial]\neg\varphi_i$ 'G$_2$'
     where $p_0, p_1, p_2 \in PV$ and $\varphi_i = p_i \wedge \bigwedge\{\neg p_j : 0 \le j \le 2, j \ne i\}$ for each $i \in \{0, 1, 2\}$

The logic KD4G$_2$ is defined analogously in the language $\mathcal{L}_{[\partial]}$ by deleting axioms 5–10.

Let $\mathcal{F} = (W, R)$ be a finite Kripke frame in which all axioms of KD4G$_2$.UC are valid, and with $W \subseteq PV$. Importantly, R may not be reflexive. But we do have:

- R is transitive.
- $R(w) \neq \emptyset$ for every $w \in W$.
- $\mathcal{F}$ is connected (using axiom C).
- $\mathcal{F}$ is 'locally 2-connected'. That is, for every $w \in W$, the frame $(R(w), R \upharpoonright R(w))$ is 2-connected (see definition 5.1) and so has at most two connected components. (This is easy to prove using validity of G$_2$ in $\mathcal{F}$.)

As earlier, define the binary relation $R^\bullet$ on W by $R^\bullet wu$ iff $Rwu \wedge \neg Ruw$. For $w \in W$, define $C(w) = \{u \in W : Rwu \wedge Ruw\}$. We say that w is a *leaf* if $R^\bullet(w) = \emptyset$. In that case, the axiom $\langle \partial \rangle \top$ and transitivity give Rww. For $w \in W$ we define $W_w = R(w) \cup \{w\}$, and define the Kripke frame

$$\mathcal{F}_w = (W_w, R \upharpoonright W_w).$$

This is connected — a connected component containing w must be W_w — and a generated subframe of $\mathcal{F}$.

Lemma 6.1. *For each connected generated subframe $\mathcal{G}$ of $\mathcal{F}$, there is a linear model $\overline{\mathcal{G}}$ based on $\mathbb{R}$, and such that for each $x \in \mathbb{R}$ and $v \in \mathcal{G}$:*

G1. *There is a unique $u \in PV$ with $\overline{\mathcal{G}}, x \models u$. Moreover, $u \in \mathcal{G}$. We will write this u as $f_{\mathcal{G}}(x)$.*

G2. *$\overline{\mathcal{G}}, x \models \langle \partial \rangle v$ iff $R(f_{\mathcal{G}}(x), v)$.*

G3. *There are $y < x < z$ in $\mathbb{R}$ with $\overline{\mathcal{G}}, y \models v$ and $\overline{\mathcal{G}}, z \models v$.*

G4. *If v is a leaf, there are linear models A, B such that $\overline{\mathcal{G}} \cong A + \overline{\mathcal{F}_v} + B$. (Reminder: linear models are nonempty.)*

Proof. We prove the lemma by complete induction on $|\mathcal{G}|$. So take a connected generated subframe $\mathcal{G}$ of $\mathcal{F}$, and inductively assume the lemma for smaller subframes than $\mathcal{G}$. There are two cases.

Case 1. Suppose that $\mathcal{G} = \mathcal{F}_w$ for some *reflexive* $w \in W$ (i.e., with Rww). Choose such a w (it need not be unique). Now, as in definition 4.1, we let

$$\overline{\mathcal{G}} = \text{Shuffle}\big(\{\widehat{w} + \overline{\mathcal{F}_u} + \widehat{w} : u \in R^\bullet(w)\} \cup \{\widehat{u} : u \in C(w)\} ; \widehat{w}\big) \quad (4)$$

Inductively, $\overline{\mathcal{F}}_u$ is defined for each $u \in R^\bullet(w)$, so as earlier, we see that this shuffle is well defined. It is easy to confirm that $\overline{\mathcal{G}}$ meets the requirements G1–G4. We leave the reader to verify G1 and G3. We briefly check G2. It holds inductively for any x in a submodel of $\overline{\mathcal{G}}$ of the form $\overline{\mathcal{F}}_u$. Any x not in such a submodel is a $\overline{\mathcal{G}}$-endpoint, so $\overline{\mathcal{G}}, x \models \langle \partial \rangle v$ iff $v \in \mathrm{supp}(\overline{\mathcal{G}})$ ($\Rightarrow$ is trivial and $\Leftarrow$ follows from lemma 3.1(1)). It follows easily from G1 and G3 that $\mathrm{supp}(\overline{\mathcal{G}}) = W_w$, so this is iff $v \in W_w$. Since w is reflexive, this is iff $R(w, v)$. But $f_{\mathcal{G}}(x) \in C(w)$, so by transitivity of R, this is iff $R(f_{\mathcal{G}}(x), v)$, as required.

We now check G4. Suppose that $v \in \mathcal{G} = \mathcal{F}_w$ is a leaf. If $R^\bullet wv$ then it is plain that G4 holds for v, since $\overline{\mathcal{F}}_v$ is an 'ingredient' of the shuffle in equation (4) defining $\overline{\mathcal{G}}$. If instead $\neg R^\bullet wv$, then $v \in C(w)$, so $\mathcal{F}_v = \mathcal{F}_w = \mathcal{G}$. Since $\overline{\mathcal{G}}$ is a shuffle, it is easily seen that $\overline{\mathcal{G}} \cong A + \overline{\mathcal{G}} + B = A + \overline{\mathcal{F}}_v + B$ for some A, B.

Case 2. Suppose otherwise. As $\mathcal{G}$ is connected and locally 2-connected, a little thought shows that there is a sequence

$$\ldots u_{-1}, d_0, u_0, d_1, u_1, d_2, \ldots$$

of elements of $\mathcal{G}$ with the following properties:

- $Ru_i d_i \wedge Ru_i d_{i+1}$ for each $i \in \mathbb{Z}$.
- $\mathcal{G} = \bigcup_{j < i} W_{u_j} = \bigcup_{k > i} W_{u_k}$ for each $i \in \mathbb{Z}$.
- $\{d_i : i \in \mathbb{Z}\}$ is the set of leaves that lie in $\mathcal{G}$.
- For each $i \in \mathbb{Z}$, let C_i and D_i be the connected components of the frame $(R(u_i), R \upharpoonright R(u_i))$ that contain d_i and d_{i+1}, respectively. Then $R(u_i) = C_i \cup D_i$.

We briefly indicate one way to choose such a sequence. As $\mathcal{G}$ is connected, there is a finite 'zigzag cycle' $d_0, u_0, \ldots, d_{n-1}, u_{n-1}, d_n$ with $d_n = d_0$, $Ru_i d_i \wedge Ru_i d_{i+1}$ for each $i < n$, every leaf in $\mathcal{G}$ is among $d_0, \ldots, d_n$, and $\mathcal{G} = \bigcup_{i < n} W_{u_i}$. For each d_i that is not a leaf, there is a leaf d_i' in $\mathcal{G}$ with $Rd_i d_i'$. We can replace d_i by d_i'. So we may assume that all the d_i are leaves. Take any $i < n$. As $(R(u_i), R \upharpoonright R(u_i))$ is 2-connected, it has connected components C, D (possibly equal) with $C \cup D = R(u_i)$. Suppose $d_i \in C$, say. If $d_{i+1} \notin D$ then choose any leaf $d \in D$ and replace the part d_i, u_i, d_{i+1} of the cycle by $d_i, u_i, d, u_i, d_{i+1}$. Do this for each $i < n$. After these insertions we obtain a cycle $d_0, u_0, \ldots, d_{m-1}, u_{m-1}, d_m$ with $d_m = d_0$, for some $m \geq n$. Now the $\mathbb{Z}$-sequence

$$\ldots, u_{m-1}, d_0, u_0, \ldots, d_{m-1}, u_{m-1}, d_0, u_0, \ldots, d_{m-1}, u_{m-1}, d_0, \ldots$$

has the required properties.

Each C_i $(i \in \mathbb{Z})$ is the domain of a connected generated subframe of $\mathcal{F}$ which we denote by $\mathcal{C}_i$, and similarly for $\mathcal{D}_i$. If $\mathcal{C}_i = \mathcal{G}$, then $u_i \in \mathcal{G} = C_i \subseteq R(u_i) \subseteq W_{u_i} \subseteq \mathcal{G}$, so u_i is reflexive and $\mathcal{G} = \mathcal{F}_{u_i}$, contradicting the case assumption. So $|\mathcal{C}_i| < |\mathcal{G}|$, and similarly, $|\mathcal{D}_i| < |\mathcal{G}|$. Let $\overline{\mathcal{C}_i}, \overline{\mathcal{D}_i}$ be the linear models given by the inductive hypothesis. As $d_i \in C_i$, $d_{i+1} \in D_i$, and they are leaves, by the inductive hypothesis there are linear models A_i, B_i, A_i', B_i' $(i \in \mathbb{Z})$ such that:

$$\overline{\mathcal{C}_i} \cong A_i + \overline{\mathcal{F}_{d_i}} + B_i$$
$$\overline{\mathcal{D}_i} \cong A_i' + \overline{\mathcal{F}_{d_{i+1}}} + B_i' \tag{5}$$

Plainly, A_i has a greatest element and no least element, B_i has a least element and no greatest element, and similarly for A_i', B_i'.

We now set

$$\overline{\mathcal{G}} = \sum_{j \in \mathbb{Z}} \left(\overline{\mathcal{F}_{d_j}} + B_j + \widehat{u}_j + A_j' \right). \tag{6}$$

In effect, $\overline{\mathcal{G}}$ is the sum

$$\cdots + \overline{\mathcal{F}_{d_0}} + B_0 + \widehat{u_0} + A_0' + \overline{\mathcal{F}_{d_1}} + B_1 + \widehat{u_1} + A_1' + \overline{\mathcal{F}_{d_2}} + B_2 + \widehat{u_2} + A_2' + \cdots$$

Clearly, the underlying order of each $\overline{\mathcal{F}_{d_j}} + B_j + \widehat{u}_j + A_j'$ is isomorphic to an interval of $\mathbb{R}$ with a greatest point but no least point. So by proposition 3.1(2), $\overline{\mathcal{G}}$ can be assumed to have domain $\mathbb{R}$.

Let us check the requirements of the lemma. Requirement G1 is proved by induction as before. For G2, suppose that $x \in \widehat{u}_j$ for some $j \in \mathbb{Z}$. Referring to equation (6), x lies just after a submodel B_j of $\mathcal{G}$ that is isomorphic to a final submodel of $\overline{\mathcal{C}_j}$. Take $y \in B_j$, so that $y < x$. Trivially, if $y < z < x$ and $\overline{\mathcal{G}}, z \models v$ then $v \in C_j$. Conversely, by G3 for $\overline{\mathcal{C}_j}$, for any $v \in C_j$ and $y < x$ there is $z \in B_j$ with $y < z < x$ and $\overline{\mathcal{G}}, z \models v$. Similarly, a copy of an initial submodel A_j' of $\overline{\mathcal{D}_j}$ can be found just after x, so all and only the elements of D_j 'occur' arbitrarily near to x on its right. Combining these two observations, we see that for any $v \in W$ we have $\overline{\mathcal{G}}, x \models \langle \partial \rangle v$ iff $v \in C_j \cup D_j = R(u_j) = R(f_{\mathcal{G}}(x))$. Condition G2 for x follows. Every other element $x \in \overline{\mathcal{G}}$ lies in an open interval of a structure $\overline{\mathcal{C}_j}$ or $\overline{\mathcal{D}_j}$ (of the form $\overline{\mathcal{F}_{d_j}} + B_j$ or $A_j' + \overline{\mathcal{F}_{d_{j+1}}}$, respectively), so G2 holds inductively for x.

For G3, let $v \in \mathcal{G}$ and $x \in \mathbb{R}$ be given. Suppose that x lies in the submodel $\overline{\mathcal{F}_{d_i}} + B_i + \widehat{u}_i + A_i'$, say, of $\overline{\mathcal{G}}$. By assumption on the u_i, there are $j, k \in \mathbb{Z}$ with $j < i < k$ and $v \in W_{u_j} \cap W_{u_k}$. Let y' be the element of $\overline{\mathcal{G}}$ in the submodel $\widehat{u}_j$ (see equation (6)). So $y' < x$. If $v = u_j$, then plainly

$\overline{\mathcal{G}}, y' \models v$. If $v \neq u_j$ then $v \in R(u_j)$, so by G2 we have $\overline{\mathcal{G}}, y' \models \langle \partial \rangle v$. Either way it is clear that $\overline{\mathcal{G}}, y \models v$ for some $y < x$. The case $z > x$ is similar, using k.

For G4, note that any leaf $v \in \mathcal{G}$ is equal to some d_j, and as equation (6) plainly shows, $\overline{\mathcal{F}_{d_j}}$ occurs as an interval in $\overline{\mathcal{G}}$. □

Now $\mathcal{F}$ is itself a connected generated subframe of $\mathcal{F}$, so by the lemma, $\overline{\mathcal{F}}$ can be found, with underlying order $\mathbb{R}$. Let $f = f_{\mathcal{F}}$. By property G3 of lemma 6.1, $f : \mathbb{R} \to W$ is surjective.

Now let $g : PV \to \wp(W)$ be an assignment into $\mathcal{F}$, and let $\mathcal{M}$ be the Kripke model (W, R, g). Define $h : PV \to \wp(\mathbb{R})$ by $h(p) = f^{-1}(g(p)) = \{x \in \mathbb{R} : \mathcal{M}, f(x) \models p\}$. This gives us a linear model $M = (\mathbb{R}, <, h)$.

Lemma 6.2. *For every $x \in \mathbb{R}$ and $\mathcal{L}_{[\partial]\forall}$-formula φ, we have $M, x \models \varphi$ iff $\mathcal{M}, f(x) \models \varphi$.*

Proof. By induction on φ. The atomic and boolean cases are easy. Assume the result for φ. Then $M, x \models \forall\varphi$ iff $M, y \models \varphi$ for every $y \in \mathbb{R}$, iff $\mathcal{M}, w \models \varphi$ for every $w \in W$ (inductively, and since f is surjective), iff $\mathcal{M}, f(x) \models \forall\varphi$. Finally, $M, x \models \langle \partial \rangle \varphi$ iff for every open interval $O \subseteq \mathbb{R}$ containing x, there is $y \in O \setminus \{x\}$ with $M, y \models \varphi$. Inductively, this holds iff for every open O containing x, there is $y \in O \setminus \{x\}$ with $\mathcal{M}, f(y) \models \varphi$. As $\mathcal{M}$ is finite, there are only finitely many values of f, so this is equivalent to saying that for some $w \in W$ with $\mathcal{M}, w \models \varphi$, every open O containing x contains a point $y \neq x$ with $f(y) = w$. This is plainly equivalent to $\overline{\mathcal{F}}, x \models \langle \partial \rangle w$ for some $w \in W$ with $\mathcal{M}, w \models \varphi$. By G2 of lemma 6.1, this holds iff $R(f(x), w)$ for some $w \in W$ with $\mathcal{M}, w \models \varphi$ — that is, iff $\mathcal{M}, f(x) \models \langle \partial \rangle \varphi$. □

Theorem 6.1 (Lucero-Bryan, 2011). *The $\mathcal{L}_{[\partial]\forall}$-logic $\mathsf{L}_{[\partial]\forall}$ of $\mathbb{R}$ is* $\mathrm{KD4G_2.UC}$.

Proof. Again we leave it to the reader to check that $\mathrm{KD4G_2.UC} \subseteq \mathsf{L}_{[\partial]\forall}$ (soundness). (Axiom $\mathrm{G_2}$ is valid over $\mathbb{R}$ because if O is an open interval of $\mathbb{R}$ and $x \in O$, then $O \setminus \{x\}$ is not the union of three pairwise disjoint nonempty open sets [19, lemma 31].) For the converse (completeness), again we take an $\mathcal{L}_{[\partial]\forall}$-formula $\varphi \notin \mathrm{KD4G_2.UC}$ and show that $\neg\varphi$ is satisfiable over $\mathbb{R}$.

By [12, corollary 5.22], $\mathrm{KD4G_2.UC}$ has the finite model property. (This is nontrivial and is proved by an unorthodox filtration in a style due to Shehtman [19]; see also [23].) So we may take a finite Kripke model $\mathcal{M} =$

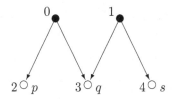

Fig. 1.   The model $\mathcal{M}$

(W, R, g) in which $\neg\varphi$ is satisfied, and such that the axioms of $KD4G_2.UC$ are valid in the frame (W, R). As usual, we may assume that (W, R) is the frame $\mathcal{F}$ studied above. Let M and f be as above. As f is surjective, we can take $x \in \mathbb{R}$ with $\mathcal{M}, f(x) \models \neg\varphi$. By lemma 6.2, $M, x \models \neg\varphi$, and so $\neg\varphi$ is satisfiable over $\mathbb{R}$ as required.   □

We leave it as an exercise to show that $KD4G_2$ is the logic of $\mathbb{R}$ in the sublanguage of $\mathcal{L}_{[\partial]\forall}$ without $\forall$.

Once again, the proof transforms a finite Kripke model of a formula effectively into a model over $\mathbb{R}$ satisfying the formula, in a way that can be applied in practical examples.

Example 6.1. The formula $\langle\partial\rangle p \wedge \langle\partial\rangle q \wedge \exists(\langle\partial\rangle q \wedge \langle\partial\rangle s \wedge [\partial]\neg p)$ is true at 0 in the Kripke model $\mathcal{M}$ shown in figure 1. Its set of worlds is $\{0, \ldots, 4\}$, and the relation R is indicated by the arrows; $0, 1$ are R-irreflexive and $2, 3, 4$ are R-reflexive. Plainly, $\mathcal{M}$ is connected and locally 2-connected. We select a sequence $\ldots, d_0, u_0, d_1, u_1, \ldots$ of elements of $\mathcal{M}$ as follows:

$$
\begin{array}{ccccccccccc}
\ldots & u_0 & u_1 & u_2 & u_3 & u_4 & u_5 & u_6 & u_7 & u_8 & u_9 & \ldots \\
\ldots & d_0 & d_1 & d_2 & d_3 & d_4 & d_5 & d_6 & d_7 & d_8 & d_9 & \ldots \\
\ldots & 2 & 0 & 3 & 1 & 4 & 1 & 3 & 0 & 2 & 0 & 3 & 1 & 4 & 1 & 3 & 0 & 2 & 0 & 3 & 1 & \ldots
\end{array}
$$

This 'loops' over all points in $\mathcal{M}$ and meets the conditions in the proof above. In the notation of the proof, we have $C_0 = \{2\}$, $D_0 = C_1 = \{3\}$, $D_1 = C_2 = \{4\}$, $D_2 = C_3 = \{3\}$, and so on. So $D_{i-1} = C_i = \{d_i\}$ for every $i \in \mathbb{Z}$.

We assume as usual that $\{0, \ldots, 4\} \subseteq PV$. Let $\mathcal{F}$ be the frame of $\mathcal{M}$, and define linear models

$$
\begin{aligned}
P &= \mathrm{Shuffle}(\{\widehat{2}\}; \widehat{2}) \cong \overline{\mathcal{F}_2} \\
Q &= \mathrm{Shuffle}(\{\widehat{3}\}; \widehat{3}) \cong \overline{\mathcal{F}_3} \\
S &= \mathrm{Shuffle}(\{\widehat{4}\}; \widehat{4}) \cong \overline{\mathcal{F}_4}
\end{aligned}
$$

As P is a shuffle, we can find a copy of it in the middle of itself, so $P \cong P_A + P + P_B$ for some suitable P_A and P_B, and similarly for Q, S. For each i, as $D_{i-1} = C_i = \{d_i\}$ we have $\mathcal{D}_{i-1} = \mathcal{C}_i = \mathcal{F}_{d_i}$, so

$$\overline{\mathcal{C}_0} = \overline{\mathcal{F}_{d_0}} = \overline{\mathcal{F}_2} = P \cong P_A + P + P_B$$
$$\overline{\mathcal{D}_0} = \overline{\mathcal{C}_1} = \overline{\mathcal{F}_{d_1}} = \overline{\mathcal{F}_3} = Q \cong Q_A + Q + Q_B$$
$$\overline{\mathcal{D}_1} = \overline{\mathcal{C}_2} = \overline{\mathcal{F}_{d_2}} = \overline{\mathcal{F}_4} = S \cong S_A + S + S_B$$
$$\overline{\mathcal{D}_2} = \overline{\mathcal{C}_3} = \overline{\mathcal{F}_{d_3}} = \overline{\mathcal{F}_3} = Q \cong Q_A + Q + Q_B$$
$$\overline{\mathcal{D}_3} = \overline{\mathcal{C}_4} = \overline{\mathcal{F}_{d_4}} = \overline{\mathcal{F}_2} = P \cong P_A + P + P_B$$
$$\overline{\mathcal{D}_4} = \overline{\mathcal{F}_{d_5}} = \overline{\mathcal{F}_3} = Q \cong Q_A + Q + Q_B$$

and so on. Equation (5) in the proof of lemma 6.1 tells us to write $\overline{\mathcal{C}_i} \cong A_i + \overline{\mathcal{F}_{d_i}} + B_i$ and $\overline{\mathcal{D}_i} \cong A_i' + \overline{\mathcal{F}_{d_{i+1}}} + B_i'$, for each i and for suitable A_i, B_i, A_i', B_i'. So we can take

$$\ldots \quad B_0 = P_B \qquad A_0' = Q_A$$
$$B_1 = Q_B \qquad A_1' = S_A$$
$$B_2 = S_B \qquad A_2' = Q_A$$
$$B_3 = Q_B \qquad A_3' = P_A$$
$$B_4 = P_B \qquad A_4' = Q_A \quad \ldots$$

According to equation (6) in the proof, we define the linear model

$$\overline{\mathcal{F}} = \sum_{j \in \mathbb{Z}} \left(\overline{\mathcal{F}_{d_j}} + B_j + \widehat{u}_j + A_j' \right)$$
$$= \cdots + \underbrace{P + P_B + \widehat{0} + Q_A}_{j=0} + \underbrace{Q + Q_B + \widehat{1} + S_A}_{j=1} + \underbrace{S + S_B + \widehat{1} + Q_A}_{j=2}$$
$$+ \underbrace{Q + Q_B + \widehat{0} + P_A}_{j=3} + \underbrace{P + P_B + \widehat{0} + Q_A}_{j=4} + \cdots$$

In this example, the expression obviously simplifies to

$$\overline{\mathcal{F}} \cong \cdots + P + \widehat{0} + Q + \widehat{1} + S + \widehat{1} + Q + \widehat{0} + P + \widehat{0} + Q + \cdots$$

and if we assign p to the set of points in copies of P, and similarly for q, s, we obtain an entirely sensible and reasonable linear model M over $\mathbb{R}$ in which $\langle \partial \rangle p \wedge \langle \partial \rangle q \wedge \exists (\langle \partial \rangle q \wedge \langle \partial \rangle s \wedge [\partial] \neg p)$ is true at any point in a copy of $\widehat{0}$.

7. Conclusion

We have proved completeness theorems for some 'spatial' logics over $\mathbb{R}$ in a fairly simple way. Spatial logic is of burgeoning interest and the methods used here may find further application. For example, there is potential for model checking a formula against a description of a model over $\mathbb{R}$ using

176

shuffles and other operators, and this has already been explored for temporal logic in [6]. Some of the theorems that we have reproved here were originally proved in more general forms, for certain topological spaces. It remains to be seen whether the methods of this paper can be adapted to apply in this generality.

Acknowledgments

The author thanks the organisers for inviting him to the conference, Rob Goldblatt for handling the paper, Valentin Shehtman for helpful bibliographic information, and the referee for finding several mistakes in the submitted paper and for very detailed comments and suggestions.

References

1. M. Aiello, J. van Benthem and G. Bezhanishvili, Reasoning about space: the modal way, *J. Logic Computat.* **13**, 889–920 (2003).
2. G. Bezhanishvili and M. Gehrke, *A new proof of completeness of S4 with respect to the real line*, Tech. Rep. PP-2002-06, ILLC, Amsterdam (2002).
3. J. P. Burgess and Y. Gurevich, The decision problem for linear temporal logic, *Notre Dame J. Formal Logic* **26**, 115–128 (1985).
4. A. Chagrov and M. Zakharyaschev, *Modal logic*, Oxford Logic Guides, Vol. 35 (Clarendon Press, Oxford, 1997).
5. K. Doets, Monadic π_1^1-theories of π_1^1-properties, *Notre Dame Journal of Formal Logic* **30**, 224–240 (1989).
6. T. French, J. McCabe-Dansted and M. Reynolds, *Synthesis and model checking for continuous time: Long version*, tech. rep., CSSE, University of Western Australia (2012), http://www.csse.uwa.edu.au/~mark/research/Online/sctm.htm.
7. Y. Gurevich, Monadic second-order theories, in *Model-Theoretic Logics*, eds. J. Barwise and S. Feferman (Springer-Verlag, New York, 1985) pp. 479–507.
8. P. Kremer, Strong completeness of S4 wrt the real line, manuscript; http://individual.utoronto.ca/philipkremer/onlinepapers/strongcompletenessR.pdf, (2012).
9. A. Kudinov, Topological modal logic of $\mathbb{R}$ with inequality, *Russian Mathematical Surveys* **63**, 163–166 (2008).
10. T. Lando and D. Sarenac, Fractal completeness techniques in topological modal logic: Koch curve, limit tree, and the real line, preprint, http://philosophy.berkeley.edu/file/698/FractalCompletenessTechniques.pdf, (2011).
11. H. Läuchli and J. Leonard, On the elementary theory of linear order, *Fundamenta Mathematicae* **59**, 109–116 (1966).
12. J. G. Lucero-Bryan, The d-logic of the real line, *J. Logic Computat.* (2011), online, doi:10.1093/logcom/exr054.

13. J. McKinsey and A. Tarski, The algebra of topology, *Annals of Mathematics* **45**, 141–191 (1944).

14. G. Mints, A completeness proof for propositional S4 in Cantor space, in *Logic at Work*, ed. E. Orłowska, Studies in Fuzziness and Soft Computing, Vol. 24 (Physica-Verlag, Heidelberg/New York, 1998) pp. 79–88. Essays dedicated to the memory of Elena Rasiowa. ISBN 3-7908-1164-5.

15. G. Mints and T. Zhang, A proof of topological completeness for S4 in $(0, 1)$, *Ann. Pure. Appl. Logic* **133**, 231–245 (2005).

16. J. M. Plotkin (ed.), *Hausdorff on Ordered Sets*, History of Mathematics, Vol. 25 (American Mathematical Society, Providence, R.I., 2005).

17. A. Rabinovich, Composition theorem for generalized sum, *Fundam. Inf.* **79**, 137–167 (January 2007).

18. J. G. Rosenstein, *Linear orderings* (Academic Press, New York, 1982).

19. V. B. Shehtman, *Derived sets in Euclidean spaces and modal logic*, Tech. Rep. X-1990-05, University of Amsterdam (1990).

20. V. B. Shehtman, 'Everywhere' and 'here', *Journal of Applied Non-classical Logics* **9**, 369–379 (1999).

21. V. B. Shehtman, Modal logic of topological spaces, habilitation thesis, (Moscow, 2000). In Russian.

22. S. Shelah, The monadic theory of order, *Annals of Mathematics* **102**, 379–419 (1975).

23. M. Zakharyaschev, A sufficient condition for the finite model property of modal logics above K4, *Logic J. IGPL* **1**, 13–21 (1993).

ON A QUESTION OF CSIMA ON COMPUTATION-TIME DOMINATION

XIA HUA[1]   JIANG LIU[2*]   GUOHUA WU[1†]

[1] *Division of Mathematical Sciences*
School of Physical and Mathematical Sciences
Nanyang Technological University, Singapore 637371

[2] *Chongqing Institute of Green and Intelligent Technology*
Chinese Academy of Sciences, Chongqing 401122, China

Keywords: Computably enumerable sets, settling-time reducibility, settling functions, Ershov hierarchy

1. Introduction

The settling-time reducibility, $<_{st}$, among the c.e. sets, was first proposed by Nabutovsky and Weinberger in [5] and Soare in [6]. For a computably enumerable (*c.e.* for short) set A with an effective enumeration $\{A_s\}_{s\in\omega}$, the *settling function* of A w.r.t. this enumeration is

$$m_A = \mu s[A_s \upharpoonright x = A \upharpoonright x],$$

where $A \upharpoonright x = \{y \leq x \mid y \in A\}$.

Definition 1.1. Let A and B be two c.e. sets, with effective enumerations $\{A_s\}_{s\in\omega}$ and $\{B_s\}_{s\in\omega}$ respectively. Say that A settling-time dominates B, denoted as $B <_{st} A$ if for any computable function f, $m_A(x) > f(m_B(x))$ is true for almost all x, i.e. m_A dominates $f \circ m_B$.

---

*Liu is partially supported by projects NSFC-61202131 and cstc2012ggB40004 from Chongqing Natural Science Foundation.

†Wu is partially supported by AcRF grants RG37/09, M52110101 and MOE2011-T2-1-071 (ARC 17/11, M45110030) from Ministry of Education of Singapore.

Nies observed that this reduction does not depend on the choice of enumerations, and hence $<_{st}$ is an ordering on *c.e.* sets. In [2], Csima considered a direct generalization of $<_{st}$ to Δ_2^0 sets.

Definition 1.2. Let A be a Δ_2^0 set with effective approximation $\{A_s\}_{s\in\omega}$. The *settling function* for A w.r.t. this enumeration, $m_A(x)$, is defined as the least stage after which the approximation does not change up to x:

$$m_A(x) = (\mu s)(\forall t \geq s)[A_t \upharpoonright x = A \upharpoonright x].$$

Csima proved that $<_{st}$ on Δ_2^0 set is not reflexive, as for each $n \geq 2$, there is a properly *n-c.e.* set A with two *n-c.e.* approximations $\{A_s\}_{s\in\omega}$ and $\{\tilde{A}_s\}_{s\in\omega}$ such that for any total computable function f, $m_{\tilde{A}}$ dominates $f(m_A)$. When $\{A_s\}_{s\in\omega}$ is an effective enumeration of a given Δ_2^0 set, the settling function of $\{A_s\}_{s\in\omega}$ is quite different from the computation function, even though they coincide when the given set is c.e.

Definition 1.3. Let A be a Δ_2^0 set with effective approximation $\{A_s\}_{s\in\omega}$. The *computation function* of A w.r.t. this enumeration, $C_A(x)$, is the least stage s such that A_s and A agree up to x. That is,

$$C_A(x) = (\mu s \geq x)[A_s \upharpoonright x = A \upharpoonright x].$$

It is a folklore that for any Δ_2^0 set A, $C_A \equiv_T A$. Based on Nabutovsky and Weinberger's notion of settling-time dominating on c.e. sets, Csima proposed the notion of computation-time dominating among Δ_2^0 sets.

Definition 1.4. Let A and B be two Δ_2^0 sets, with effective approximation $\{A_s\}_{s\in\omega}$ and $\{B_s\}_{s\in\omega}$ respectively. Say that A computation-time dominates B w.r.t. to these enumerations, if for all total computable function f, $C_A(x) > f(C_B(x))$ is true for almost all x, i.e. C_A dominates $f \circ C_B$.

In [2], Csima showed that Nies' observation that settling-time dominating does not depend on the enumerations is not true for computation-time dominating, when Δ_2^0 enumerations are involved. What Csima proved is the existence of c.e. set A with a c.e. enumeration $\{A_s\}_{s\in\omega}$ and a 3-c.e. approximation $\{\tilde{A}_s\}_{s\in\omega}$ such that the computation function of $\{A_s\}_{s\in\omega}$ dominates $f \circ C_{\tilde{A}}$ for any total computable function f. In the paper, Csima asked whether this kind of computation-time domination ordering depending on approximations is also true for 2-c.e. sets. In this paper, we give a confirmative answer to Csima's question.

Theorem 1.1. *There exists a 2-c.e. set A with two different 2-c.e. approximations $\{A_s\}_{s \in \omega}$ and $\{\tilde{A}_s\}_{s \in \omega}$ such that for every total computable function f, such that for almost all x,*

$$C_{\tilde{A}}(x) \geq f(C_A(x)).$$

Our notations and terminology are standard and generally follow Soare [7].

2. Requirements and basic strategy

We will construct a 2-c.e. set A with two different 2-c.e. approximations $\{A_s\}_{s \in \omega}$ and $\left\{\tilde{A}_s\right\}_{s \in \omega}$ such that for each e, the following requirements are satisfied:

$$\mathcal{R}_e : \varphi_e \text{ is total} \Rightarrow (\forall^\infty x)[C_{\tilde{A}}(x) \geq \varphi_e(C_A(x))],$$

where φ_e is the eth partial computable function.

Csima's idea of satisfying a single $\mathcal{R}_e$ requirement is direct. That is, first partition ω into infinitely many nonempty blocks, B_n, $n \in \omega$. To satisfy $\mathcal{R}_e$, it is enough to show that for each $n \geq e$, for all numbers x in B_n, $C_{\tilde{A}}(x) \geq \varphi_e(C_A(x))$. Here is the idea. Let x_n be an element in B_n, and put x_n into A_s at stage s, remove it at stage $s + 1$. According to [2], this is called a tempting process. Now wait for $\varphi_e(s)$ to converge. If it does not converge, then $\mathcal{R}_e$ is satisfied. Otherwise, let $t > s$ be a stage at which $\varphi_e(s)$ converges, and we do at stage t is to reenumerate x_n into A_t, and also into $\tilde{A}_t$. Note that $A_t \restriction (x_n + 1) = A_s \restriction (x_n + 1)$, and so $C_A(x_n + 1) = s$. By $\varphi_e(s) < t$, we have

$$C_{\tilde{A}}(x_n + 1) = t > \varphi_e(s) = \varphi_e(C_A(x_n + 1)).$$

Csima's idea works well to show that for all numbers x in B_n, $C_{\tilde{A}}(x) \geq \varphi_e(C_A(x))$ and hence $\mathcal{R}_e$ is satisfied. In Csima's argument, $\cup \tilde{A}_s$ is a c.e. set, while $\{A_s : s \in \omega\}$ is a 3-c.e. approximation of this set. It seems that the use of one element for temptation of each block makes a 3-c.e. approximation quite necessary.

Our construction is a variant of Csima's original idea, and is able to make approximations 2-c.e. Instead of using one element for the temptation, we use two-element. That is, we start the temptation by enumerating a number $x + 1$ into A_s, and then enumerating x at stage $s + 1$. At stage $s + 1$, we also enumerate both x and $x + 1$ into $\widetilde{A}_{s+1}$ (we do this to ensure that $\lim_s A_s$ and $\lim_s \widetilde{A}_s$ are equal). Now if $\varphi_e(s)$ converges at stage t, then $\varphi_e(s) < t$, and we remove x from both A_t and $\widetilde{A}_t$. Then $A_t \upharpoonright (x + 2) = A_s \upharpoonright (x + 2)$ and $C_A(x + 2) = s$. As $C_{\widetilde{A}}(x + 2) = t$, $C_{\widetilde{A}}(x) \geq \varphi_e(C_A(x))$. Again, we can show that $\mathcal{R}_e$ is satisfied.

We now consider the general case. As in [2], we partition ω into infinitely many consecutive blocks B_0, B_1, $\cdots$, such that each block B_n is composed of $n^3 + 1$ subblocks, $B_{n,0}$, $B_{n,1}$, $\cdots$, B_{n,n^3}, and each subblock contains $2n + 2$ numbers. Here the numbers of $B_{n,i}$ are less than the numbers in $B_{n,j}$, if i is less than j, and we will make sure that for each $e < n$, if φ_e is total, then $C_{\widetilde{A}}(x) \geq \varphi_e(C_A(x))$ is true for all x in B_n. It can be done on a single subblock, and we let B_n have $n^3 + 1$ subblocks because whenever we do actions on some block B_m with $m < n$, we need to restart our work on B_n on a new subblock. We will see that $n^3 + 1$ blocks is enough to deal with the actions of these B_m's.

Say that we activate block B_n at stage s via an unused subblock $B_{n,k} = \{b_0, b_1, \cdots, b_{2n+1}\}$ with $b_0 < b_1 < \cdots < b_{2n+1}$, if $B_{n,k}$ is the largest unused subblock of B_n, and our actions at stage s and the following stages are:

- at stage s, enumerate $b_{2n+1}, \cdots, b_3, b_1$, into A_s;
- at stage $s + 1$, enumerate b_{2n} into A_{s+1}
- at stage $s + 2$, enumerate b_{2n-2} into A_{s+2},

   $\cdots$

- at stage $s + n + 1$, enumerate b_0 into A_{s+n+1}.
  We use c_n to denote $s + n + 1$, for convenience.

Following [2], we call this a *temptation* process. We say that B_n is fully activated at stage $s + n + 1$ via subblock $B_{n,k}$ if this process is done, and from this stage, B_n continues to be active, until we do actions on some block $B_m, m < n$ (if so, we say that block B_n is deactivated, and consequently, c_n is undefined automatically), and B_n needs to be activated later via a(nother) unused subblock (a new c_n will be defined in this case), which means that we abandon the subblock $B_{n,k}$ used just now, and will activate B_n via $B_{n,k-1}$. In the construction, at any stage, for each block B_n, at most one subblock is active, and the first subblock of B_n being activated will be the greatest one, i.e. B_{n,n^3}. A subblock $B_{n,k}, k < n^3$, will be activated only after $B_{n,k+1}$, the previously activated subblock, is deactivated by some action on a smaller block.

In the construction, the temptation process can skip several stages, so when we say that a computation becomes convergent at stage s, we mean this computation has not been convergent previously, except those stages being skipped in some temptation process.

We say that a block B_n requires attention at stage s if one of the following occurs:

- B_n is not active (so c_n is undefined), or
- There is some $e \leq n$ such that $\varphi_e(c_n)$ becomes convergent at stage s.

3. Construction

We now provide the construction of the 2-c.e. set, and two different 2-c.e. approximations. First, we partition ω into infinitely many blocks $B_n, n \in \omega$, such that each block B_n is composed of $n^3 + 1$ subblocks, $B_{n,0}, B_{n,1}, \cdots, B_{n,n^3}$, and each subblock contains $2n + 2$ numbers.

Stage 0: Do nothing.

Stage $s > 0$: Activate block B_s first via the greatest subblock $B_{n,n\cdot2^n} = \{b_0, b_1, \cdots, b_{2n+1}\}$ and do the temptation process as follows:

- at stage s, enumerate $b_{2s+1}, \cdots, b_3, b_1$, into A_s;
- at stage $s + 1$, enumerate b_{2s} into A_{s+1}
- at stage $s + 2$, enumerate b_{2s-2} into A_{s+2},
  $\cdots$
- at stage $2s + 1$, enumerate b_0 into A_{2s+1}; and enumerate $b_0, b_1, \cdots, b_{2n+1}$ into $\widetilde{A}_{2s+1}$. Set $c_n = 2s + 1$.

Now we consider whether there are some n and $e \leq n$ such that B_n is active via a subblock $B_{n,k}$ and $\varphi_e(c_n)$ becomes convergent at stage s. Find the largest $i \leq n$ such that $b_{2i} \in A_s \cap B_{n,k}$, and remove b_{2i} from both A and $\widetilde{A}$. We say that we do actions on B_n at stage s. For $m > n$, if B_m is active via a subblock $B_{m,l}$, deactivate $B_{m,l}$, do the temptation process for subblock $B_{m,l-1}$, and declare that B_m is active via a subblock $B_{m,l-1}$. Redefine $c_m = s + m + 1$.

Go to stage $2s + 2$. We skip the stages from $s + 1$ to $2s + 1$ as we do the temptation during these stages.

This completes the construction.

4. Verification

We now verify that the constructed approximations $\{A_s\}_{s \in \omega}$ and $\{\widetilde{A}_s\}_{s \in \omega}$ satisfying the requirements. Obviously, $\lim_s A_s = \lim_s \widetilde{A}_s$, and both $\{A_s\}_{s \in \omega}$ and $\{\widetilde{A}_s\}_{s \in \omega}$ are 2-c.e. approximations.

Now we check that each block B_n has enough subblocks to carry out the temptations.

Lemma 4.1. *For each block B_n, $n \in \omega$, there is a subblock $B_{n,k}$ and a stage s such that after stage s, B_n keeps active via $B_{n,k}$ forever.*

Proof. By the construction, the active subblock of B_n can be shifted to a smaller one only when some number in B_m, $m < n$, is removed from A and $\widetilde{A}$. So this kind of shifting can happen at most

$$1 \cdot 2^1 + 2 \cdot 2^2 + \cdots + n \cdots 2^n < n \cdot 2^n$$

many times, and $n \cdot 2^n$ many subblocks in B_n enables us to find an unused subblock in B_n whenever needed. As a consequence, we will activate a subblock $B_{n,k}$ of B_n at some stage in the construction, and from then on B_n will not be deactivated afterwards. That is, B_n keeps active via subblock $B_{n,k}$ forever. □

The following lemma shows that each requirement is satisfied.

Lemma 4.2. *For each $e \in \omega$, if φ_e is total, then for each $n \geq e$, and for all $x \in \cup_{n \geq e} B_n$, $C_{\widetilde{A}}(x) \geq \varphi_e(C_A(x))$. $\mathcal{R}_e$ is satisfied.*

Proof. First we prove by induction on n that we do actions on each B_n at most finitely often. Suppose that it is true for B_j, $j < n$. Let s_n be the last stage we do actions for these B_js. Then after stage s_n, after the temptation for B_n (the last one) has been done, as c_n is now fixed, we do actions on block B_n, because $\varphi_e(c_n)$ becomes convergent for some $e \leq n$. Obviously, we do such actions on block B_k at most $n + 1$ many times. It completes the induction.

Now we show that for each e, if $n \geq e$, then for all $x \in B_n$, $C_{\widetilde{A}}(x) \geq \varphi_e(C_A(x))$. By the statement above, let s, k be the numbers such that after stage s, $B_{n,k}$ keeps active forever. This means that c_n becomes fixed.

Let y be the least number in $B_{n,k}$ being removed from A and $\widetilde{A}$. Obviously, y is the last number being removed from block B_n, and hence for

$x \geq y, x \in B_n$, $C_{\tilde{A}}(x) = C_{\tilde{A}}(y) = t$. Suppose that y is removed at stage t. As $C_A(x) \leq c_n$, if φ_e, $e < n$, is total, then $\varphi_e(c_n)$ converges by stage t; otherwise, when it converges later, we have to do action on B_n again, contradicting our assumption on y. Therefore $\varphi_e(C_A(x)) \downarrow [t] < t = C_{\tilde{A}}(x)$ by the convention that $\varphi_e(n) \downarrow [t]$ implies $\varphi_e(m) \downarrow [t]$ for all $m < n$.

Now, we consider $x < y$ in B_n. Because a fresh subblock of B_n is always activated when we remove number from B_{n-1}. It follows that $C_{\tilde{A}}(x) = C_{\tilde{A}}(\max B_{n-1})$, $C_A(x) = C_A(\max B_{n-1})$ for all $x \in B_n$ less than y. Since $e < n$, φ_e would have been considered in the block B_{n-1}. Using the previous argument, we have $C_{\tilde{A}}(\max B_{n-1}) > \varphi_e(C_A(\max B_{n-1}))$. Therefore, $C_{\tilde{A}}(x) > \varphi_e(C_A(x))$. $\qquad \square$

This complete the proof of Theorem 1.1.

References

1. B. F. Csima. *Comparing c.e. sets based on their settling-times*, Computability in Europe 2007, *Lecture Notes in Computer Science* 4497 (2007), 196-204.
2. Csima: *The settling time reducibility ordering and Δ_2^0 sets*, J. Logic Comp. 19 (2009), 145-150.
3. B. F. Csima and R. Shore. *The settling-time reducibility ordering*, *J. Symbolic Logic* 72 (2007), 1055-1071.
4. B. F. Csima and R. I. Soare. *Computability results used in differential geometry*, *J. Symbolic Logic* 71 (2006), 1394-1410.
5. A. Nabutovsky and S. Weinberger. The fractal nature of Riem/Diff I. *Geometrica Dedicata* 101 (2003), 1-54.
6. R. I. Soare. Computability theory and diferential geometry. *Bull. Symbolic Logic* 10(2004), 457-486.
7. R. I. Soare. *Recursively enumerable sets and degrees*, Springer-Verlag, Heidelberg, 1987.

A GENERALIZATION OF BETH MODEL TO FUNCTIONALS OF HIGH TYPES

F. KACHAPOVA

*School of Computing and Mathematical Sciences, Auckland University of Technology,
Auckland 1142, New Zealand
E-mail farida.kachapova@aut.ac.nz*

Beth model is one of the tools in intuitionistic proof theory. Van Dalen constructed a Beth model for intuitionistic analysis with choice sequences. Bernini and Wendel investigated intuitionistic theories with many types of choice sequences, n-functionals, and their relation with classical type theory. In this paper we construct a Beth model for a basic intuitionistic theory with many types of functionals using recursive approach to define nodes in the Beth model. This model is a tool for justifying consistency of intuitionistic principles for functionals of high types. The principles that hold in our Beth model can be added to the initial basic theory in order to develop it into a relatively strong intuitionistic theory, where a significant part of classical type theory can be interpreted, with the purpose of contributing to the programme of justifying classical mathematics from the intuitionistic point of view. In this paper we show that Kripke's schema for n-functionals and standard axioms for lawless n-functionals hold in our Beth model.

Keywords: Beth model; forcing; choice sequence; functional; lawless; intuitionistic principle; Kripke's schema

1. Introduction

Intuitionistic analysis studies choice sequences, which are freely proceeding sequences of natural numbers. While analysis/ second-order arithmetic is the most important part of set theory, sets of high types and functionals of high types are also needed for developing some parts of mathematics.

Here we consider a basic axiomatic theory L with functionals of type n, or n-functionals, for all $n = 1, 2, \ldots$. We interpret 1-functionals as choice sequences, and $(n + 1)$-functionals as sequences of n-functionals; they are generalizations of choice sequences. The language of L is a modification of the language of a very strong intuitionistic theory HL introduced by Bernini [1]. The theory HL was shown to be inconsistent by Wendel [2]. The theory HL was modified by Wendel [2] and Bernini [3] but without proofs of consistency. Kashapova [4] modified HL and created a Beth-Kripke model as a proof of consistency of the modified theory. In this paper a new, improved model for functionals of many types is constructed and it is a Beth model.

Van Dalen [5] constructed a Beth model for choice sequences and showed that several principles of intuitionistic analysis hold in this model. We generalize this model to a Beth model with many types of functionals. Then we prove the consistency of the Beth model for the fragment of L with functionals of types $1, \ldots, s$ (for any $s \geqslant 1$). In this paper we also show that Kripke's schema and standard axioms for lawless n-functionals hold in our Beth model.

The paper starts a reverse engineering project: testing intuitionistic principles with n-functionals in the Beth model and extending the theory L by the ones that hold in the model. This project (which will be continued in next paper) aims to construct a rich language for expressing mathematical theorems and a strong, consistent intuitionistic theory with this language where a significant part of classical type theory can be interpreted; this is intended as a contribution to the programme of justifying classical mathematics from the intuitionistic point of view.

In this paper we will use the same terminology as in the books [6] and [7] by Kolmogorov and Dragalin; in particular, this relates to the following concepts of a logical-mathematical language and an axiomatic theory.

A *logical-mathematical language of first order* (or *logical-mathematical language*, or just *language* in short) is defined as $\Omega = \langle Srt, Cnst, Fn, Pr \rangle$, where

Srt is a non-empty set of sorts of objects, and for each sort $\pi \in Srt$ there is a countable collection of variables of this sort;

$Cnst$ is the set of all constants of the language;

Fn is the set of all functional symbols of the language;

Pr is the set of all predicate symbols of the language.

In the language Ω we can construct terms, atomic formulas and formulas.

Terms of different sorts are constructed recursively from constants and variables using functional symbols. The complexity of a term t is the number of occurrences of functional symbols in t.

For each predicate symbol $P(x_1, \ldots, x_k)$ and terms $t_1, \ldots, t_k$ of corresponding sorts we construct an atomic formula $P(t_1, \ldots, t_k)$.

Formulas are constructed from atomic formulas and logical constant $\perp$

using logical connectives $\wedge$ (conjunction), $\vee$ (disjunction) and $\supset$ (implication), and quantifiers $\forall$ and $\exists$. The logical connective $\equiv$ (equivalence) is defined as follows: $\varphi \equiv \psi$ means $(\varphi \supset \psi) \wedge (\psi \supset \varphi)$. The complexity of a formula φ is the number of occurrences of logical symbols (connectives and quantifiers) in φ.

Next we define an *axiomatic theory* (or just *theory* in short) as $Th = \langle \Omega, l, A \rangle$, where

Ω is a logical-mathematical language;

l is the logic of the theory; we will consider only two well-known logics: the classical logic CPC (classical predicate calculus) and the intuitionistic logic HPC (Heyting's predicate calculus); their detailed description can be found, for example, in [7];

A is some set of closed formulas of the language Ω called the set of non-logical axioms of Th. When axioms are stated as non-closed formulas, it means that they must be closed by universal quantifiers over all parameters.

The notation $Th \vdash \varphi$ (formula φ is derivable in the theory Th) means that φ is derivable in the logic l from a finite subset of the axiom set A. The theory Th is consistent if it is not true that $Th \vdash \perp$.

2. Definitions

2.1. *Definition of Beth model*

The following algebraic definition of Beth model is a slight modification of the definition in [5].

Definition 2.1. A Beth model $\mathcal{B}$ for a logical-mathematical language Ω is a quadruple of objects $\langle M, \leq, D, Val \rangle$ described as follows.

$\langle M, \leq \rangle$ is a partially ordered set, called the domain of the model; elements of M are denoted $\alpha, \beta, \gamma, \ldots$.

D defines domains for variables of the language: for each sort π, D_π is the domain for objects of the sort π.

We define an *evaluated formula* as a formula of the language Ω, in which all parameters are replaced by objects from suitable domains. Another way to describe an evaluated formula is given in [5]: we assume that the language Ω is extended by constants for all elements of D_π for each sort π ; then an evaluated formula is defined as a closed formula of the extended language.

Val is a valuation mapping that assigns T or $\perp$ to any couple (α, φ), where $\alpha \in M$ and φ is an evaluated formula, in such a way that this monotonicity condition holds:

$$(\beta \le \alpha \ \& \ Val(\alpha, \varphi) = T) \Rightarrow Val(\beta, \varphi) = T.$$

Elements of M are interpreted as moments or worlds, the relation $<$ on M is interpreted as "later in time". As in [5], a path in M through α is defined as a maximal linearly ordered subset of M containing α. A path in M is defined as a maximal linearly ordered subset of M.

A topological definition of Beth model can be found in [7].

Definition 2.2. In the Beth model a **forcing** relation $\alpha \Vdash \varphi$ is defined for any $\alpha \in M$ and evaluated formula φ by induction on the complexity of φ.

For an atomic formula φ, $\alpha \Vdash \varphi$ iff for any path S in M through α, there is $\beta \in S$ such that $Val(\beta, \varphi) = T$;

$\alpha \not\Vdash \bot$ (falsity is never forced);

$\alpha \Vdash \psi \wedge \eta$ iff ($\alpha \Vdash \psi \ \& \ \alpha \Vdash \eta$);

$\alpha \Vdash \psi \vee \eta$ iff for any path S through α, there is $\beta \in S$ such that $\beta \Vdash \psi$ or $\beta \Vdash \eta$;

$\alpha \Vdash \psi \supset \eta$ iff for any $\beta \le \alpha, (\beta \Vdash \psi \Rightarrow \beta \Vdash \eta)$;

$\alpha \Vdash \forall x \psi(x)$ iff for any element d of the appropriate domain, $\alpha \Vdash \psi(d)$;

$\alpha \Vdash \exists x \psi(x)$ iff for any path S through α, there is $\beta \in S$ and an element d of the appropriate domain, such that $\beta \Vdash \psi(d)$.

We will assume that $\langle M, \le \rangle$ is a tree, which is the usual case.

For the Beth model $\mathcal{B}$, $\mathcal{B} \Vdash \varphi$ (an evaluated formula φ holds in $\mathcal{B}$) iff $\varepsilon \Vdash \varphi$, where ε is the root of the tree M.

2.2. Facts about Beth models

The following lemma and other facts about Beth models are explained in more detail in [5].

Lemma 2.1. *A Beth model $\mathcal{B}$ for a logical-mathematical language Ω has the following properties.*

(1) Monotonicity: $(\alpha \Vdash \varphi \wedge \beta \le \alpha) \Rightarrow \beta \Vdash \varphi.$

(2) $\alpha \Vdash \varphi \Leftrightarrow$ *for any path S in M through α there is $\beta \in S$ with $\beta \Vdash \varphi$.*

(3) Soundness theorem. For any closed formula φ of the language Ω,

$$HPC \vdash \varphi \ \Rightarrow \ \mathcal{B} \Vdash \varphi.$$

We use classical logic in metamathematics and in this lemma, in particular.

Van Dalen [5] constructed a Beth model for the language of intuitionistic analysis FIM, which has two types of variables: over natural numbers and over choice sequences. We generalize this model to a Beth model for a language with many types of functionals.

3. Axiomatic Theories L and L_s

The language of theory L has the following variables:

$x, y, z, \ldots$ over natural numbers (variables of type 0),
and variables of type n $(n \geq 1)$:

$F^n, G^n, H^n, \ldots$ over n-functionals (functionals of type n);
$A^n, B^n, C^n, \ldots$ over lawlike n-functionals;
$\mathcal{F}^n, \mathcal{G}^n, \mathcal{H}^n, \ldots$ over lawless n-functionals.

Constants: 0 of type 0 and for each $n \geq 1$ a constant K^n (an analog of 0 for type n).

Functional symbols: $N^n, Ap^n (n \geq 1)$ and a functional symbol for each primitive recursive function, including S for successor.

Predicate symbols: $=_n$ for each $n \geq 0$.

Terms and n-functionals are defined recursively as follows.

Every numerical variable is a term.
Constant 0 is a term.
Every variable of type n is an n-functional.
Constant K^n is an n-functional.
If f is a functional symbol for a primitive recursive function with k numerical arguments and m functional arguments, $t_1, \ldots, t_k$ are terms and $Z_1, \ldots, Z_m$ are 1-functionals, then $f(t_1, \ldots, t_k, Z_1, \ldots, Z_m)$ is a term.
If Z is an n-functional, then $N^n(Z)$ is an n-functional (a successor of Z).
If Z is a 1-functional, t is a term, then $Ap^1(Z, t)$ is a term.
If Z is an $(n+1)$-functional, t is a term, then $Ap^{n+1}(Z, t)$ is an n-functional.

$Ap^n(Z, t)$ is interpreted as the result of application of functional Z to term t. Denote $Ap^n(Z, t)$ as $Z(t)$.

Here 1-functional is interpreted as a function from natural numbers to natural numbers, and $(n + 1)$-functional is interpreted as a function from natural numbers to n-functionals.

As usual, lawlike (or constructive) functionals are determined by laws or algorithms. Lawless functionals are opposite extreme: they could not be determined by any law. At each moment only an initial segment of a lawless functional is known, and there are no restrictions on its future values.

Atomic formulas:

$t =_0 \tau$, where t and τ are terms,

$Z =_n V$, where Z and V are n-functionals ($n \geqslant 1$).

Formulas are constructed from atomic formulas using logical connectives and quantifiers.

For a formula φ, $sort(\varphi)$ is the maximal type of parameters in φ; it is 0 if φ has no parameters.

The theory L has intuitionistic predicate logic HPC with equality axioms and the following non-logical axioms.

(1) $Sx \neq 0$, $Sx = Sy \supset x = y$.

(2) Defining relations for all primitive recursive functions.

(3) Induction for natural numbers: $\varphi(0) \wedge \forall x \, (\varphi(x) \supset \varphi(Sx)) \supset \forall x \varphi(x)$.

(4) $K^{n+1}(x) = K^n$, $\neg (N^n (F^n) = K^n)$.

(5) $N^{n+1} \left(F^{n+1}\right)(x) = N^n (F(x))$, $N^n (F^n) = N^n (G^n) \supset F = G$.

The first three axioms define arithmetic at the bottom level. The axioms 4 and 5 describe K^n and N^n as analogs of zero and successor function, respectively, on level n.

Denote L_s the fragment of the theory L with types not greater than s.

The theory L is a modification of Bernini's theory HL [1]. The language of L_1 has one type of functionals and is essentially the language of intuitionistic analysis FIM.

4. Beth Model for Language L_s

Any formal proof in the theory L is finite and therefore it is a proof in some fragment L_s. So we generalize the model of van Dalen [5] to a Beth model $\mathcal{B}_s$ for the fragment L_s for any $s \geqslant 1$, instead of constructing a model for the entire language L with infinitely many types of functionals.

4.1. *Notations*

A sequence of elements $x_0, x_1, \ldots, x_n$ is denoted $x = \langle x_0, x_1, \ldots, x_n \rangle$, and its length is denoted $lth(x)$. Symbol $*$ denotes the concatenation function:

$$\langle x_0, x_1, \ldots, x_n \rangle * \langle y_0, y_1, \ldots, y_k \rangle = \langle x_0, x_1, \ldots, x_n, y_0, y_1, \ldots, y_k \rangle.$$

For a sequence x, $\langle x \rangle_n$ denotes its nth element and $\overline{x}(n)$ denotes the initial segment of x of length n assuming that $n < lth(x)$.

For a function f on natural numbers, $\overline{f}(n)$ denotes the initial segment of f of length n, that is the sequence $\langle f(0), f(1), \ldots, f(n-1) \rangle$.

The notation $f : a \dashrightarrow b$ means that f is a partial function from set a to set b.

For a function f of two variables denote $f^{[x]} = \lambda y. f(x, y)$.

The notation $Z \downarrow$ means that the object Z is defined.

Next we introduce a few notations for a fixed set b.

$b^{(n)}$ is the set of all sequences of elements of b that have length n.

b^* is the set of all finite sequences of elements of b.

On the set b^* a partial order $\leqslant$ is defined by:

$$y \leqslant x \text{ iff } x \text{ is an initial segment of } y.$$

With this order b^* is a tree growing down. Its root is the empty sequence $\langle \, \rangle$.

Denote $\Delta(b)$ the set of all partial functions $f : (b^* \times \omega) \dashrightarrow b$ that satisfy the following two conditions:

(i) for any $x, y \in b^*, (y \leqslant x \Rightarrow f^{[x]} \subseteq f^{[y]})$ (monotonicity);

(ii) for any path S in b^*, $\bigcup \{ f^{[x]} \mid x \in S \}$ is a total function on ω (completeness along each path).

4.2. Definition of Beth model $\mathcal{B}_s$

Definition 4.1. We fix $s \geqslant 1$ and define a Beth model $\mathcal{B}_s$ for the language L_s:

$$\mathcal{B}_s = \langle M, \preccurlyeq, D, Val \rangle,$$

where each component is described as follows.

1) Domain for natural numbers is $a_0 = \omega$.

For $k \geqslant 1$, domain for k-functionals is $a_k = \Delta(a_{k-1})$. Thus,

$$a_k = \{ f \mid f : (a_{k-1}^* \times \omega) \dashrightarrow a_{k-1}, \ f \text{ is monotonic \& complete along each path} \}.$$

Domain for lawlike k-functionals:

$$b_k = \{ f \in a_k \mid f^{[<>]} \text{ is a total function} \}.$$

Before defining a domain for lawless functionals we introduce an auxiliary set $c_{k+1}(k \geqslant 0)$:

$$c_{k+1} = \{ \xi|\ \xi : (\omega \times a_k) \to a_k \ \&\ \forall n \ (\xi^{[n]} \ is \ a \ bijection \ on \ a_k) \}.$$

Elements of c_{k+1} are called k-permutations. For any k-permutation ξ we define $\nu(\xi)$ as the function $f \in a_{k+1}$ such that for any $n \in \omega$, $x \in a_k^*$:

$$f(x,n) = \begin{cases} \xi^{[n]}(< x >_n) & \text{if } n < lth(x), \\ \text{undefined} & \text{otherwise}, \end{cases}$$

Domain for lawless for k-functionals is $l_k = \{\nu(\xi) \mid \xi \in c_k\}, k \geqslant 1$.

2) The domain of the model is $M = \displaystyle\bigcup_{m=0}^{\infty} \left(a_0^{(m)} \times \ldots \times a_{s-1}^{(m)} \right)$.

Elements of M are denoted $\alpha, \beta, \gamma, \ldots$. We denote $lh(\alpha) = m$ if $\alpha \in a_0^{(m)} \times \ldots \times a_{s-1}^{(m)}$.

3) Order on M is defined by:

$$\beta \preceq \alpha \text{ iff } < \beta >_i \leqslant < \alpha >_i \text{ for each } i < s.$$

With this order M is a subtree of the direct product of trees $a_0^*, \ldots, a_{s-1}^*$. M is a non-countable tree growing down. Its root is

$$\varepsilon = \langle\ \underbrace{\langle\ \rangle, \ldots, \langle\ \rangle}_{s}\ \rangle.$$

4) It remains to define a valuation mapping Val. The definition consists of steps (a)-(d).

(a) For each constant Q of the language L we define its interpretation $\widehat{Q}$.

i) $\widehat{0} = 0$.

ii) The interpretation of constant K^k $(k = 1, \ldots, s)$ is given by:

$\widehat{K}^1(x, n) = 0$ for any $n \in \omega$ and x;

$\widehat{K}^{k+1}(x, n) = \widehat{K}^k$ for any $n \in \omega$ and x.

(b) For any $\alpha \in M$ and functional symbol h of the language L_s we define its interpretation $\widehat{h}^{[\alpha]}$ as follows.

i) If h is a symbol for a primitive recursive function f, then $\widehat{h}^{[\alpha]} = f$.

ii) Next we define a successor function S^k of type k $(k = 0, 1, \ldots, s)$: $S^0(x) = S(x) = x + 1$, where S is a successor function for natural numbers;

$S^{n+1}(f) = S^n \circ f.$

Functional symbol N^k is interpreted by: $\widehat{N}^{k[\alpha]} = S^k$.

iii) Functional symbol Ap^k is interpreted by: $\widehat{Ap}^{k[\alpha]}(f,n) = f(\langle\alpha\rangle_{k-1}, n)$ for any $n \in \omega, f \in a_k, k = 1, \ldots, s$.

Thus, a k−functional f does not depend on the entire node α but only on its $(k-1)^{th}$ component.

We define an *evaluated term* as a term of the language L_s, in which all parameters are replaced by objects from suitable domains. We define an *evaluated k-functional* as a k-functional of the language L_s, in which all parameters are replaced by objects from suitable domains. These definitions are similar to the definition of an evaluated formula.

(c) Suppose Z is an evaluated term or n-functional. We define its interpretation $Z^{[\alpha]}$ at node α by induction on the complexity of Z.

i) If Z is an evaluated variable, then $Z^{[\alpha]} = Z$.

ii) If Z is a constant, then $Z^{[\alpha]} = \widehat{Z}$.

iii) If $Z = h(V_1, \ldots, V_m)$, where h is a functional symbol, then

$$Z^{[\alpha]} = \widehat{h}^{[\alpha]}(V_1^{[\alpha]}, \ldots, V_m^{[\alpha]}).$$

(d) The valuation mapping Val is defined by:

$$Val(\alpha, Z = V) = T \text{ iff } Z^{[\alpha]} \downarrow \ \& \ V^{[\alpha]} \downarrow \ \& \ Z^{[\alpha]} = V^{[\alpha]}.$$

This completes the definition of $\mathcal{B}_s = \langle M, \preccurlyeq, D, Val \rangle$.

Clearly, $b_k \subseteq a_k$ and $l_k \subseteq a_k, k = 1, \ldots, s$. For $\alpha \in M, lh(\alpha) = lth(\langle\alpha\rangle_k)$ for any $k < s$.

Next lemma states natural properties of the application predicate.

Lemma 4.1. *Suppose* $k = 1, \ldots, s$, $f \in a_k$, $n \in \omega$ *and* $\widehat{Ap}^{k[\alpha]}(f,n) \downarrow$. *Then*

(1) $\widehat{Ap}^{k\,[\alpha]}(f,n) \in a_{k-1}$;

(2) $\beta \preccurlyeq \alpha \Rightarrow \widehat{Ap}^{k\,[\beta]}(f,n) \downarrow \ \& \ \widehat{Ap}^{k\,[\beta]}(f,n) = \widehat{Ap}^{k\,[\alpha]}(f,n).$

Proof.

(1) Since $\langle\alpha\rangle_{k-1} \in a_{k-1}^*$, then $\widehat{Ap}^{k\,[\alpha]}(f,n) = f(\langle\alpha\rangle_{k-1}, n) \in a_{k-1}$ by the definition of a_k.

(2) Suppose $\beta \preccurlyeq \alpha$. Then $\langle \beta \rangle_{k-1} \leqslant \langle \alpha \rangle_{k-1}$. By monotonicity of f, $f(\langle \beta \rangle_{k-1}, n) \downarrow$ and $f(\langle \beta \rangle_{k-1}, n) = f(\langle \alpha \rangle_{k-1}, n)$, which implies the conclusion of part (2). $\qquad \square$

In the next lemma we show the validity of the interpretations of terms and functionals.

Lemma 4.2. *Suppose* $\alpha \in M$, *and* Z, V *are evaluated k-functionals ($k = 1, \ldots, s$) or terms of the language L_s ($k = 0$ if Z, V are terms). Then the following holds.*

(1) $Z^{[\alpha]} \downarrow \Rightarrow Z^{[\alpha]} \in a_k$.

(2) $Z^{[\alpha]} \downarrow \& \beta \preccurlyeq \alpha \Rightarrow Z^{[\beta]} \downarrow \& Z^{[\beta]} = Z^{[\alpha]}$.

(3) For any path S in M there is $\gamma \in S$ such that $Z^{[\gamma]} \downarrow$.

(4) $[Val(\alpha, Z = V) = T] \& \beta \preccurlyeq \alpha \Rightarrow [Val(\beta, Z = V) = T]$.

Proof.

(1) is proven by induction on the complexity of Z. The case $Z = Ap^{k+1}(V, t)$ follows from Lemma 4.1.1), other cases are trivial.

(2) is proven by induction on the complexity of Z. The case $Z = Ap^{k+1}(V, t)$ follows from Lemma 4.1.2), other cases are trivial.

(3) is proven by induction on the complexity of Z. We consider only the non-trivial case $Z = Ap^{k+1}(V, t)$. Consider a path S in M. By the inductive assumption there are $\alpha, \beta \in S$ such that $V^{[\alpha]} \downarrow$ and $t^{[\beta]} \downarrow$.

Denote $f = V^{[\alpha]}, n = t^{[\beta]}$ and $S' = \{ \langle \delta \rangle_k \mid \delta \in S \}$. Then S' is a path in a_k^*, $f \in a_{k+1}$ and $n \in \omega$ by part 1. The function f is complete along each path in a_k^*, so there is $x \in S'$ with $f^{[x]}(n) \downarrow$.

For some $\delta \in S$, $x = \langle \delta \rangle_k$. Denote $\gamma = min\{\alpha, \beta, \delta\}$. Then $V^{[\gamma]} = V^{[\alpha]} = f$, $t^{[\gamma]} = t^{[\beta]} = n$ and $\langle \gamma \rangle_k \leqslant \langle \delta \rangle_k = x$, since $\gamma \preccurlyeq \delta$.

So $Z^{[\gamma]} = V^{[\gamma]}(\langle \gamma \rangle_k, t^{[\gamma]}) = f(\langle \gamma \rangle_k, n) = f(x, n)$, since f is monotonic; hence $Z^{[\gamma]} \downarrow$.

(4) follows from (2). $\qquad \square$

Lemma 4.2.4) implies that $\mathcal{B}_s$ is a Beth model. Clearly $\mathcal{B}_s$ is embedded in $\mathcal{B}_{s+1}$ and $\mathcal{B}_1$ is essentially the van Dalen's model [5]. Our interpretation of a k-functional is similar to the interpretation of a choice sequence in [5]; the difference is in the recursive use of the tree a_{k-1}^* instead of the tree ω^*.

4.3. Properties of the model $\mathcal{B}_s$

Lemma 4.3. *Suppose $\alpha \in M$, Z is an evaluated term or functional, and φ is an evaluated formula of the language L_s. Then*

$$Z^{[\alpha]} \downarrow \Rightarrow [\alpha \Vdash \varphi(Z) \Leftrightarrow \alpha \Vdash \varphi(Z^{[\alpha]})].$$

Proof. It follows from statement 4.5.1 on pg. 85 of [7]. □

Lemma 4.4. *Suppose $\alpha \in M, f \in l_k, n \in \omega$. Then*

$$f(n)^{[\alpha]} \downarrow \Leftrightarrow n < lh(\alpha).$$

Proof. It follows from the interpretation of a lawless functional. □

The following are some examples of k-functionals in the model $\mathcal{B}_s$, $k = 1, \ldots, s$.

(1) $\widehat{K}^k$ is a lawlike k-functional; it is defined at the root of M and it is the same at every node; $\widehat{K}^k \in b_k$.

(2) For $x \in a_{k-1}^*, n \in \omega$ define:

$$f(x,n) = \begin{cases} <x>_n & \text{if } n < lth(x), \\ undefined & \text{if } n \geqslant lth(x). \end{cases}$$

Clearly $f = \nu(\xi)$, where ξ is an element of c_k given by: $\xi(n,x) = x$. So $f \in l_k$; f is a lawless k-functional whose values are determined by the path.

(3) For $x \in a_{k-1}^*, n \in \omega$ define:

$$g(x,n) = \begin{cases} undefined & \text{if } lth(x) = 0, \\ <x>_0 & \text{if } lth(x) > 0. \end{cases}$$

The k-functional g is neither lawlike, nor lawless.

Theorem 4.1 (Soundness for L_s).

$$L_s \vdash \varphi \;\Rightarrow\; \mathcal{B}_s \Vdash \bar{\varphi},$$

where $\bar{\varphi}$ is the closure of the formula φ, that is the formula φ with universal quantifiers over all its parameters.

Proof. The proof is by induction of the length of the derivation of φ. For axioms and derivation rules of the intuitionistic logic HPC it follows from the Soundness theorem for Beth models (Lemma 2.1.3). The axioms for equality hold in the model due to the definition of $Val(\alpha, Z = V)$. The proof for axioms 1-3 is obvious. The proof for axioms 4 and 5 follows from the interpretations of constant K^n and functional symbol N^n. $\square$

In the rest of the paper we will show that some intuitionistic principles hold in the model $\mathcal{B}_s$ and therefore can be consistently added to L_s.

5. Principle of Primitive Recursive Completeness of Lawlike Functions in the Model $\mathcal{B}_s$

This principle is stated in the language of L as follows.

$$(\text{LK}) \qquad \exists A^1 \forall x \, (A(x) = t) \,,$$

where t is any term that does not contain variables other than variables of type 0 and variables over lawlike 1-functionals.

Denote (LK_s) the formula (LK) with types not greater than s.

It is quite obvious that (LK_s) holds in the model $\mathcal{B}_s$ but the proof involves several steps. First we introduce a set $\tilde{b}_n (n = 0, 1, \dots, s)$ recursively:

$$\tilde{b}_0 = \omega, \qquad \tilde{b}_{k+1} = \left\{ f \in b_{k+1} \mid f^{[<>]} : \omega \to \tilde{b}_k \right\}.$$

Clearly, $\tilde{b}_n \subseteq b_n \subseteq a_n$ and $\tilde{b}_1 = b_1$.

Lemma 5.1. $f \in \tilde{b}_n \quad \Rightarrow \quad S^n(f) \in \tilde{b}_n$.

Proof. The proof is by induction on n. The case $n = 0$ is obvious. Assume it is true for n. For $f \in \tilde{b}_{n+1}$ and any $x \in \omega, f(<>, x) \in \tilde{b}_n$, so

$$S^{n+1}(f)(<>, x) = S^n \circ f(<>, x) = S^n(f(<>, x)) \in \tilde{b}_n$$

by the inductive assumption.

Hence $S^{n+1}(f)^{[<>]} : \omega \to \tilde{b}_n$ and $S^{n+1}(f) \in \tilde{b}_{n+1}$. $\square$

Lemma 5.2. *Suppose $\bar{X}$ is a list of variables $X_1, \dots, X_r$ and $\bar{f}$ is a list $f_1, \dots, f_r$ of objects from corresponding domains. Suppose the list $\bar{X}$ can contain only variables of type 0 and variables over lawlike 1-functionals. Suppose $Z(\bar{X})$ is an n-functional ($n = 1, \dots, s$) or a term (then $n = 0$), and all its parameters are in the list $\bar{X}$. Then*

$$Z^{[\varepsilon]} \downarrow \quad \& \quad Z^{[\varepsilon]} \in \tilde{b}_n.$$

Proof. The proof is by induction on the complexity of Z. We consider only three non-trivial cases.

If $Z(\bar{X})$ is a variable X_1, then X_1 is a variable of type 0 or a variable over lawlike 1-functionals. In the first case $Z(\bar{f}) = f_1 \in \omega = \tilde{b}_0$ and in the second case $Z(\bar{f}) = f_1 \in b_1 = \tilde{b}_1$.

Case $Z = N^n(V)$. By the inductive assumption, $V^{[\varepsilon]} \in \tilde{b}_n$. Then $Z^{[\varepsilon]} = S^n\left(V^{[\varepsilon]}\right) \in \tilde{b}_n$ by the previous lemma.

Case $Z = Ap^{n+1}(V, t)$. By the inductive assumption, $V^{[\varepsilon]} \in \tilde{b}_{n+1}$ and $t^{[\varepsilon]} \in \omega$. Then $Z^{[\varepsilon]} = V^{[\varepsilon]}\left(<>, t^{[\varepsilon]}\right) \in \tilde{b}_n$ by the definition of $\tilde{b}_{n+1}$. $\square$

Theorem 5.1. *The closure of the formula (LK$_s$) holds in the model* $\mathcal{B}_s$.

Proof. Suppose $\bar{X}$ is a list of variables $X_1, \ldots, X_r$ and $\bar{f}$ is a list $f_1, \ldots, f_r$ of objects from corresponding domains. Suppose the list $\bar{X}$ can contain only variables of type 0 and variables over lawlike 1-functionals, and $t(x, \bar{X})$ is a term with all its parameters in the list $x, \bar{X}$.

Define a function g by the following: for any $x \in \omega$ and any u,

$$g(u, x) = t\left(x, \bar{f}\right)^{[\varepsilon]}.$$

By the previous lemma, for any $x \in \omega$ and any u, $g(u, x) \downarrow$ and $g(u, x) \in \omega$. So $g \in b_1$ and $\varepsilon \Vdash \forall x \left[g(x) = t(x, \bar{f})\right]$, which proves the theorem. $\square$

6. Permutations in the Model $\mathcal{B}_s$

This technical section contains some facts about the model $\mathcal{B}_s$ necessary for the rest of the paper. We defined a k-permutation as an element of the set c_{k+1}, see Definition 4.1.

In this section we fix permutations $\xi_0 \in c_1, \ldots, \xi_{s-1} \in c_s$. They generate a few mappings introduced in the following definition.

Definition 6.1.

(1) Define $\tilde{\xi}_k$ by the following: for any $m \geqslant 1$,

$$\tilde{\xi}_k(x_0, \ldots, x_{m-1}) = \langle\, \xi_k^{[0]}(x_0), \ldots, \xi_k^{[m-1]}(x_{m-1})\, \rangle.$$

(2) Define $\tilde{\xi}$ by the following:

$$\tilde{\xi}(y_0, \ldots, y_{s-1}) = \langle\tilde{\xi}_0(y_0), \ldots, \tilde{\xi}_{s-1}(y_{s-1})\rangle.$$

(3) Define η_k by the following:

$$\eta_k(x, n) = (\xi_k^{[n]})^{-1}(x).$$

(4) Define Λ_k by induction on k.

$\Lambda_0(n) = n$.

Assume $k \geqslant 1$ and Λ_{k-1} is defined. Define $\Lambda_k(f) = g$, where

$$g(x, n) = \Lambda_{k-1}\left(f\left(\tilde{\xi}_{k-1}(x), n\right)\right).$$

Clearly if f is a lawlike k-functional, then $\Lambda_k(f) = f$.

The following technical lemma gives reasons for introducing the aforementioned mappings.

Lemma 6.1. *Suppose* $k = 0, \ldots, s - 1$.

(1) $\eta_k \in c_{k+1}$.

(2) $\tilde{\xi}_k$ *is a bijection on* a_k^* *and* $\tilde{\eta}_k$ *is the inverse of* $\tilde{\xi}_k$.

(3) $\tilde{\xi}$ *is a bijection on* M *and* $\tilde{\eta}$ *is the inverse of* $\tilde{\xi}$.

(4) $\tilde{\xi}_k$ *is an isomorphism of the partially ordered set* a_k^*, *i.e. a bijection preserving the order:*

$$y \leqslant x \Leftrightarrow \tilde{\xi}_k(y) \leqslant \tilde{\xi}_k(x).$$

(5) $\tilde{\xi}$ *is an isomorphism of the partially ordered set* M, *so*

$$\beta \preccurlyeq \alpha \Leftrightarrow \tilde{\xi}(\beta) \preccurlyeq \tilde{\xi}(\alpha).$$

(6) *Suppose* S *is a path in* a_k^* *through* x *and* $S' = \left\{\tilde{\xi}_k(y) \mid y \in S\right\}$. *Then* S' *is a path in* a_k^* *through* $\tilde{\xi}_k(x)$.

(7) *Suppose* S *is a path in* M *through* α *and* $S' = \left\{\tilde{\xi}(\beta) \mid \beta \in S\right\}$. *Then* S' *is a path in* M *through* $\tilde{\xi}(\alpha)$.

(8) *For* $k = 0, 1, \ldots, s$,

$$f \in a_k \Leftrightarrow \Lambda_k(f) \in a_k.$$

Proof. The first five statements follow from the definitions. Part 6 follows from parts 1, 2 and 4. Part 7 follows from parts 1, 3 and 5.

Part 8. $\Rightarrow$ The proof is by induction on k. The basis step is obvious. We will consider the inductive step. We assume that the statement holds for k and prove it for $k + 1$.

Suppose $f \in a_{k+1}$ and $g = \Lambda_{k+1}(f)$. To show that $g \in a_{k+1}$ we check three conditions.

(i) We need to prove that $g : (a_k^* \times \omega) \dashrightarrow a_k$.
Suppose $x \in a_k^*$, $n \in \omega$ and $y = g(x, n)$ is defined. Then $\tilde{\xi}_k(x) \in a_k^*$ and $y = \Lambda_k \left(f\left(\tilde{\xi}_k(x), n \right) \right)$ is defined. So $f\left(\tilde{\xi}_k(x), n \right) \in a_k$ and by the inductive assumption $y = \Lambda_k \left(f\left(\tilde{\xi}_k(x), n \right) \right) \in a_k$.

(ii) Monotonicity.
Suppose $y \leqslant x$ and $g(x, n) \downarrow$. Then $f\left(\tilde{\xi}_k(x), n \right) \downarrow$ and $\tilde{\xi}_k(y) \leqslant \tilde{\xi}_k(x)$ by part 4. By monotonicity of f, $f\left(\tilde{\xi}_k(y), n \right) \downarrow$ and $f\left(\tilde{\xi}_k(y), n \right) = f\left(\tilde{\xi}_k(x), n \right)$, so $g(y, n) \downarrow$ and $g(y, n) = g(x, n)$.

(iii) Completeness along each path.
Suppose S is a path in a_k^* and $n \in \omega$. Denote $S' = \left\{ \tilde{\xi}_k(z) \mid z \in S \right\}$. By part 6, S' is a path in a_k^*. There is $y \in S'$ such that $f(y, n) \downarrow$. Denote $x = \tilde{\eta}_k(y)$. By part 2, $y = \tilde{\xi}_k(x)$. So $x \in S$, $f\left(\tilde{\xi}_k(x), n \right)$ is defined and belongs to a_k. By the inductive assumption $g(x, n) = \Lambda_k \left(f\left(\tilde{\xi}_k(x), n \right) \right) \in a_k$ and $g(x, n)$ is defined.

$\Leftarrow$ This follows from the previous if we use $\eta_0, \ldots, \eta_{s-1}$ instead of $\xi_0, \ldots, \xi_{s-1}$. $\qquad\square$

The following lemma describes a relation between permutations and forcing.

Lemma 6.2. *Suppose $\bar{X}$ is a list of variables $X_1, \ldots, X_r$ of types $k_1, \ldots, k_r$, respectively. Suppose $\bar{f}$ is a list $f_1, \ldots, f_r$ of objects from corresponding domains and $\bar{g}$ is the list $g_1, \ldots, g_r$, where $g_i = \Lambda_{k_i}(f_i)$. Suppose $\beta = \tilde{\xi}(\alpha)$.*

(1) Suppose $Z(\bar{X})$ is an n-functional ($n = 1, \ldots, s$) or a term (then $n = 0$), and all its parameters are in the list $\bar{X}$. Then

i) $Z(\bar{f})^{[\beta]} \downarrow \Rightarrow Z(\bar{g})^{[\alpha]} \downarrow \,\&\, Z(\bar{g})^{[\alpha]} = \Lambda_n \left(Z(\bar{f})^{[\beta]} \right).$

ii) $Z(\bar{f})^{[\beta]} \downarrow \Leftrightarrow Z(\bar{g})^{[\alpha]} \downarrow.$

iii) Suppose $V(\bar{X})$ is also an n-functional or a term (if $n = 0$), and all its parameters are in the list $\bar{X}$. Then

$$\left[Val\left(\beta,\ Z(\bar{f}) = V(\bar{f}) \right) = T \right] \Leftrightarrow \left[Val\left(\alpha,\ Z(\bar{g}) = V(\bar{g}) \right) = T \right].$$

(2) Suppose $\varphi(\bar{X})$ is a formula of L_s and all its parameters are in the list $\bar{X}$. Then

$$\beta \Vdash \varphi(\bar{f}) \Leftrightarrow \alpha \Vdash \varphi(\bar{g}).$$

Proof.

(1) i) It is proven by induction on the complexity of Z. We consider the case $Z = Ap^{n+1}(V, t)$. Other cases are straightforward. Suppose $Z(\bar{f})^{[\beta]} \downarrow$. Then

$$V(\bar{f})^{[\beta]} \downarrow \ \& \ t(\bar{f})^{[\beta]} \downarrow \ \& \ V(\bar{f})^{[\beta]} \left(\langle \beta \rangle_n, t(\bar{f})^{[\beta]} \right) \downarrow.$$

By the inductive assumption

$$V(\bar{g})^{[\alpha]} \downarrow \ \& \ t(\bar{g})^{[\alpha]} \downarrow \ \& \ V(\bar{g})^{[\alpha]} = \Lambda_{n+1} \left(V(\bar{f})^{[\beta]} \right) \ \& \ t(\bar{g})^{[\alpha]} = t(\bar{f})^{[\beta]}.$$

$$Z(\bar{g})^{[\alpha]} = V(\bar{g})^{[\alpha]} \left(\langle \alpha \rangle_n, t(\bar{g})^{[\alpha]} \right) = \Lambda_{n+1} \left(V(\bar{f})^{[\beta]} \right) \left(\langle \alpha \rangle_n, t(\bar{f})^{[\beta]} \right)$$

$$= \Lambda_n \left[V(\bar{f})^{[\beta]} \left(\tilde{\xi}_n \left(\langle \alpha \rangle_n \right), t(\bar{f})^{[\beta]} \right) \right] = \Lambda_n \left[V(\bar{f})^{[\beta]} \left(\langle \beta \rangle_n, t(\bar{f})^{[\beta]} \right) \right]$$

$$= \Lambda_n \left(Z(\bar{f})^{[\beta]} \right).$$

So $Z(\bar{g})^{[\alpha]} \downarrow$.

ii) The implication from right to left follows from the previous part if we use $\eta_0, \ldots, \eta_{s-1}$ instead of $\xi_0, \ldots, \xi_{s-1}$.

iii) The implication from left to right follows from part i). The implication from right to left is proven using $\eta_0, \ldots, \eta_{s-1}$ instead of $\xi_0, \ldots, \xi_{s-1}$.

(2) The proof is by induction on the complexity of φ using Definition 2.2 of forcing.

(a) φ is an atomic formula $Z(\bar{X}) = V(\bar{X})$.

$\Rightarrow$ Suppose $\beta \Vdash \varphi(\bar{f})$. Consider an arbitrary path S in M through α. Denote $S' = \left\{ \tilde{\xi}(\gamma) \mid \gamma \in S \right\}$. By Lemma 6.1.7), S' is a path in M through β. So there is $\beta' \in S'$ such that $Val \left(\beta', Z(\bar{f}) = V(\bar{f}) \right) = T$. For some $\alpha' \in S, \beta' = \tilde{\xi}(\alpha')$ and by part 1.iii), $Val \left(\alpha', Z(\bar{g}) = V(\bar{g}) \right) = T$. So $\alpha \Vdash \varphi(\bar{g})$.

$\Leftarrow$ follows from the previous if we take $\eta_0, \ldots, \eta_{s-1}$ instead of $\xi_0, \ldots, \xi_{s-1}$.

(b) The case $\psi \vee \eta$ is similar to the previous one.

In next two cases we prove only the implication from left to right; the implication from right to left follows if we take $\eta_0, \ldots, \eta_{s-1}$ instead of $\xi_0, \ldots, \xi_{s-1}$.

(c) φ is $\psi \supset \eta$.

Suppose $\beta \Vdash \varphi(\bar{f})$, $\gamma \preccurlyeq \alpha$ and $\gamma \Vdash \psi(\bar{g})$. Denote $\gamma' = \tilde{\xi}(\gamma)$. Then $\gamma' \preccurlyeq \beta$, by the inductive assumption $\gamma' \Vdash \psi(\bar{f})$, so $\gamma' \Vdash \eta(\bar{f})$ and $\gamma \Vdash \eta(\bar{g})$ by the inductive assumption.

(d) φ is $\exists Y \psi(\bar{X}, Y)$.

Suppose $\beta \Vdash \varphi(\bar{f})$ and S is a path in M through α. Denote $S' = \left\{ \tilde{\xi}(\gamma) \mid \gamma \in S \right\}$. Then S' is a path through β. There is $\beta' \in S'$ and y from the appropriate domain such that $\beta' \Vdash \psi(\bar{f}, y)$. For some $\alpha' \in S, \beta' = \tilde{\xi}(\alpha')$, so by the inductive assumption $\alpha' \Vdash \psi(\bar{g}, y)$. Hence $\alpha \Vdash \varphi(\bar{g})$.

Other cases are straightforward. $\qquad\qquad\qquad\qquad\qquad\qquad\square$

The following lemma states some properties of forcing in the model $\mathcal{B}_s$.

Lemma 6.3. *Suppose $\bar{X}$ is a list of variables $X_1, \ldots, X_r$ of types $k_1, \ldots, k_r$, respectively and $\bar{f}$ is a list $f_1, \ldots, f_r$ of objects from corresponding domains.*

(1) Suppose $\varphi(\bar{X})$ is a formula of L_s with all its parameters in the list $\bar{X}$ and $sort(\varphi) \leqslant m$, where $m \geqslant 1$. Suppose $\alpha, \beta \in M$ and

$$\bigwedge_{i=0}^{m-1} (\langle \alpha \rangle_i = \langle \beta \rangle_i).$$

Then

$$\alpha \Vdash \varphi(\bar{f}) \iff \beta \Vdash \varphi(\bar{f}).$$

(2) Suppose $\varphi(\mathcal{H}^p, \bar{X})$ is a formula of L_s with all its parameters in the list $\mathcal{H}^p, \bar{X}$; $sort(\varphi) \leqslant p$ and φ does not have non-lawlike parameters of type p other than $\mathcal{H}^p$. Suppose

$$\alpha \in M \ \& \ g, h \in l_p \ \& \ (\forall n < lh(\alpha)) \left(g(n)^{[\alpha]} = h(n)^{[\alpha]} \right).$$

Then

$$\alpha \Vdash \varphi(g, \bar{f}) \iff \alpha \Vdash \varphi(h, \bar{f}).$$

Proof.

(1) Denote $l = lh(\alpha)$. Since $\langle\alpha\rangle_0 = \langle\beta\rangle_0$, we have $lh(\beta) = lh(\alpha) = l$. For any $k = 0, \ldots, s - 1$ define ξ_k by the following: for $n \in \omega, x \in a_k$

$$\xi_k(n, x) = \begin{cases} \langle\langle\beta\rangle_k\rangle_n & \text{if } (n < l \;\&\; x = \langle\langle\alpha\rangle_k\rangle_n), \\ \langle\langle\alpha\rangle_k\rangle_n & \text{if } (n < l \;\&\; x = \langle\langle\beta\rangle_k\rangle_n), \\ x & \text{otherwise .} \end{cases}$$

Then each $\xi_k \in c_{k+1}$ and $\beta = \tilde{\xi}(\alpha)$. Denote $g_i = \tilde{\xi}_{k_i}(f_i)$ for $i = 1, \ldots, r$. By Lemma 6.2.2),

$$\beta \Vdash \varphi(\bar{f}) \;\Leftrightarrow\; \alpha \Vdash \varphi(\bar{g}). \tag{1}$$

If $k < m$, then $\langle\alpha\rangle_k = \langle\beta\rangle_k$ and

$$(\forall n \in \omega)(\forall x \in a_k)\left(\xi_k^{[n]}(x) = x\right),$$

$$(\forall y \in a_k^*)\left(\tilde{\xi}_k(y) = y\right),$$

so for any $j \leqslant m$,

$$(\forall z \in a_j)\left(\Lambda_j(z) = z\right).$$

Since $sort(\varphi) \leqslant m$, for each $i = 1, \ldots, r$, $k_i \leqslant m$ and $g_i = \Lambda_{k_i}(f_i) = f_i$. So by formula (1)

$$\beta \Vdash \varphi(\bar{f}) \;\Leftrightarrow\; \alpha \Vdash \varphi(\bar{f}).$$

(2) For some $\chi_1, \chi_2 \in c_p$, $g = \nu(\chi_1)$ and $h = \nu(\chi_2)$. Clearly $p \geqslant 1$. Define permutations $\xi_0, \ldots, \xi_{s-1}$:

$$\xi_{p-1}^{[n]} = \left(\chi_2^{[n]}\right)^{-1} \circ \chi_1^{[n]}; \quad \xi_k^{[n]}(x) = x \text{ for } k \neq p - 1.$$

By Lemma 6.2.2), it is sufficient to prove the following:

$$\tilde{\xi}(\alpha) = \alpha; \tag{2}$$

$$\Lambda_{k_i}(f_i) = f_i, \; i = 1, \ldots, r; \tag{3}$$

$$\Lambda_p(h) = g. \tag{4}$$

Proof of formula (2). Denote $\alpha = \langle\alpha_0, \ldots, \alpha_{s-1}\rangle$, then $\tilde{\xi}(\alpha) = \langle\tilde{\xi}_0(\alpha_0), \ldots, \tilde{\xi}_{s-1}(\alpha_{s-1})\rangle$. Clearly $\tilde{\xi}_k(\alpha_k) = \alpha_k$ for $k \neq p - 1$.

Now consider $k = p - 1$. Fix $n < lh(\alpha)$. Since $g = \nu(\chi_1)$ and $h = \nu(\chi_2)$,

$$\chi_1^{[n]}\left(\langle \alpha_{p-1}\rangle n\right) = g(\alpha_{p-1}, n) = g(n)^{[\alpha]} = h(n)^{[\alpha]} = h(\alpha_{p-1}, n) = \chi_2^{[n]}\left(\langle \alpha_{p-1}\rangle n\right).$$

Thus, $\chi_1^{[n]}\left(\langle \alpha_{p-1}\rangle n\right) = \chi_2^{[n]}\left(\langle \alpha_{p-1}\rangle n\right)$ and

$$\xi_{p-1}^{[n]}\left(\langle \alpha_{p-1}\rangle n\right) = \left(\chi_2^{[n]}\right)^{-1}\left(\chi_1^{[n]}\left(\langle \alpha_{p-1}\rangle n\right)\right) = \langle \alpha_{p-1}\rangle n.$$

So for any $n < lh(\alpha)$, $\xi_{p-1}^{[n]}\left(\langle \alpha_{p-1}\rangle n\right) = \langle \alpha_{p-1}\rangle n$ and $\tilde{\xi}_{p-1}(\alpha_{p-1}) = \alpha_{p-1}$.

Proof of formula (3). For $j < p - 1$, $\tilde{\xi}_j(x) = x$ for any $x \in a_j^*$ and

$$(\forall y \in a_j)\left(\Lambda_j(y) = y\right), j < p. \tag{5}$$

If f_i is a domain object for a lawlike k_i-functional, then $\Lambda_{k_i}(f_i) = f_i$; otherwise $k_i < p$ and $\Lambda_{k_i}(f_i) = f_i$ by (5).

Proof of formula (4). For any $n \in \omega, x \in a_{p-1}^*$,

$$\Lambda_p(h)(x, n) = \Lambda_{p-1}\left(h\left(\tilde{\xi}_{p-1}(x), n\right)\right) = h\left(\tilde{\xi}_{p-1}(x), n\right) \text{ by (5)}.$$

$g = \nu(\chi_1)$ and $h = \nu(\chi_2)$,

$$g(x, n) = \begin{cases} \chi_1^{[n]}\left(\langle x\rangle n\right) & \text{if } n < lth(x), \\ \text{undefined} & \text{otherwise}. \end{cases}$$

Since $\tilde{\xi}_{p-1}(x)$ has the same length as x, it is sufficient to prove:

$$h\left(\tilde{\xi}_{p-1}(x), n\right) = g(x, n) \text{ for any } n < lth(x).$$

Indeed, $h\left(\tilde{\xi}_{p-1}(x), n\right) = \chi_2^{[n]}\left(\langle \tilde{\xi}_{p-1}(x)\rangle n\right) = \chi_2^{[n]}\left(\xi_{p-1}^{[n]}\left(\langle x\rangle n\right)\right)$

$$= \chi_2^{[n]}\left[\left(\chi_2^{[n]}\right)^{-1} \circ \chi_1^{[n]}\left(\langle x\rangle n\right)\right] = \chi_1^{[n]}\left(\langle x\rangle n\right) = g(x, n). \qquad \square$$

7. Lawless Functionals in the Model $\mathcal{B}_s$

According to the Definition 4.1, a lawless k-functional is an element of l_k; a lawless k-functional can be seen as a labeled tree a_{k-1}^*, where each node is assigned an element of a_{k-1} in such a way that immediately after each node all elements of a_{k-1} occur.

7.1. Axioms for lawless functionals in the model $\mathcal{B}_s$

The following are modifications for the language L of the axioms for lawless sequences in [5] and the axioms for lawless functionals in [1].

(LL1) $\exists \mathcal{H}^n (\forall y \leq x)(\mathcal{H}(y) = F^n(y))$.

(LL2) $\mathcal{H}^n = \mathcal{G}^n \vee \mathcal{H}^n \neq \mathcal{G}^n$.

(LL3) $\varphi(\mathcal{H}^n) \supset \exists x \forall \mathcal{G}^n [(\forall y < x)\, (\mathcal{G}(y) = \mathcal{H}(y)) \supset \varphi(\mathcal{G})]$,

where $sort(\varphi) \leq n$ and φ does not have non-lawlike parameters of type n other than $\mathcal{H}^n$.

The first formula is the axiom of existence of lawless functionals, and the third one is the principle of open data.

Denote (LL) a conjunction of the closures of (LL1), (LL2) and (LL3); denote (LL_s) the formula (LL) with types not greater than s.

Theorem 7.1. $\mathcal{B}_s \Vdash (LL_s)$.

Proof.

(LL1). Fix $f \in a_n$ and a path S in M. By induction on x we will prove:

$$(\exists \alpha \in S)(\forall y \leq x) f(y)^{[\alpha]} \downarrow .$$

For $x = 0$ it follows from Lemma 4.2.3).

Assume the formula holds for x: for some $\beta \in S$, $(\forall y \leq x) f(y)^{[\beta]} \downarrow$. By Lemma 4.2.3), there is $\gamma \in S$ such that $f(x+1)^{[\gamma]} \downarrow$. For $\alpha = min\{\beta, \gamma\}$ we have $(\forall y \leq x+1) f(y)^{[\alpha]} \downarrow$.

Now let us prove that (LL1) holds in the model. Fix an arbitrary $x \in \omega$ and $\alpha \in S$ such that $(\forall y \leq x) f(y)^{[\alpha]} \downarrow$.

The path $S = \{\langle \bar{g}_0(k), \ldots, \bar{g}_{s-1}(k) \rangle \mid k \in \omega\}$ for some functions $g_0 : \omega \to a_0, \ldots, g_{s-1} : \omega \to a_{s-1}$. Denote $\beta = \langle \bar{g}_0(x+1), \ldots, \bar{g}_{s-1}(x+1) \rangle$ and $\gamma = min\{\alpha, \beta\}$. Then $\beta, \gamma \in S$ and $(\forall y \leq x) f(y)^{[\gamma]} \downarrow$.

Define ξ by:

$$\xi(y, z) = \begin{cases} f(y)^{[\gamma]} & \text{if } y \leq x \ \& \ z = g_{n-1}(y), \\ g_{n-1}(y) & \text{if } y \leq x \ \& \ z = f(y)^{[\gamma]}, \\ z & \text{otherwise} . \end{cases}$$

For any $y \leq x$ and $z \in a_{n-1}$ we have $g_{n-1}(y) \in a_{n-1}$ and $f(y)^{[\gamma]} \in a_{n-1}$, so $\xi : (\omega \times a_{n-1}) \to a_{n-1}$ and for any $y \in \omega$, $\xi^{[y]}$ is a bijection on a_{n-1}. Therefore $\xi \in c_n$ and $h = \nu(\xi) \in l_n$. It remains to prove:

$$\gamma \Vdash (\forall y \leq x) [h(y) = f(y)] .$$

For any $y \leqslant x$, $h(y)^{[\gamma]} = h\left(\langle\gamma\rangle_{n-1}, y\right) = \xi^{[y]}\left(\langle\langle\gamma\rangle_{n-1}\rangle_y\right)$. Since $\gamma \preccurlyeq \beta$,

we have $\langle\gamma\rangle_{n-1} \leqslant \langle\beta\rangle_{n-1}$ and $\langle\langle\gamma\rangle_{n-1}\rangle_y = g_{n-1}(y)$. So
$h(y)^{[\gamma]} = \xi^{[y]}(g_{n-1}(y)) = f(y)^{[\gamma]}$ by the definition of ξ.

For (LL2) the proof follows from the interpretation of equality.

(LL3). Suppose $\bar{X}$ is a list of variables $X_1, \ldots, X_r$ and $\bar{f}$ is a list
$f_1, \ldots, f_r$ of objects from corresponding domains. Suppose all parameters
of φ are in the list $\mathcal{H}^n, \bar{X}$. Suppose $\alpha \in M, h \in l_n$ and

$$\alpha \Vdash \varphi(h, \bar{f}). \tag{6}$$

Take $x = lh(\alpha)$. It is sufficient to prove that for any $g \in l_n$,

$$\alpha \Vdash (\forall y < x)\,(g(y) = h(y)) \supset \varphi(g, \bar{f}).$$

Suppose $\beta \preccurlyeq \alpha$ and $(\forall y < x)\,\beta \Vdash (g(y) = h(y))$. Since
$x = lh(\alpha)$, by Lemma 4.4, $(\forall y < x)\,\left[g(y)^{[\alpha]} \downarrow\ \&\ h(y)^{[\alpha]} \downarrow\right]$.
So $(\forall y < x)\,\left(g(y)^{[\alpha]} = h(y)^{[\alpha]}\right)$. By (6) and Lemma 6.3.2), $\alpha \Vdash \varphi(g, \bar{f})$. $\square$

8. Kripke's Schema in the Model $\mathcal{B}_s$

For an n-functional F of language L denote $F(0)^n$ the term $F\underbrace{(0)\ldots(0)}_{n}$.

In the language L Kripke's schema is the following formula:

(KS) $\exists G^k \left[\varphi \equiv \exists x \left(G(x)(0)^{k-1} \neq 0\right)\right]$,

where G is not a parameter of φ and $k \geqslant max\,\{1, sort(\varphi)\}$.

Denote (KS$_s$) the formula (KS) with types not greater than s.

Theorem 8.1. *The closure of the formula (KS$_s$) holds in the model $\mathcal{B}_s$.*

Proof.
In case $s = 1$ the proof is the same as in [5], since the model described
there is essentially $\mathcal{B}_1$. We will assume $s \geqslant 2$.

Denote ψ the formula φ, in which all parameters are replaced by objects
from corresponding domains.

First we decribe the proof for case $k = 2$ (other cases will be considered
later). Then $sort(\varphi) \leqslant 2$. We need the following notations:

$$\tilde{M} = \bigcup_{m=0}^{\infty} \left(a_0^{(m)} \times a_1^{(m)} \right);$$

$$contraction(\alpha) = \bar{\alpha}(2);$$

$$extension(a) = a * \langle \underbrace{\langle \widehat{K}^2, \dots, \widehat{K}^2 \rangle}_{m}, \dots, \underbrace{\langle \widehat{K}^{s-1}, \dots, \widehat{K}^{s-1} \rangle}_{m} \rangle \text{ for } a \in a_0^{(m)} \times a_1^{(m)}.$$

This introduces a smaller domain $\tilde{M}$. Clearly,

$$\alpha \in M \; \Rightarrow \; contraction(\alpha) \in \tilde{M};$$

$$a \in \tilde{M} \; \Rightarrow \; extension(a) \in M.$$

The operations of contraction and extension preserve the length of a node.

$$\text{For } a \in \tilde{M}, contraction(extension(a)) = a. \tag{7}$$

By Lemma 6.3.1), for any formula χ with $sort(\chi) \leqslant 2$ and its evaluation $\tilde{\chi}$:

$$contraction(\alpha) = contraction(\beta) \; \Rightarrow \; (\alpha \Vdash \tilde{\chi} \Leftrightarrow \beta \Vdash \tilde{\chi}). \tag{8}$$

Define a function g by the following:

$$g(v, n) = \begin{cases} q_{v,n} & \text{if } lth(v) \geqslant n, \\ \text{undefined} & \text{otherwise}, \end{cases}$$

where

$$q_{v,n}(u, m) = \begin{cases} 1 & \text{if } lth(u) \geqslant n \; \& \; extension\,(\bar{u}(n), \bar{v}(n)) \Vdash \psi, \\ 0 & \text{if } lth(u) \geqslant n \; \& \; extension\,(\bar{u}(n), \bar{v}(n)) \not\Vdash \psi, \\ \text{undefined} & \text{otherwise}. \end{cases}$$

It is sufficient to prove:

$$g \in a_2 \tag{9}$$

and

$$\varepsilon \Vdash [\psi \equiv \exists x \, (g(x)(0) \neq 0)],$$

which means that for any $\alpha \in M$,

$$\alpha \Vdash \psi \Leftrightarrow \alpha \Vdash \exists x \, [g(x)(0) \neq 0] \,. \tag{10}$$

Proof of formula (9) .

(i) To prove $g : (a_1^* \times \omega) \dashrightarrow a_1$ suppose $v \in a_1^*, n \in \omega$ and $g(v, n) \downarrow$. Then $lth(v) \geqslant n$ and $g(v, n) = q_{v,n}$. We need to show that $q_{v,n} \in a_1$. Clearly, $q_{v,n} : (a_0^* \times \omega) \dashrightarrow a_0$.
To prove the monotonicity of $q_{v,n}$ suppose $q_{v,n}(u, m) \downarrow$ and $u' \leqslant u$. Then $lth(u') \geqslant lth(u) \geqslant n, \bar{u}'(n) = \bar{u}(n)$ and $q_{v,n}(u', m) = q_{v,n}(u, m)$.
To prove completeness of $q_{v,n}$ suppose S_0 is a path in a_0^* and $m \in \omega$. Denote u the element of S_0 with length n.
If $extension \, (\bar{u}(n), \bar{v}(n)) \Vdash \psi$, then $q_{v,n}(u, m)$ equals 1, otherwise it equals 0; in both cases it is defined.
(ii) Monotonicity of g.
Suppose $g(v, n) \downarrow$ and $v' \leqslant v$. Then $lth(v') \geqslant lth(v) \geqslant n$, so $g(v', n) = q_{v',n}$ is defined. Since $\bar{v}'(n) = \bar{v}(n)$, then $q_{v',n} = q_{v,n}$.
(iii) Completeness of g.
Consider a path S_1 in a_1^*. Denote v the element of S_1 with length n. Then $g(v, n)$ is defined.

Proof of formula (10) .
$\Rightarrow$ Suppose $\alpha \Vdash \psi$. Denote $a = contraction(\alpha)$, $u = \langle \alpha \rangle_0$ and $v = \langle \alpha \rangle_1$. By (7-8), $extension(a) \Vdash \psi$. Take $n = lh(\alpha)$. Then $a = \langle \bar{u}(n), \bar{v}(n) \rangle$ and

$$g(n)(0)^{[\alpha]} = g\left(\langle \alpha \rangle_1, n\right)\left(\langle \alpha \rangle_0, 0\right) = g(v, n)(u, 0) = q_{v,n}(u, 0) = 1.$$

So $\alpha \Vdash [g(n)(0) = 1]$, which implies $\alpha \Vdash \exists x \, [g(x)(0) \neq 0]$.
$\Leftarrow$ Suppose

$$\alpha \Vdash \exists x \, [g(x)(0) \neq 0] \,. \tag{11}$$

Consider an arbitrary path S in M through α. According to Lemma 2.1.2), it is sufficient to prove: $(\exists \beta \in S)\beta \Vdash \psi$. By (11), there are $n \in \omega$ and $\gamma \in S$ such that $\gamma \Vdash g(n)(0) \neq 0$. Since $g \in a_2$, there is $\delta \in S$, for

which $g(n)(0)^{[\delta]} \downarrow$. Take $\beta = min\{\gamma, \delta\}$. Then $\beta \in S$, $g(n)(0)^{[\beta]} = 1$ and $lh(\beta) \geqslant n$.
Denote $b = contraction(\beta)$, $u = \langle \beta \rangle_0, v = \langle \beta \rangle_1$ and $c = \langle \bar{u}(n), \bar{v}(n) \rangle$.

Since $1 = g(n)(0)^{[\beta]} = g\left(\langle \beta \rangle_2, n\right)\left(\langle \beta \rangle_1, 0\right) = g(v, n)(u, 0)$, we have $extension(c) \Vdash \psi$.

Since $extension(b) \preccurlyeq extension(c)$, then $extension(b) \Vdash \psi$ and by (7-8), $\beta \Vdash \psi$. This completes the proof of formula (10) and the proof for case $k = 2$.

Other cases are similar and use similar notations. In case $k = 1$ the functional g is defined by:

$$g(u, n) = \begin{cases} 1 & \text{if } lth(u) \geqslant n \text{ \& } extension\,(\bar{u}(n)) \Vdash \psi, \\ 0 & \text{if } lth(u) \geqslant n \text{ \& } extension\,(\bar{u}(n)) \nVdash \psi, \\ \text{undefined} & \text{otherwise .} \end{cases}$$

In case $k \geqslant 3$ the functional g is defined by:

$$g(u_k, n) = \begin{cases} q_{u_k,n}^{(k-1)} & \text{if } lth(u_k) \geqslant n, \\ \text{undefined} & \text{otherwise ,} \end{cases}$$

where

$$q_{u_k,n}^{(k-1)}(u_{k-1}, m_{k-1}) = \begin{cases} q_{u_k,n,u_{k-1},m_{k-1}}^{(k-2)} & \text{if } lth(u_{k-1}) \geqslant n, \\ \text{undefined} & \text{otherwise ,} \end{cases}$$

where

$$\ldots \ldots \ldots$$

$$q_{u_k,n,\ldots,u_3,m_3}^{(2)}(u_2, m_2) = \begin{cases} q_{u_k,n,\ldots,u_2,m_2}^{(1)} & \text{if } lth(u_2) \geqslant n, \\ \text{undefined} & \text{otherwise ,} \end{cases}$$

where

$$q_{u_k,n,\ldots,u_2,m_2}^{(1)}(u_1, m_1) =$$

$$= \begin{cases} 1 & \text{if } lth(u_1) \geqslant n \text{ \& } extension\,(\bar{u}_1(n),\ldots,\bar{u}_k(n)) \Vdash \psi, \\ 0 & \text{if } lth(u) \geqslant n \text{ \& } extension\,(\bar{u}_1(n),\ldots,\bar{u}_k(n)) \nVdash \psi, \\ \text{undefined} & \text{otherwise .} \end{cases}$$

Similarly to case $k = 2$, we have: $g \in a_k$ and

$$\varepsilon \Vdash \left[\psi \equiv \exists x \left(g(x)(0)^{k-1} \right) \neq 0 \right]. \qquad \square$$

9. Discussion

In this paper we constructed Beth model $\mathcal{B}_s$ for the basic intuitionistic theory L_s with many types of functionals. We proved that Kripke's schema with n-functionals and axioms for lawless n-functionals hold in this model. Next we are planning to formalize all the proofs within classical type theory, to investigate whether the theory of the creating subject and other intuitionistic principles hold in the model and to generalize the model $\mathcal{B}_s$ for the fragment L_s to a Beth model for the full theory L with infinitely many types of functionals.

The intuitionistic principles that hold in the Beth model can be added to L_s to create a strong and consistent intuitionistic theory with n-functionals, where a significant part of classical type theory can be interpreted. The described Beth model is a first step in this direction.

Acknowledgements

The author thanks the referee and the editor for their valuable comments and suggestions that helped to improve this paper.

References

1. S. Bernini, A very strong intuitionistic theory, *Studia Logica* **35**, 377 (1976).
2. N. Wendel, The inconsistency of Bernini's very strong intuitionistic theory, *Studia Logica* **37**, 341 (1978).
3. S. Bernini, A note on my paper "A very strong intuitionistic theory", *Studia Logica* **37**, 349 (1978).
4. F. Kashapova, Intuitionistic theory of functionals of higher type, *Math. Notes* **45**, 66 (1989).
5. D. van Dalen, An interpretation of intuitionistic analysis, *Ann. Math. Logic* **13**, 1 (1978).
6. A. N. Kolmogorov and A. G. Dragalin, *Introduction to Mathematical Logic* (Moscow University, Moscow, 1982).
7. A. Dragalin, *Mathematical Intuitionism. Introduction to Proof Theory* (American Mathematical Society, USA, 1987).

A Computational Framework for the Study of Partition Functions and Graph Polynomials

T. Kotek[1] and J.A. Makowsky[2] and E.V. Ravve[3]

[1] *Institut für Informationssysteme*
Technische Universität Wien, Vienna, Austria
E-mail: tomuko@gmail.com

[2] *Faculty of Computer Science*
Technion–Israel Institute of Technology, Haifa, Israel
E-mail: janos@cs.technion.ac.il

[3] *Department of Software Engineering*
ORT Braude College, Karmiel, Israel
Email: cselena@braude.ac.il

Partition functions and graph polynomials have found many applications in combinatorics, physics, biology and even the mathematics of finance. Studying their complexity poses some problems. To capture the complexity of their combinatorial nature, the Turing model of computation and Valiant's notion of counting complexity classes seem most natural. To capture the algebraic and numeric nature of partition functions as real or complex valued functions, the Blum-Shub-Smale (BSS) model of computation seems more natural. As a result many papers use a naive hybrid approach in discussing their complexity or restrict their considerations to sub-fields of $\mathbb{C}$ which can be coded in a way to allow dealing with Turing computability.

In this paper we propose a unified natural framework for the study of computability and complexity of partition functions and graph polynomials and show how classical results can be cast in this framework.

Keywords: Graph polynomials, partition functions, models of computation, complexity of computation, counting problems

1. Introduction

The study of graph polynomials and partition functions has become a focal point in the research linking discrete mathematics to applied mathematics. Initiated 100 years ago with Birkhoff's paper on the chromatic polynomial, [1], and extended in W. Tutte's paper of 1954, [2], graph polynomials remained for long an exotic subject. However, physicists and chemists independently were led to study similar mathematical abstractions, [3,4]. For recent survey, see [5]. Since the 1980ties many versions of partition functions (aka Potts models, Ising models, Jones polynomial) found applications in fields as diverse as statistical mechanics, chemical graph theory, knot theory, biology and the mathematics of finance.

In their landmark paper [6] F. Jaeger, D.L. Vertigan and D.J.A. Welsh analyzed the complexity of the Tutte and Jones polynomials. Given a point in the complex plane $\bar{a}$ they looked at the complex valued graph parameter $T_{\bar{a}}(-)$ defined by the evaluation $T(-;\bar{a})$ of the Tutte polynomial at the point $\bar{a}$.

The Tutte polynomial is a special case of a *graph polynomial*, a functor which associates with a graph $G = (V, E)$ a polynomial in a polynomial ring $\mathcal{R}$ which is invariant under graph isomorphisms.

Let $\mathcal{R}$ be a sub-field of the complex numbers $\mathbb{C}$. Let $P(G; \overline{X})$ be a graph polynomial in the indeterminates $X_1, \ldots, X_n$ with coefficients in $\mathcal{R}$. For $\bar{a} \in \mathcal{R}^n$, $P(-;\bar{a})$ is a graph invariant taking values in $\mathcal{R}$. We could restrict the graphs to be from a class (graph property) $\mathcal{C}$ of graphs.

We are interested in the complexity of computing $P(-;\bar{a})$ for graphs from $\mathcal{C}$. If for all graphs $G \in \mathcal{C}$ the value of $P(-;\bar{a})$ is a graph invariant taking values in $\mathbb{N}$, we can work in the *Turing model of computation*. Otherwise we identify the graph G with its adjacency matrix M_G, and we work in the *Blum-Shub-Smale (BSS) model of computation*, mostly over the complex numbers $\mathbb{C}$.

We take our inspiration from the classical result of F. Jaeger and D.L. Vertigan and D.J.A. Welsh on the complexity of evaluations of the Tutte polynomial, [6]. They show:

- either evaluation at a point $(a, b) \in \mathbb{C}^2$ is polynomial time computable in the Turing model, and (a, b) lies in some quasi-algebraic set of dimension 1,
- or some $\sharp$**P**-complete problem is reducible to the evaluation at $(a, b) \in \mathbb{C}^2$.
- To stay in the Turing model of computation, they assume that (a, b) is in some finite dimensional extension of the field $\mathbb{Q}$.

The proof of the second part is *hybrid* in its nature: The reduction is more naturally placed in the BSS *model* of computation, However, although there are various analogues for $\sharp$**P** in BSS, cf. [7–10] there seemed to be no counterpart for $\sharp$**P**-completeness in the BSS model suitable for graph polynomials. In [7,8] what is proposed as $\sharp$**P** for BSS counts the zeros of a polynomial, in case this is finite. Nevertheless, we think it *more natural* to work entirely in the BSS model of computation, and we will propose a new framework based on evaluations of polynomials rather than on counting zeros.

This paper is a first step towards a more general theory. First we formulate the situation described above for the Tutte polynomial entirely in the BSS model of computation over the complex numbers $\mathbb{C}$.

We use the framework of SOL-polynomials introduced in [11]. SOL-polynomials form a large class of graph polynomials, which are in a precise sense definable in Second Order Logic. These are then used to define the class $\mathrm{SOLEVAL}_{\mathbb{C}}$ of graph parameters which are evaluations of SOL-polynomials in $\mathbb{C}$. The class $\mathrm{SOLEVAL}_{\mathbb{C}}$ will serve as the previously missing counterpart for $\sharp\mathbf{P}$ in the BSS model. We then investigate to what extent the Tutte polynomial is typical and formulate several versions of "Difficult Evaluation Properties" and "Difficult Point Properties" (DPP). We examine many cases from the literature where these "Difficult Evaluation Properties" hold. Finally, we formulate several conjectures.

The value of the paper is mostly conceptual. It puts complexity questions of graph polynomials into a uniform framework which allows to compare results scattered in the literature. But last but not least, it opens new avenues of research.

We assume the reader is vaguely familiar with the BSS-model of computation, cf. [12], and with the basic of complexity of counting, cf. [13,14].

2. The complexity of graph parameters with values in $\mathbb{C}$

2.1. *Valiant's counting functions and their Turing complexity*

L. Valiant in [15] introduced the counting complexity class $\sharp\mathbf{P}$ which has complete problems with respect to polynomial time Turing reductions. Typical $\sharp\mathbf{P}$-complete problems are the number of 3-colorings of a graph or the number of perfect matchings. Degrees are equivalence classes of decision problems or counting problems with respect to $\mathbf{P}$-time Turing reducibility. Ladner's Theorem asserts that, assuming $\mathbf{P} \neq \mathbf{NP}$, for every degree $[g] \in \mathbf{NP} - \mathbf{P}$ there is $[g'] \in \mathbf{NP} - \mathbf{P}$ with $[g'] < [g]$, [16–18]. It seems to be folklore that the same holds also, if we replace $\mathbf{NP}$ by $\sharp\mathbf{P}$ and $\mathbf{P}$ by $\mathbf{FP}$.

2.2. *Graph parameters in the BSS-model*

We define the framework for graph parameters in the BSS model over some sub-field $\mathcal{R}$ of the complex numbers $\mathbb{C}$. When we get more specific $\mathcal{R}$ will be

$\mathbb{C}$. A *graph invariant* or *graph parameter* is a function $f : \bigcup_n \{0,1\}^{n \times n} \to \mathcal{R}$ which is invariant under permutations of columns and rows of the input adjacency matrix. Graph invariants include decision problems. A *graph transformation* is a function $T : \bigcup_n \{0,1\}^{n \times n} \to \bigcup_n \{0,1\}^{n \times n}$ which is invariant under permutations of columns and rows of the input adjacency matrix. The BSS-*P-time computable functions* over $\mathcal{R}$, $\mathrm{FP}_{\mathcal{R}}$ are the functions $f : \{0,1\}^{n \times n} \to \mathcal{R}$ BSS-computable in time $O(n^c)$ for some fixed $c \in \mathbb{N}$. Let f_1, f_2 be graph invariants. f_1 *is* BSS-*P-time reducible to* f_2, $f_1 \leq_P f_2$ if there are BSS-P-time computable functions T and F such that

(i) T is a graph transformation ;
(ii) For all graphs G with adjacency matrix M_G we have

$$f_1(M_G) = F(f_2(T(M_G)))$$

Two graph invariants f_1, f_2 are BSS-*P-time equivalent over* $\mathcal{R}$, $f_1 \sim_{\mathcal{R}} f_2$, if $f_1 \leq_{\mathcal{R}} f_2$ and $f_2 \leq_{\mathcal{R}} f_1$.

2.3. Cones and degrees

The BSS model over a ring $\mathcal{R}$ deals traditionally with *decision problems* where the input is an $\mathcal{R}$-vector. A function f maps $\mathcal{R}$-vectors into $\mathcal{R}$. There is a decision problem associated with f: Given $\overline{X}$ is it true that $f(\overline{X}) = a$. In the study of graph polynomials decision problems and functions have as input $(0,1)$-matrices and the decision problems and functions have to be *graph invariants*.

We denote by $\mathrm{FEXP}_{\mathbb{C}}$ the set of functions computable in time $2^{O(n^c)}$ in the BSS-model over $\mathbb{C}$. If $c = 1$ we write $\mathrm{FE}_{\mathbb{C}}$ and speak of simple exponential time. $\mathrm{FP}_{\mathbb{C}}$ are the functions computable in polynomial time. Let g, g' be two graph parameters in $\mathrm{FEXP}_{\mathbb{C}}$. We denote by $[g]_{\mathcal{R}}$ the equivalence class (BSS-degree) of all graph parameters $g' \in EXP_{\mathcal{R}}$ under the equivalence relation $\sim_{\mathcal{R}}$. Analogously, $[g]_T$ denotes the corresponding equivalence class in the Turing model of computations. We denote by $\langle g \rangle_{\mathcal{R}}$ the class (BSS-cone) $\{g' \in \mathrm{FEXP}_{\mathcal{R}} : g \leq_{\mathcal{R}} g'\}$. There are BSS-NP-complete problems for all $\mathcal{R}$ under consideration here, and instead of specifying them, we consider the complete problems in $\mathbf{NP}_{\mathcal{R}}$ to be a degree (which may vary with the choice of the Ring $\mathcal{R}$). The cone of an $\mathbf{NP}$-complete problem forms the $\mathbf{NP}$-hard problems. There is no well developed theory of degrees and cones for functions, especially for graph parameters, in the BSS model.

G. Malajovich and K. Meer proved an analogue of Ladner's Theorem for decision problems in the BSS-model over $\mathbb{C}$. Iterating their argument, one gets:

Theorem 2.1. (G. Malajovich and K. Meer) *Assume* $\mathbf{P}_\mathbb{C} \neq \mathbf{NP}_\mathbb{C}$. *Then for every degree* $[g] \in \mathbf{NP}_\mathbb{C} - \mathbf{P}_\mathbb{C}$ *there is* $[g'] \in \mathbf{NP}_\mathbb{C} - \mathbf{P}_\mathbb{C}$ *with* $[g'] < [g]$.

Note that the corresponding result over the reals $\mathbb{R}$ or any other field $\mathcal{R}$ is not known to hold, cf. [19].

2.4. *Discrete counting in* BSS

K. Meer in [7] introduced an analogue of $\sharp\mathbf{P}$ for discrete counting in the BSS model over $\mathbb{R}$. In [20] a first attempt of studying graph polynomials in the BSS model is discussed.

The definitions from [7] can easily be extended to the BSS model over $\mathbb{C}$. Let us look at *counting functions*, i.e., functions $f : \mathbb{C}^\infty \to \mathbb{N} \cup \{\infty\}$. For a complexity class of functions FC we denote by FC^{count} the class of counting functions in FC. Such a function f is in $\sharp\mathbf{P}_\mathbb{C}$ if there exists a polynomial time BSS-machine over $\mathbb{C}$ and a polynomial q such that

$$f(y) = |\{z \in \mathbb{C}^{q(size(y))} : M(y,z) \text{ accepts }\}|$$

It is not difficult to see, cf. [7], that

$$FP_\mathcal{R}^{count} \subseteq \sharp\mathbf{P}_\mathcal{R} \subseteq FE_\mathcal{R}$$

for every sub-field $\mathcal{R}$ of $\mathbb{C}$. Typical examples from [7] over the reals $\mathbb{R}$ are counting zeroes of multivariate polynomials of degree at most 4 ($\sharp 4 - \text{FEAS}$) or counting the number of sign changes of a sequence of real numbers ($\sharp$SC).

Over the complex numbers also the number of k-colorings of a graph is in $\sharp\mathbf{P}_\mathbb{C}$ for fixed k. To see this, we associate with a graph $G = (V, E)$ with $V = \{1, \dots, n\} = [n]$ the following set $\mathcal{E}^k_{color}(G)$ of equations:

(i) $x_i^k - 1 = 0, i \in [n]$
(ii) $\sum_{d=0}^{k-1} x_i^{k-1-d} x_j^d = 0$, for all $(i, j) \in E$.

Clearly, $\mathcal{E}^k_{color}(G)$ has at most k^n many complex solutions. Now, D.A. Beyer in his Ph.D. thesis, [21], observed that

Proposition 2.1. (D. Beyer) *A graph G is k-colorable iff $\mathcal{E}^k_{color}(G)$ has a complex solution, and each solution corresponds exactly to proper k-coloring of G.*

This also shows that for fixed k deciding k-colorability is in $\mathbf{NP_C}$. However, it seems unlikely that it is $\mathbf{NP_C}$-hard, because of the very special form of the equations involved. For further discussion, cf. [22].

In S. Margulies' Ph.D. thesis, [23, Chapter 2] the following is shown:

Theorem 2.2. (S. Margulies) *Every decision problem in* $\mathbf{NP}$ *(in the Turing model) can be encoded as solvability problem of sets of equations over* $\mathbb{C}$.

Using the fact that $\sharp$SAT is $\sharp\mathbf{P}$ complete we get:

Theorem 2.3. *Every function* $f \in \sharp\mathbf{P}$ *(in the Turing model) has an encoding in* $\sharp\mathbf{P_C}$ *(in the* BSS *model).*

In particular we can place the permanent and Hamiltonian functions in the BSS model over $\mathbb{C}$. These functions are usually studied in Valiant's theory of algebraic circuits, cf. [24].

Corollary 2.1. *The functions* per *and* ham *of* $(0,1)$ *matrices are in* $\sharp\mathbf{P_C}$.

2.5. The difficult counting hypothesis (DCH)

There are very few explicit $\sharp\mathbf{P_C}$-complete problems in the literature. The paper [25] shows that the computation of the Euler characteristic of an affine or projective complex variety is complete in this class for Turing reductions in the BSS-model of computation. But there are no explicit $\sharp\mathbf{P_C}$-complete problems in graph theory which correspond to problems which are in $\sharp\mathbf{P}$-complete problems in the Turing model of computation. This is due to the fact that counting discrete solution sets of polynomial equations does not correspond to solvability in a parsimonious way. However, some problems in $\sharp\mathbf{P_C}$ are $\mathbf{NP_C}$-hard, because a set of polynomial equations is solvable if the number of solutions in the above sense is different from 0 but may be ∞. Solvability of systems of polynomial equations is $\mathbf{NP_C}$-complete. It might not be too difficult to construct artificilly problems which are $\sharp\mathbf{P_C}$-complete.

However, k-colorability as expressed by a set of equations is *unlikely* to be $\mathbf{NP_C}$-hard, because of the special form of the equations. Therefore, also counting the number of colorings is not known to be $\mathbf{NP_C}$-hard in the BSS-model. On the other hand, it would be *truly surprising* if counting the number of k-colorings were in $\mathrm{FP}_{\mathbb{C}}^{count}$.

We therefore formulate the following two complexity hypotheses for the BSS model over $\mathbb{C}$:

Strong difficult counting hypothesis (SDCH)
Every counting function in $f \in \sharp P_\mathbb{C}$ with discrete input which is $\sharp P$-hard in the Turing model is $NP_\mathbb{C}$-hard in the BSS model over $\mathbb{C}$.

and

Weak difficult counting hypothesis (WDCH)
A counting function in $f \in \sharp P_\mathbb{C}$ which is $\sharp P$-hard in the Turing model cannot be in $FP_\mathbb{C}^{count}$.

The following is easy to see:

Proposition 2.2. *Assume* $P_\mathbb{C} \neq NP_\mathbb{C}$. *Then* **SDCH** *implies* **WDCH**.

2.6. *Evaluations of graph polynomials over* $\mathbb{C}$

Let $\mathcal{C}$ be a graph property. Let $P(G; \overline{X})$ be a graph polynomial with indeterminates $X_1, \ldots, X_m$. We define

$$\mathrm{EASY}_\mathbb{C}(P, \mathcal{C}) = \{\overline{a} \in \mathbb{C}^n : P(-; \overline{a}) \in \mathrm{FP}_\mathbb{C}\}$$

and

$$\mathrm{HARD}_\mathbb{C}(P, \mathcal{C}) = \{\overline{a} \in \mathbb{C}^n : P(-; \overline{a}) \text{ is } NP_\mathbb{C} - \mathrm{hard}\}$$

Let f be a counting function not in $\mathrm{FP}_\mathbb{C}^{count}$ or a decision problem not in $P_\mathbb{C}$.

$$f - \mathrm{HARD}_\mathbb{C}(P, \mathcal{C}) = \{\overline{a} \in \mathbb{C}^n : P(-; \overline{a}) \in \langle f \rangle_\mathbb{C}\}$$

We omit $\mathcal{C}$ if $\mathcal{C}$ is the class of all finite graphs.

Clearly, if **SDCH** is true and f is $\sharp P$-complete in the Turing model then

$$f - \mathrm{HARD}_\mathbb{C}(P, \mathcal{C}) = \mathrm{HARD}_\mathbb{C}(P, \mathcal{C}).$$

How can we describe $\mathrm{EASY}_\mathbb{C}(P, \mathcal{C})$ and $\mathrm{HARD}_\mathbb{C}(P, \mathcal{C})$?

3. Case studies: Three graph polynomials

3.1. *The chromatic polynomial and its complexity*

Let $G = (V(G), E(G))$ be a graph, and $\lambda \in \mathbb{N}$.

A *proper* λ-*vertex-coloring* is a map $c : V(G) \rightarrow [\lambda]$ such that $(u, v) \in E(G)$ implies that $c(u) \neq c(v)$. Let $\chi(G, \lambda)$ be the number of proper λ-vertex-colorings. Hundred years ago in 1912, G. Birkhoff showed that $\chi(G, \lambda)$ is a *polynomial in* $\mathbb{Z}[\lambda]$. Henceforth we treat $\chi(G, \lambda)$ as a polynomial over $\mathbb{C}$ and consider also evaluations of λ for arbitrary complex numbers. $\chi(G, \lambda)$ is called the *chromatic polynomial of G*. In 1973, R. Stanley showed that for simple graphs G, $\mid \chi(G, -1) \mid$ counts the number of *acyclic orientations* of G. These classical results can be found in many monographs, e.g., [26–28].

For fixed $a \in \mathbb{C}$ we define the graph parameter $\chi_a(G)$ and look at its complexity as a function of G. We follow [29]. There are three values $a = 0, 1, 2$ for which $\chi_a(G)$ can be computed in polynomial time both in the Turing model and in the BSS-model of computation. Furthermore, for $\chi_3(G)$ and $\chi_{-1}(G)$ are $\sharp$P-complete in the Turing model of computation.

Let $G_1 \bowtie G_2$ denote the join of two graphs. We observe that

$$\chi(G \bowtie K_n, \lambda) = (\lambda)^{\underline{n}} \cdot \chi(G, \lambda - n) \qquad (\star)$$

From this we get

(i) $\chi(G \bowtie K_1, 4) = 4 \cdot \chi(G, 3)$
(ii) $\chi(G \bowtie K_n, 3 + n) = (n + 3)^{\underline{n}} \cdot \chi(G, 3)$, hence for $n \in \mathbb{N}$ with $n \geq 3$ it is $\sharp$P-complete.

Here $x^{\underline{n}} = x \cdot (x - 1) \cdot \ldots \cdot (x - n + 1)$. These reductions work in the Turing model for λ in some Turing-computable field extending $\mathbb{Q}$. The reductions also work in BSS if performed directly on the graph parameters, and not on the equivalent problem of solvability of the equations $\mathcal{E}^k_{color}(G)$.

If we have an oracle for some $q \in \mathbb{Q} - \mathbb{N}$ which allows us to compute $\chi_q(G)$ we can compute $\chi(G, q')$ for any $q' \in \mathbb{Q}$ as follows:

Algorithm $A(q, q', \mid V(G) \mid)$:

(i) Given G the degree of $\chi(G, q)$ is at most $n = \mid V(G) \mid$.
(ii) Use the oracle and $(\star)$ to compute $n + 1$ values of $\chi(G, \lambda)$.
(iii) Using Lagrange interpolation we can compute $\chi(G, q')$ in polynomial time.

We note that this algorithm is purely algebraic and works for all graphs G, $q \in (F) - \mathbb{N}$ and $q' \in F$ for any field F extending $\mathbb{Q}$.

Hence we get that for all $q_1, q_2 \in \mathbb{C} - \mathbb{N}$ the graph parameters are *polynomially reducible to each other*.

Furthermore, for $3 \leq i \leq j \in \mathbb{N}$, $\chi(G, i)$ is reducible to $\chi(G, j)$. This now works in the BSS-model over $\mathbb{C}$.

3.2. *Dichotomy of the difficulty of evaluations*

We summarize the situation for the chromatic polynomial as follows:

(i) $\mathrm{EASY}_{\mathbb{C}}(\chi) = \{0,1,2\}$

(ii) The remaining cases can be divided into those points $a \in \mathbb{C}$ where $\chi_a(G)$ is both in $\sharp\mathbf{P}$ and $\sharp\mathbf{P}_{\mathbb{C}}$, and into those points a for which there is no counting interpretation.

(iii) In the Turing model the degrees of $\chi_3(G)$ and $\chi_{-1}(G)$ are the same, because both are in $\sharp\mathbf{P}$ and they are $\sharp\mathbf{P}$-complete.

(iv) In BSS over $\mathbb{C}$ we only get that $[\chi_3] \leq [\chi_{-1}]$

(v) If we had $[\chi_3] = [\chi_{-1}]$ then $\chi_3\mathrm{HARD}$ would consist of all the χ_a with $a \neq 0,1,2$. Under SDCH these would indeed be $\mathbf{NP}_{\mathbb{C}}$-hard. Under WDCH it would be different from $\mathrm{EASY}_{\mathbb{C}}(\chi)$.

(vi) Among the graph parameters $\chi_a(G)$: $a \in \mathbb{C}$, either all are in $\mathrm{EASY}_{\mathbb{C}}(\chi)$ or there is smallest degree, namely $[\chi_3(G)]$ which is not easy. In other words, Ladner's theorem does not hold for the evaluations of the chromatic polynomial, and, assuming that $[\chi_3(G)] \notin \mathrm{FP}_{\mathbb{C}}$, $[\chi_3(G)]$ is a minimal degree.

(vii) It is conceivable that in BSS over $\mathbb{C}$ we have

$$[\chi_3(G)] < [\chi_4(G)] < \ldots < [\chi_j(G)] < \ldots < [\chi_{-1}(G)] = [\chi_a(G) : a \in \mathbb{C} - \mathbb{N}]$$

We have a *Dichotomy Theorem* for the evaluations of $\chi(-,\lambda)$:

(i) $\mathrm{EASY}_{\mathbb{C}}(\chi) = \{0,1,2\}$ is a *quasi-algebraic set* (a finite Boolean combination of algebraic sets) of *dimension* 0.

(ii) All other evaluations are at least as difficult as $\chi_3(G)$, which is a *quasi-algebraic* set of *dimension* 1.

(iii) Under the assumption of **SDCH** we get the dichotomy that all evaluations of $\chi(G,x)$ are either in $\mathrm{EASY}_{\mathbb{C}}(\chi)$ or in $\mathrm{HARD}_{\mathbb{C}}(\chi)$.

(iv) Under the assumption of **WDCH** we get the dichotomy that all evaluations of $\chi(G,x)$ are either in $\mathrm{EASY}_{\mathbb{C}}(\chi)$ or in $\chi_3 - \mathrm{HARD}_{\mathbb{C}}(\chi)$.

3.3. The complexity of the Tutte and the cover polynomial

The Tutte polynomial $T(G,X,Y)$ is a bivariate polynomial and $\chi(G,\lambda) \leq_P T(G,1-\lambda,0)$. For our discussion here the exact definition of the Tutte polynomial is not needed. A good reference is [30, Chapter 10].

We have the following dichotomy theorem:

(i) $\mathrm{EASY}_{\mathbb{C}}(T) = \mathrm{H} \cup \mathrm{Except}$, where $H = \{(x,y) \in \mathbb{C}^2 : (x-1)(y-1) = 1\}$ and Except contains the points $(0,0), (1,1), (-1,-1), (0,-1)$, $(-1,0), (i,-i), (-i,i), (j,j^2), (j^2,j)$ where $j = e^{\frac{2\pi i}{3}}$. This is a quasi-algebraic set of dimension 1.

(ii) For all other points $a \in \mathbb{C}^2$ evaluating the Tutte polynomial is at least as hard as $T(G, -2, 0)$. This is a quasi-algebraic set of dimension 2.

(iii) Furthermore there is a most difficult evaluation point in $\overline{A}_{hard} \in \mathbb{C}^2$ and most evaluation points are in the degree of $\overline{A}_{hard}$.

(iv) As in the case of the chromatic polynomial, one can use the **WDCH** or **SDCH** to sharpen the dichotomy.

The cover polynomial $C(D; X, Y)$ is a bivariate graph polynomial for digraphs D which was introduced by F.R.K. Chung and R.L. Graham, [31]. Its complexity was studied in [32] and again follows the same pattern.

(i) $\text{EASY}_{\mathbb{C}}(C) = \{(0,0), (0,-1), (1,-1)\}$. This is a quasi-algebraic set of dimension 0.

(ii) For all other points $a \in \mathbb{C}^2$ evaluating the cover polynomial is at least as hard as $C(D; 0, 1)$ or $C(D; 1, 0)$. This is a quasi-algebraic set of dimension 2.

(iii) Again there is also a hardest evaluation point.

Note that $C(D, 0, 1)$ is the permanent of the adjacency matrix of D and $C(D, 1, 0)$ counts the number of Hamiltonian paths of D. Both these evaluations are $\sharp$**P**-complete in the Turing model. At the moment it is not clear to us which of them is reducible to the other in BSS over $\mathbb{C}$.

3.4. *Lessons learned from the case study*

In the introduction we proposed to study the complexity of graph parameters in the framework of BSS. Graph parameters are often integer valued, but there are many cases which may have values in $\mathbb{Q}, \mathbb{R}$ or $\mathbb{C}$, in particular graph parameters of weighted graphs. To include all the cases from the literature we chose to deal with case of complex valued graph parameters. In particular our graph parameters are functions in FEXP$_{\mathbb{C}}$. We adapted the notions of **P**-time reducibility of BSS to functions in FEXP$_{\mathbb{C}}$ and to graph parameters in particular. The degree structure of FEXP$_{\mathbb{C}}$ under **P**-time reducibility is not well understood. At the lowest end we have the functions FP$_{\mathbb{C}}$. A typical **NP**$_{\mathbb{C}}$-hard problem is the solvability of polynomial equations, which can be coded as a graph parameter of a hyper-edge weighted hyper-graph which reflects the structure of the polynomial equations. Therefore some graph parameters are **NP**$_{\mathbb{C}}$-hard. However, the exact complexity of most graph parameters in the BSS model over $\mathbb{C}$ is not known, even under the Difficult Counting Hypotheses **SDCH** and **WDCH**.

We have shown that $\sharp$**P**$_{\mathbb{C}}$ does capture surprisingly well the classical counting problems ♯SAT and the like. However, it has the disadvantage of allowing only values in $\mathbb{N}$ or the value ∞. What we need is an extension of ♯P of the Turing model in BSS over $\mathbb{C}$ which takes arbitrary values in $\mathbb{C}$. S. Toda's Theorem states that decision problems in $\mathbf{P}^{\sharp\mathbf{P}}$ contain the whole polynomial hierarchy $\mathbf{PH}$, cf. [14], or equivalently all the decision problems definable in Second Order Logic, cf. [33]. These consideration lead us to the introduction of the class SOLEVAL$_{\mathbb{C}}$ of graph parameters as a substitute for ♯P of the Turing model.

4. The complexity class SOLEVAL

The class of functions in ♯$\mathbf{P}_{\mathbb{C}}$ has the disadvantage that it counts only the number of solutions of polynomial equations provided this number is finite. Evaluating the chromatic polynomial at an irrational point is therefore not expected to be in ♯$\mathbf{P}_{\mathbb{C}}$. We now propose a complexity class for graph parameters, SOLEVAL$_{\mathcal{R}}$ which is better suited to study the complexity of graph parameters with real or complex values, than generalizations of counting complexity classes.

Our class SOLEVAL$_{\mathcal{R}}$ contains virtually all graph parameters from standard graph theory books. But it is not difficult to come up with natural candidates for counting problems not in SOLEVAL$_{\mathcal{R}}$: The game of HEX played on graphs with two disjoint unary predicates on vertices comes to mind. Deciding winning positions is **PSpace**-complete, cf. [34,35]. Therefore counting the winning strategies is not in SOLEVAL unless **PSpace** coincides with the polynomial hierarchy in the Turing model.

In [11] the class of graph parameters definable in Second Order Logic (SOL) was introduced. Roughly speaking these are graph polynomials where summation and products are allowed to range over first order or second order variables of formulas in SOL. For the full definition we refer to [11] but we give a few illustrative examples.

Example 4.1. *(The independence polynomial)* Let $ind(G, i)$ denote the number of independent sets of size i of a graph G. The graph polynomial $ind(G, X) = \sum_i ind(G, i) \cdot X^i$, can be written also as

$$ind(G, X) = \sum_{I \subseteq V(G)} \prod_{v \in I} X$$

where I ranges over all independent sets of G and v ranges over all elements in I. To be an independent set is definable by a formula of SOL in the vocabulary of graphs.

Example 4.2. *(The chromatic number)* The chromatic number $\chi(G)$ of a graph G is the smallest $k \in \mathbb{N}$ such that $\chi(G; k) \neq 0$. We can represent $\chi(G)$ as the evaluation of the polynomial $\sum_{v \in C : \phi(v, C)} X$ where $\phi(v, C)$ is the formula

$$C(v) \wedge \psi_{coloring}(C) \wedge (\forall C'(\psi_{coloring}(C') \to \psi_{inject}(C, C')))$$

where

(i) $\psi_{coloring}(C)$ says that "there is a function $f : V(G) \to C$ which is a proper coloring".

(ii) $\psi_{inject}(C, C')$ says that "there is an injective function $s : C \to C'$"

Here the vocabulary of graphs is augmented by a unary predicate C, but the polynomial is independent of the the the interpretation of C.

A *simple* SOL-*polynomial* $p(G, X)$ is a polynomial of the form

$$p(G, X) = \sum_{A : A \subseteq V(G) : \phi(A)} \prod_{v : \psi(v)} X$$

where A ranges over all subsets of $V(G)$ satisfying $\phi(A)$ and v ranges over all elements of $V(G)$ satisfying ψ. Both formulas ϕ and ψ are SOL-formulas, but summation be be over first and second order variables, while products are only over first order variables.

For the general case

- One allows several indeterminates $X_1, \ldots, X_t$.
- One allows summation over relations $S \subseteq V(G)^k$ of any fixed arity k rather than over sets.
- Instead of the standard monomials other bases of monomials can be used. For example, in the case of one indeterminate X^n could be replace by $\binom{X}{n}$ or the falling factorials $X^{\underline{n}}$.
- One gives an inductive definition.
- One allows arbitrary finite relational vocabularies and is not restricted to the vocabulary of graphs.
- In some applications one looks at graphs with a linear order on the vertices, but one requires the definition to be *invariant under the ordering*, i.e., different orderings still give the same polynomial.

The general case includes the chromatic polynomial, the Tutte polynomial and its variations, the cover polynomial, and virtually all graph polynomials from the literature. We shall see more examples in the sequel.

Proposition 4.1. *For every field* $\mathcal{R}$ *we have* *(i)* SOLEVAL$_\mathcal{R}$ $\subseteq$ FEXP$_\mathcal{R}$.

(ii) If $f \in \sharp\mathbf{P}$ in the Turing model, then $f \in$ SOLEVAL$_\mathcal{R}$.

Sketch. (i) is shown by induction on the defining formulas. (ii) follows from Fagin's characterization of **NP** in the Turing model as the problems definable in existential SOL, cf. [33]. □

5. Degrees of evaluations

Ultimately one wants to study the structure of the degrees in SOLEVAL$_\mathcal{R}$ for various sub-fields of $\mathbb{C}$. Is there a maximal degree in SOLEVAL$_\mathcal{R}$, in other words, does it have complete graph parameters? Does it contain interesting sub-hierarchies?

Here we set ourselves a more moderate task. We start a general investigation of the degree structure of the evaluations of a fixed (set of) graph polynomial(s).

Let $\mathfrak{P}$ be a family of SOL-polynomials. If $\mathfrak{P} = \{F\}$ is a singleton, we omit the set brackets. For a sub-field $\mathcal{R} \subseteq \mathbb{C}$ we denote by EVAL$_\mathcal{R}(\mathfrak{P})$ the set of graph parameters which are evaluations of polynomials in $\mathfrak{P}$ over $\mathcal{R}$. If $\mathfrak{P}$ consists of all SOL-polynomials, we denote it by SOLEVAL$_\mathcal{R}$.

5.1. *The partial order of the degrees*

If $F(G : \overline{X})$ is a a SOL-polynomial in n indeterminates $\overline{X} = (X_1, \ldots, X_n)$ we can partially pre-order its evaluations points $\overline{a} \in \mathbb{C}^n$ by the partial pre-order of the degrees of evaluating $F(G, \overline{a})$. So we write $[\overline{a}]_F$ for the degree of $F(G, \overline{a})$. What interest us is the partial order one gets by taking the quotient of the pre-order with respect to the equivalence relation of polynomial time bi-reducibility.

Example 5.1. The quotient order on EVAL(χ) of the chromatic polynomial is discrete and linear with FP$_\mathbb{C}$ as its first element and $[\chi(G : -1)]$ as its last element.

Example 5.2. The quotient order on EVAL(T) of the Tutte polynomial consists of one element, or it has a minimal element above FP$_\mathbb{C}$ and a maximal element. The case of the cover polynomial is similar.

5.2. *Difficult points are dense*

Another way of studying the complexity of EVAL(F) for a particular SOL-polynomial in n indeterminates is topological.

Theorem 5.1. *Let $\mathcal{R}$ be a sub-field of $\mathbb{C}$ which is a metric space. Let $F(G;\overline{X})$ be a SOL-polynomial such that for some $\overline{a} \in \mathbb{C}^n$ the graph parameter $F(G;\overline{a}) \notin \mathrm{FP}_\mathcal{R}$. Then the set of difficult evaluation points $\mathrm{EVAL}(F)_\mathcal{R} - \mathrm{FP}_\mathcal{R}$ is dense in $\mathcal{R}$.*

Sketch. Assume there is a neighborhood U where all evaluation points are in $\mathrm{FP}_\mathcal{R}$. So we can compute in polynomial time enough values of $F(G,x)$ to use multi-dimensional Lagrange interpolation and compute the coefficients of $F(G,x)$. But then all evaluations of $F(G,a)$ are in $\mathrm{FP}_\mathcal{R}$. $\quad\square$

In the examples so far more is true:

Example 5.3. The evaluations of the maximal degree of $\mathrm{EVAL}(\chi)_\mathcal{R}$ are dense in $\mathcal{R}$.

Example 5.4. The evaluations of the maximal degree of $\mathrm{EVAL}(T)_\mathcal{R}$ are dense in $\mathcal{R}^2$.

Theorem 5.2. *Let F be a SOL-polynomial in n indeterminates. Assume there is an open set $U \subseteq \mathbb{C}^n$ and a point $\overline{A} \in \mathbb{C}^n$ such that for all $\overline{b} \in U$ the degree $[\overline{b}]_F \leq [\overline{a}]_F$. Then $[\overline{a}]_F$ is a maximal degree in $\mathrm{EVAL}(F)_\mathbb{C}$.*

Sketch. The previous proof still works, since $[\overline{a}]_F$ is closed under polynomial reductions and computations. $\quad\square$

6. The difficult point property

Given a graph polynomial $F(G, \overline{X})$ in n indeterminates $X_1, \ldots, X_n$ we are interested in the set $\mathrm{EASY}_\mathbb{C}(F)$.

We say that F has the *weak difficult point property* (**WDPP**) if

(i) there is a quasi-algebraic subset $A \subset \mathbb{C}^n$ of dimension $\leq n - 1$ which contains $\mathrm{EASY}_\mathbb{C}(F)$, and
(ii) there exists finitely many $\overline{a}_i : i \leq \alpha \in \mathbb{N}$ such that for each $\overline{b} \in \mathbb{C}^n - A$ the evaluation $F(G, \overline{b})$ is $F(G, \overline{a}_i) - \mathrm{HARD}_\mathbb{C}$ for some $\overline{a}_i \in \mathbb{C}^n$ and for all $i \leq \alpha$ the evaluation $F(G, \overline{a}_i)$ is not in $\mathrm{FP}_\mathbb{C}$.

We say that F has the *strong difficult point property* (**SDPP**) if **WDPP** holds for F with $A = \mathrm{EASY}_\mathbb{C}(F)$.

Without the requirement that A has a small dimension the **SDPP** is a *dichotomy property*, in the sense that every evaluation point is either easy or at least as hard as one of the evaluations at $\overline{a}_i, i \leq \alpha$. The two versions,

WDPP and **SDPP**, have a *quantitative aspect*: The set of easy points is rare in a strong sense: they are in a quasi-algebraic set of lower dimension.

Again the Difficult Counting Hypotheses **WDCH** and **SDCH** can be used to sharpen both Difficult Point Properties.

7. Examples for the WDPP and SDPP

In the discussion of the examples in this section we assume the Weak Difficult Counting Hypothesis **WDCaH**. We have seen in the case study, that the chromatic polynomial $\chi(G; \lambda)$ and the Tutte polynomial $T(G; X, Y)$ both have the **SDPP** with $\alpha = 1$ and the cover polynomial $C(D; X, Y)$ has the **SDPP** with $\alpha = 2$.

7.1. The Bollobás-Riordan polynomials

The Bollobás-Riordan polynomial is a generalization of the Tutte polynomial for graphs with colored edges colored with k colors, [36]. It has $4k$ many indeterminates. Its complexity was studied in [37,38], where it was shown that it satisfies **WDPP**.

7.2. The interlace polynomials

The interlace polynomial was introduced and intensively studied in [39–43]. It is a polynomial in two indeterminates. It complexity was studied in [44,45] where it was shown that it satisfies the **WDPP**.

7.3. Counting weighted homomorphisms aka partition functions

Let $A \in \mathbb{C}^{n \times n}$ be a complex, symmetric matrix, and let G be a graph. Let

$$Z_A(G) = \sum_{\sigma: V(G) \to [n]} \prod_{(v,w) \in E(G)} A_{\sigma(v), \sigma(w)}$$

Z_A is called a *partition function*.

If $\mathbf{X}$ is the matrix $(X_{i,j})_{i,j \le n}$ of indeterminates, then $Z_{\mathbf{X}}$ is a graph polynomial in n^2 indeterminates, and Z_A is an evaluation of $Z_{\mathbf{X}}$. i

Partition functions have their origin in statistical mechanics and have a very rich literature. In [46] a characterization is given of all multiplicative graph parameters which can be presented as partition functions. The complexity of partition functions was studied in a series of papers, [47–50].

Jin-yi Cai, Xi Chen and Pinyan Lu, [50], building on [47] proved a dichotomy theorem for $Z_\mathbf{X}$ where $\mathcal{R} = \mathbb{C}$. Analyzing their proofs reveals that $Z_\mathbf{X}$ satisfies the **SDPP** for $\mathcal{R} = \mathbb{C}$.

There are various generalizations of this to Hermitian matrices, cf. [51], and beyond.

7.4. *Generalized colorings*

Let $f : V(G) \to [k]$ be a coloring of the vertices of $G = (V(G), E(G))$. We look at variations of coloring properties which have been defined in the literature, cf. [11].

(i) f is *proper* if $(uv) \in E(G)$ implies that $f(u) \neq f(v)$. In other words if for every $i \in [k]$ the counter-image $[f^{-1}(i)]$ induces an independent set. These are the usual colorings.

(ii) f is *convex* if for every $i \in [k]$ the counter-image $[f^{-1}(i)]$ induces a connected graph, cf. [52].

(iii) f is *t-improper* if for every $i \in [k]$ the counter-image $[f^{-1}(i)]$ induces a graph of maximal degree t. For its origins and history cf. [53].

(iv) f is *H-free* if for every $i \in [k]$ the counter-image $[f^{-1}(i)]$ induces an H-free graph. For its origins and history cf. [54].

(v) f is *acyclic* if for every $i, j \in [k]$ the union $[f^{-1}(i)] \cup [f^{-1}(i)]$ induces an acyclic graph, [55].

The following was shown in [11,56]:

Theorem 7.1. *For all the above properties, counting the number of colorings is a polynomial in k.*

7.5. *More cases where SDPP holds*

SDPP was verified for

(i) the cover polynomial $C(G; x, y)$ introduced in [31] in [57,58].

(ii) the bivariate matching polynomial for multi-graphs defined first in [4] in [59,60].

(iii) The first two authors have also verified it for the graph polynomials for convex colorings, for t-improper colorings (for multi-graphs), for acyclic colorings, and the bivariate chromatic polynomial introduced by K. Döhmen, A. Pönitz and P. Tittman in [61].

(iv) More cases can be found in the Ph.D. Theses of I. Averbouch and the first author [62,63] and in [64,65].

C. Hoffmann's PhD thesis [66] contains a general *sufficient criterion* which allows to establish the **WDPP** for a wide class of (artificially defined) graph polynomials. Unfortunately, this method does not apply in most concrete cases of generalized chromatic polynomials discovered using the methods of [11].

Problem 7.1. *Characterize the graphs H for which the evaluation of the graph polynomial of H-free colorings with k colors satisfies **SDPP** or **WDPP**.*

8. Conclusion, conjectures and open problems

The purpose of this paper was to promote the study of graph parameters in BSS over $\mathbb{C}$ by formulating and illustrating the framework and formulating some conjectures. In particular we want to renew interest in Meer's approach to counting problems over some field. In [7,67,68] the authors studied definability questions but did not study complexity and the degrees of polynomial time reducibility.

It turns out that Meer's definition of $\sharp \mathbf{P}$ in BSS adapted to the complex case, is richer than originally assumed. By translating graph properties into polynomial equations we showed that every problem in $\sharp \mathbf{P}$ in the Turing model is also in $\sharp \mathbf{P}_{\mathbb{C}}$. However, the degree structure of $\sharp \mathbf{P}_{\mathbb{C}}$ remains unclear.

Problem 8.1. *Does $\sharp \mathbf{P}_{\mathbb{C}}$ have complete problems?*

Even the complexity of classically difficult problems seems unresolved. For example it is not clear whether 3-colorability is $\mathbf{NP}_{\mathbb{C}}$-hard or whether evaluating the chromatic polynomial at the point -1 is really more difficult than evaluating at the point 3. Similarly, it is not known whether counting the number of perfect matchings or of Hamiltonian paths are of the same difficulty or even comparable. The latter is particularly unsettling, since in Valiant's theory of algebraic circuits computing the Hamiltonian of a matrix and computing the permanent are of the same difficulty (namely **VNP**-complete).

Problem 8.2. *What can we say in BSS over $\mathbb{C}$ about the four graph parameters*

(i) $\chi_3(G)$, $\chi_{-1}(G)$,
(ii) $\sharp \mathrm{pm}(G)$, which counts the number of perfect matchings, and
(iii) $\sharp \mathrm{ham}(G)$ which counts the number of Hamiltonian paths?

Are they $\mathbf{NP}_{\mathbb{C}}$ *hard? Are they mutually equally hard?*

As a more suitable complexity class for the study of graph polynomials and graph parameters we introduced the class SOLEVAL$_{\mathbb{C}}$.

Problem 8.3. *Does* SOLEVAL$_{\mathbb{C}}$ *have complete problems? More generally what is the structure of the* $\mathbf{P}_{\mathbb{C}}$-*degrees of* SOLEVAL$_{\mathbb{C}}$ *?*

By translating the results on the complexity of the Tutte polynomial into the BSS model over $\mathbb{C}$ we were able to identify the open problems the solution of which are needed to draw a complete picture.

Let us conclude with a few conjectures in BSS over $\mathbb{C}$:

Conjecture 8.1. (Difficulty of Counting)

(i) *WDCH is true.*
(ii) *SDCH is false*

Conjecture 8.2. (Difficulty Point Property)
For every SOL *polynomial* F *in* n *indeterminates the following holds:*

(i) *There is a maximal degree* $[a]_{max}$ *in* EVAL$(F)_{\mathbb{C}}$. *Furthermore, the evaluation points in* $[a]_{max}$ *form a quasi-algebraic set of dimension* n.
(ii) *Either* EVAL$(F)_{\mathbb{C}} - \text{FP}_{\mathbb{C}} = \emptyset$ *or it has a minimal degree.*
(iii) EVAL$(F)_{\mathbb{C}} \cap \text{FP}_{\mathbb{C}}$ *is quasi-algebraic of dimension* $\leq n - 1$.

This conjecture is stronger that **SDPP**. In **SDPP** we stipulate that there are finitely many minimal degrees in EVAL$(F)_{\mathbb{C}} - \text{FP}_C$. In Conjecture 8.2 we require that there is only one such degree. The Strong Difficult Counting Hypothesis **SDCH** would imply that this minimal degree is at least $\mathbf{NP}_{\mathbb{C}}$-hard. We could actually strengthen Conjecture 8.2 by requiring that there is exactly one degree in EVAL$(F)_{\mathbb{C}} - \text{FP}_C$. But in the light of conjecture 8.1 this seems like going too far.

Acknowledgment

The authors would like to thank I. Averbouch, P. Bürgisser, K. Meer, P. Koiran, and an anonymous referee for useful comments on preliminary versions of this paper. An extended abstract of this paper appeared as [69].

T. Kotek was partially supported by the Fein Foundation and the Graduate School of the Technion–Israel Institute of Technology.

J.A. Makowsky and E.V. Ravve were partially supported by the Israel Science Foundation for the project "Model Theoretic Interpretations of Counting Functions" (2007-2011) and the Grant for Promotion of Research by the Technion–Israel Institute of Technology.

228

References

1. G. Birkhoff, *Annals of Mathematics* **14**, 42 (1912).
2. W. Tutte, *Canadian Journal of Mathematics* **6**, 80 (1954).
3. R. Potts, *Mathematical Proceedings* **48**, 106 (1952).
4. C. Heilmann and E. Lieb, *Comm. Math. Phys* **25**, 190 (1972).
5. A. Sokal, The multivariate Tutte polynomial (alias Potts model) for graphs and matroids, in *Survey in Combinatorics, 2005*, , London Mathematical Society Lecture Notes Vol. 3272005.
6. F. Jaeger, D. Vertigan and D. Welsh, *Math. Proc. Camb. Phil. Soc.* **108**, 35 (1990).
7. K. Meer, *Theoretical Computer Science* **242**, 41 (2000).
8. P. Bürgisser and F. Cucker, *J. Complexity* **22**, 147 (2006).
9. S. Basu and T. Zell, *Foundations of Computational Mathematics* **10**, 429 (2010).
10. S. Basu, *Foundations of Computational Mathematics* **12**, 327 (2012).
11. T. Kotek, J. Makowsky and B. Zilber, On counting generalized colorings, in *Model Theoretic Methods in Finite Combinatorics*, eds. M. Grohe and J. Makowsky, Contemporary Mathematics, Vol. 558 (American Mathematical Society, 2011) pp. 207–242.
12. L. Blum, F. Cucker, M. Shub and S. Smale, *Complexity and Real Computation* (Springer Verlag, 1998).
13. C. Papadimitriou, *Computational Complexity* (Addison Wesley, 1994).
14. S. Arora and B. Barak, *Computational Complexity: A modern approach* (Cambridge University Press, 2009).
15. L. Valiant, *SIAM Journal on Computing* **8**, 410 (1979).
16. R. Ladner, *J. ACM* **22**, 155 (1975).
17. U. Schöning, *Theor. Comput. Sci.* **18**, 95 (1982).
18. K. Ambos-Spies, *Theor. Comput. Sci.* **63**, 43 (1989).
19. S. Ben-David, K. Meer and C. Michaux, *J. Complexity* **16**, 324 (2000).
20. J. Makowsky and K. Meer, On the complexity of combinatorial and metafinite generating functions of graph properties in the computational model of Blum, Shub and Smale, in *CSL'00*, , Lecture Notes in Computer Science Vol. 1862 (Springer, 2000).
21. D. Bayer, The division algorithm and the Hilbert scheme, PhD thesis, Harvard University1982.
22. J. A. D. Loera, J. Lee, S. Margulies and S. Onn, *Combinatorics, Probability & Computing* **18**, 551 (2009).
23. S. Margulies, Computer algebra, combinatorics, and complexity: Hilbert's nullstellensatz and NP-complete problems, PhD thesis, University of California, Davis2008.
24. P. Bürgisser, *Completeness and Reduction in Algebraic Complexity*, Algorithms and Computation in Mathematics, Vol. 7 (Springer, 2000).
25. P. Bürgisser, F. Cucker and M. Lotz, *Found. Comput. Math.* **5**, 351 (2005).
26. N. Biggs, *Algebraic Graph Theory, 2nd edition* (Cambridge University Press, 1993).
27. C. Godsil and G. Royle, *Algebraic Graph Theory*, Graduate Texts in Mathe-

matics, Graduate Texts in Mathematics (Springer, 2001).

28. R. Diestel, *Graph Theory*Graduate Texts in Mathematics, Graduate Texts in Mathematics, 3 edn. (Springer, 2005).

29. M. Linial, *SIAM Journal of Algebraic and Discrete Methods* **7**, 331 (1986).

30. B. Bollobás, *Modern Graph Theory* (Springer, 1999).

31. F. Chung and R. Graham, *Journal of Combinatorial Theory, Ser. B* **65**, 273 (1995).

32. M. Bläser and H. Dell, Complexity of the cover polynomial, in *Automata, Languages and Programming, ICALP 2007*, eds. L. Arge, C. Cachin, T. Jurdziński and A. Tarlecki, Lecture Notes in Computer Science, Vol. 4596 (Springer, 2007).

33. L. Libkin, *Elements of Finite Model Theory* (Springer, 2004).

34. S. Even and R. Tarjan, *Journal of ACM* **23**, 710 (1976).

35. M. Garey and D. Johnson, *Computers and Intractability*Mathematical Series, Mathematical Series (W.H. Freeman and Company, 1979).

36. B. Bollobás and O. Riordan, *Combinatorics, Probability and Computing* **8**, 45 (1999).

37. M. Bläser, H. Dell and J. Makowsky, Complexity of the Bollobás-Riordan polynomial. exceptional points and uniform reductions, in *Computer Science–Theory and Applications, Third International Computer Science Symposium in Russia*, eds. E. A. Hirsch, A. A. Razborov, A. Semenov and A. Slissenko, Lecture Notes in Computer Science, Vol. 5010 (Springer, 2008).

38. M. Bläser, H. Dell and J. A. Makowsky, *Theory Comput. Syst.* **46**, 690 (2010).

39. R. Arratia, B. Bollobas and G. Sorkin, The interlace polynomial: A new graph polynomial, IBM Research Report, RC 21813 (98165), 31 July 2000.

40. R. Arratia, B. Bollobás and G. Sorkin, *Combinatorica* **24.4**, 567 (2004).

41. R. Arratia, B. Bollobás and G. Sorkin, *Journal of Combinatorial Theory, Series B* **92**, 199 (2004).

42. M. Aigner and H. van der Holst, *Linear Algebra and Applications* **377**, 11 (2004).

43. B. Courcelle, *The Electronic Journal of Combinatorics* **15**, p. R69 (2008).

44. M. Bläser and C. Hoffmann, On the complexity of the interlace polynomial, arXive 0707.4565, (2007).

45. M. Bläser and C. Hoffmann, On the complexity of the interlace polynomial, in *STACS*, 2008.

46. M. Freedman, L. Lovász and A. Schrijver, *Journal of AMS* **20**, 37 (2007).

47. A. Bulatov and M. Grohe, *Theoretical Computer Science* **348**, 148 (2005).

48. L. A. Goldberg, M. Grohe, M. Jerrum and M. Thurley, *SIAM J. Comput.* **39**, 3336 (2010).

49. M. Grohe and M. Thurley, Counting homomorphisms and partition functions, in *Model Theoretic Methods in Finite Combinatorics*, eds. M. Grohe and J. Makowsky, Contemporary Mathematics, Vol. 558 (American Mathematical Society, 2011) pp. 243–292.

50. J. Cai, X. Chen and P. Lu, Graph homomorphisms with complex values: A dichotomy theorem, in *ICALP (1)*, 2010.

51. M. Thurley, *CoRR* **abs/1004.0992** (2010).

52. S. Moran and S. Snir, *Journal of Computer and System Sciences* **73.7**, 1078 (2007).

53. R. J. Kang and C. McDiarmid, *Combinatorics, Probability & Computing* **19**, 87 (2010).

54. D. Achlioptas, *Discrete Math* **165**, 21 (1997).

55. N. Alon, C. McDiarmid and B. Reed, *Random Structures and Algorithms* **2.3**, 277 (1991).

56. T. Kotek, J. Makowsky and B. Zilber, On counting generalized colorings, in *Computer Science Logic, CSL'08*, , Lecture Notes in Computer Science Vol. 52132008.

57. M. Bläser and H. Dell, Complexity of the cover polynomial, in *ICALP*, 2007.

58. M. Bläser, H. Dell and M. Fouz, *Computational Complexity* **xx**, xx (2012), accepted for publication.

59. I. Averbouch and J. Makowsky, The complexity of multivariate matching polynomials, Preprint, January 2007.

60. C. Hoffmann, *Fundamenta Informaticae* **98.4**, 373 (2010).

61. K. Dohmen, A. Pönitz and P. Tittmann, *Discrete Mathematics and Theoretical Computer Science* **6**, 69 (2003).

62. I. Averbouch, Completeness and universality properties of graph invariants and graph polynomials, PhD thesis, Technion - Israel Institute of Technology, Haifa, Israel2011.

63. T. Kotek, Definability of combinatorial functions, PhD thesis, Technion - Israel Institute of Technology, Haifa, IsraelMarch 2012. Submitted.

64. P. Tittmann, I. Averbouch and J. Makowsky, *European Journal of Combinatorics* **32**, 954 (2011).

65. I. Averbouch, B. Godlin and J. Makowsky, *European Journal of Combinatorics* **31**, 1 (2010).

66. C. Hoffmann, Computational complexity of graph polynomials, PhD thesis, Naturwissenschaftlich-Technische Fakultät der Universität des Saarlandes, Saarbrücken, Germany2010.

67. E. Grädel and K. Meer, *Lectures in Applied Mathematics* **32**, 381 (1996), A preliminary version has been presented at the 27th ACM-Symposium on Theory of Computing, Las Vegas, 1995.

68. E. Grädel and Y. Gurevich, *Information and Computation* **140**, 26 (1998).

69. T. Kotek, J. Makowsky and E. Ravve, A computational framework for the study of partition functions and graph polynomials (abstract), in *SYNASC 2012*, eds. V. Negru and et al. Proceedings of the International Symposium on Symbolic and Numeric Algorithms for Scientific Computing, SYNASC, Timisoara, Romania (IEEE Computer Society, 2013).

RELATION ALGEBRAS AND R

T. Kowalski

La Trobe University, Department of Mathematics and Statistics,
Bundoora, Victoria 3086, Australia
E-mail: t.kowalski@latrobe.edu.au

It is shown that the relevant logic **R** is complete with respect to square-increasing, commutative, integral relation algebras. This contrasts with the representable case, where completeness is known to fail.

Keywords: Relevant logics, relation algebras, completeness.

1. Introduction

In recent years there has been a steady flow, or a continuous trickle at the very least, of articles investigating the connection between *relevant logics* and *relation algebras*. It began in 2007, when Maddux observed (see [1]) that relevant logics have a natural interpretation in the language of relation algebras, with respect to which certain standard relevant logics are sound. He asked whether this interpretation was complete as well. In 2008, Mikulás proved in [2] that the relevant logic **R** was not complete with respect to square-increasing, commutative, representable relation algebras, answering thereby the original Maddux' question in the negative. Rather surprisingly, Maddux was able to show in [3] that another well-known relevant logic, namely **RM**, is complete with respect to idempotent, commutative, representable relation algebras. For several other results in similar vein, the reader is referred to Bimbó *et al.* [4] and Hirsch and Mikulás [5]. These results are mostly negative, however, there are some positive results for certain positive fragments of relevant logics (pardon the pun). These results all have a common focus on representable relation algebras.

In this article I would like to broaden the perspective a little and begin to investigate the connection between relevant logics and relation algebras, without requiring representability. Specifically, it will be shown that **R** is complete with respect to square-increasing, commutative relation algebras.

I was encouraged to investigate the non-representable case by a surprisingly wide attention one little result of mine has received. This result (see [6]), motivated by Meyer's study [7] of the logic **B**, deals with *weakly-associative* relation algebras. Since these are non-representable, giving up on representability seemed natural.

Some familiarity with lattice theory and a modicum of universal algebra is assumed. In particular, familiarity with the standard representation of distributive lattices as Priestley spaces (see [8]) will help, although we will only need its non-topological part. Other facts and notions will be recalled as needed.

For the section dealing with De Morgan monoids, some acquaintance with *residuated lattices* will be very useful, as certain basic facts from their theory are used rather quickly and with little explanation. Galatos *et al.* [9] has more on residuated lattices.

2. Relation algebras

We recall definitions and some basic facts on relation algebras that we will use later on. For a comprehensive reference book we recommend Maddux [10] or Hirsch and Hodkinson [11], preferably both.

A *relation algebra* is an algebra $\mathbf{A} = \langle A, \vee, \wedge, \cdot, \neg, \breve{\ }, e, 0, 1 \rangle$ such that

- $(A, \wedge, \vee, \neg, 0, 1)$ is a Boolean algebra,
- $(A, \cdot, \breve{\ }, e)$ is an involutive monoid, whose operations are *additive normal operators* with respect to the Boolean part.
- $x \wedge y \cdot z = 0$ iff $y \wedge x \cdot z^{\breve{\ }} = 0$ iff $z \wedge y^{\breve{\ }} \cdot x = 0$, known as *Peirce's* or *triangle* laws, hold.

The conditions above can be expressed by means of identities, namely

- some finite base of identities for Boolean algebras
- $xe = x = ex$, $x(yz) = (xy)z$, $x^{\breve{\ }\breve{\ }} = x$, $(xy)^{\breve{\ }} = y^{\breve{\ }}x^{\breve{\ }}$
  (involutive monoid identities)
- $(x \vee y)^{\breve{\ }} = x^{\breve{\ }} \vee y^{\breve{\ }}$, $x(y \vee z) = xy \vee xz$, $(x \vee y)z = xz \vee yz$
  (additivity of $\breve{\ }$ and $\cdot$)
- $0^{\breve{\ }} = 0$, $x0 = 0 = 0x$
  (normality of $\breve{\ }$ and $\cdot$)
- $x^{\breve{\ }} \cdot \neg(xy) \wedge y = 0$
  (equivalent in presence of the others to triangle laws)

and so the class RA of all relation algebras is a finitely based variety. For any set U the algebra of all binary relations over U, whose operations are set-theoretical union, intersection and complement, relation composition and converse, and the constants are the identity relation, $U \times U$ and $\emptyset$, is a relation algebra. Such relation algebras are called *set relation algebras*. A relation algebra $\mathbf{A}$ is *representable* if $\mathbf{A}$ can be embedded into a direct product of set relation algebras. The class RRA of all representable relation algebras is a proper subclass of RA; in fact RRA is a non-finitely based variety.

A relation algebra $\mathbf{A}$ is *integral* if $ab = 0$ implies $a = 0$ or $b = 0$ (so there are no "zero divisors"). Integrality is equivalent to the property that the identity element e is an atom of the underlying Boolean algebra. All integral relation algebras are simple, but not conversely.

3. Relevant logic R

We define the relevant logic $\mathbf{R}$ as a Hilbert system, following the presentation from Dunn and Restall [12] (see also the classical Anderson *et al.*[13,14] and Routley and Meyer [15] for much more on relevant logics). As usual, *formulas* are built recursively from a countable set or *propositional variables*, by means of *connectives* $\wedge$, $\vee$, $\rightarrow$ and $\sim$. The axioms are:

(1) $\alpha \rightarrow \alpha$
(2) $(\alpha \rightarrow \beta) \rightarrow ((\gamma \rightarrow \alpha) \rightarrow (\gamma \rightarrow \beta))$
(3) $((\alpha \rightarrow (\alpha \rightarrow \beta)) \rightarrow (\alpha \rightarrow \beta)$
(4) $(\alpha \rightarrow (\beta \rightarrow \gamma)) \rightarrow (\beta \rightarrow (\alpha \rightarrow \gamma))$
(5) $(\alpha \wedge \beta) \rightarrow \alpha$, $(\alpha \wedge \beta) \rightarrow \beta$
(6) $((\alpha \rightarrow \beta) \wedge (\alpha \rightarrow \gamma)) \rightarrow (\alpha \rightarrow (\beta \wedge \gamma))$
(7) $\alpha \rightarrow (\alpha \vee \beta)$, $\beta \rightarrow (\alpha \vee \beta)$
(8) $((\alpha \rightarrow \gamma) \wedge (\beta \rightarrow \gamma)) \rightarrow ((\alpha \vee \beta) \rightarrow \gamma)$
(9) $(\alpha \wedge (\beta \vee \gamma)) \rightarrow ((\alpha \wedge \beta) \vee \gamma)$
(10) $(\alpha \rightarrow \sim\alpha) \rightarrow \sim\alpha$
(11) $(\alpha \rightarrow \sim\beta) \rightarrow (\beta \rightarrow \sim\alpha)$
(12) $\sim\sim\alpha \rightarrow \alpha$

And the rules of inference are:

$$\frac{\alpha \quad \alpha \rightarrow \beta}{\beta} \qquad \frac{\alpha \quad \beta}{\alpha \wedge \beta}$$

Then, $\mathbf{R}$ is defined as the smallest set of formulas containing all the axioms and closed under the rules of inference. A "multiplication" connective, called *fusion*, can be defined in $\mathbf{R}$ putting

$$\alpha \cdot \beta \leftrightarrow \sim(\alpha \rightarrow \sim\beta)$$

Since adding fusion to **R** is *conservative*, we assume that fusion is present from the outset, and count the defining formula above as one of the axioms of **R**.

Sometimes **R** is further extended by adding an *Ackermann constant*, perhaps more informatively called *truth constant*, and requiring that it satisfies the axioms

- t
- $t \to (\alpha \to \alpha)$

Although extending **R** by the Ackermann constant is conservative as well, we will not assume the presence of t. In fact, we will assume the opposite, and on rare occasions where t is present, we will use $\mathbf{R}_t$ to denote the extension.

4. De Morgan monoids

The usual algebraic semantics for **R** and $\mathbf{R}_t$ (see Dunn [16], Routley and Meyer [17], but also Maximova [18,19] where they were studied independently) is provided by the class of *De Morgan monoids*. Out of a number of equivalent presentations of De Morgan monoids we choose the following. A De Morgan monoid is an algebra $\mathbf{A} = \langle A, \vee, \wedge, \cdot, \to, 1, 0 \rangle$ such that:

(1) $\langle A, \vee, \wedge \rangle$ is a distributive lattice,
(2) $\langle A, \cdot, 1 \rangle$ is a commutative monoid,
(3) the equivalence $a \cdot b \leq c$ iff $b \leq a \to c$ holds for all $a, b, c \in A$,
(4) the following identities hold in **A**:

    (a) $a \leq a \cdot a$
    (b) $(a \to 0) \to 0 = a$

In the terminology of residuated lattices (see [9]), a De Morgan monoid is an involutive, square-increasing, distributive $\mathbf{FL_e}$-algebra. It is customary and useful to define a unary operation $\sim$ (called *negation*) on A, putting $\sim a = a \to 0$. Since the residuation equivalence is itself equivalent to the pair of inequalities

$$a((a \to b) \wedge c) \leq b \leq a \to ((ab) \vee c)$$

and inequalities can be rendered as equations in the standard lattice-order way, De Morgan monoids form a finitely based variety, which we will denote DM. Notice that we dropped the multiplication sign, which we will do often from now on. We also assume the following partial order of binding strength

$\{\sim\} \leq \{\cdot, \rightarrow\} \leq \{\vee, \wedge\}$ among the basic operations. This ordering may seem odd for a logician (implication binds stronger than lattice connectives), but we adopt it for consistency: we want multiplication to bind stronger than lattice connectives, and the residual, being a division-like connective, belongs to the multiplication family.

Lemma 4.1. *Let* **M** *be a De Morgan monoid,* $a \in M$ *and* $X, Y \subseteq M$. *Then, the following infinitary laws hold:*

(1) $\bigvee X \cdot \bigvee Y = \bigvee \{xy \colon x \in X, \ y \in Y\}$
(2) $\bigvee X \rightarrow a = \bigwedge \{x \rightarrow a \colon x \in X\}$
(3) $a \rightarrow \bigwedge X = \bigwedge \{a \rightarrow x \colon x \in X\}$
(4) $\sim \bigvee X = \bigwedge \{\sim x \colon x \in X\}$
(5) $\sim \bigwedge X = \bigvee \{\sim x \colon x \in X\}$

where the equalities mean that if one side exists, so does the other, and they are equal.

It follows in particular that $\sim$ is a dual lattice isomorphism on any De Morgan monoid **M**. It is thus natural to define on any De Morgan monoid a binary operation $+$ putting $x + y = \sim(\sim x \cdot \sim y)$. The next lemma is straightforward to prove.

Lemma 4.2. *Let* **M** *be a De Morgan monoid. The operation* $+$ *defined as above is associative and commutative on* M. *Moreover, for any* $a, b, c \in M$, *the inequality* $a + a \leq a$ *holds, and so does the following equivalence*

$$a + b \geq c \quad \text{iff} \quad b \geq c - a$$

where $c - a$ *stands for* $\sim a \cdot c$.

It follows that for a De Morgan monoid $\mathbf{M} = \langle M, \vee, \wedge, \cdot, \rightarrow, 1, 0 \rangle$, the algebra $\widetilde{\mathbf{M}} = \langle M, \wedge, \vee, +, -, 0, 1 \rangle$ is a De Morgan monoid as well.

Lemma 4.3. *Let* **M** *be a De Morgan monoid. Then* $\sim$ *establishes an isomorphism between* **M** *and* $\widetilde{\mathbf{M}}$.

We call $\widetilde{\mathbf{M}}$ the *dual* of **M**. We will use the isomorphism between **M** and its dual in transferring results about filters of **M** to analogous results about ideals of **M**.

Lemma 4.4. *The following hold in all De Morgan monoids:*

(1) $\sim\sim x = x$
(2) $\sim(\sim(x \vee y) \vee z) = \sim(\sim x \vee z) \vee \sim(\sim y \vee z)$

(3) $x(y \lor z) = xy \lor xz$

(4) $xy \leq z$ iff $x \cdot \sim z \leq \sim y$

(5) $x \land y = \sim(\sim x \lor \sim y)$

(6) $x \to y = \sim(x \cdot \sim y) = \sim x + y$

(7) $0 = \sim 1$

(8) $x \land y \leq xy$

(9) $x + y \leq x \lor y$

(10) $x \leq y$ iff $1 \leq \sim x + y$

(11) $xy \leq z + u$ iff $1 \leq \sim x + \sim y + z + u$

(12) $1 \leq x \lor \sim x$

(13) if $\sim x \land 1 \leq x \land 1$, then $1 \leq x$

(14) if $x, y \leq 1$, then $xy = x \land y$

(15) $y \leq \sim x + xy$

(16) $\sim x(x + y) \leq y$

Proof. Most of these are well-known, so we will only prove (11), (12) and (13). For (11) we have

$$\begin{aligned} xy \leq z + u \quad &\text{iff} \quad 1 \leq xy \to (z + u) \\ &\text{iff} \quad 1 \leq \sim(xy) + (z + u) \\ &\text{iff} \quad 1 \leq (\sim x + \sim y) + (z + u) \\ &\text{iff} \quad 1 \leq \sim x + \sim y + z + u \end{aligned}$$

where the first line follows by residuation, the second by (6) above, the third by definition of $+$, and the fourth by associativity of $+$, which in turn follows from Lemma 4.3.

For (12), using (8) above we calculate $x \land \sim x \leq x \cdot \sim x = x(x \to 0) \leq 0$, and so $1 \leq x \lor \sim x$ as required.

Finally, for (13), suppose $\sim x \land 1 \leq x \land 1$. Then, $x \land 1 = (\sim x \land 1) \lor (x \land 1) = (\sim x \lor x) \land 1 = 1$, where the last equality follows by (12). Thus, $x \geq 1$ as claimed. $\qquad \square$

An alternative term-equivalent presentation of De Morgan monoids takes them to be algebras $\langle S, \cdot, \lor, \sim, 1 \rangle$ such that $\langle S, \cdot, 1 \rangle$ is a commutative monoid, $\langle S, \lor, 1 \rangle$ is a (join) semilattice, and the conditions (1) through (4) above hold. The term-equivalence is established by taking (5) through (7) as definitions.

There is yet another common presentation of De Morgan monoids, obtained by dropping the constant(s) from any of the above presentations, and requiring instead that there exists an element e such that $ea = a$ for any a. For instance Slaney [20] defines De Morgan monoids this way. But it is algebraically inconvenient, because De Morgan monoids so defined do not form a variety (closure under subalgebras fails). De Morgan monoids without constants are however essential for the embedding result, so departing from the standard terminology we will refer to them as *bare* De Morgan monoids. To be precise, we define bare De Morgan monoids as $\{\vee, \wedge, \cdot, \rightarrow\}$-reducts of De Morgan monoids. The next lemma characterises subdirectly irreducible De Morgan monoids. Before we state it, let us recall two lattice-theoretical notions. Let $\mathbf{L}$ be a lattice. An element $a \in L$ is *join irreducible* if $a = b \vee c$ implies $a = b$ or $a = c$. An element $a \in L$ is *join prime* if $a \leq b \vee c$ implies $a \leq b$ or $a \leq c$. Obviously, a join prime element is join irreducible; the converse is not true in general but it holds in distributive lattices.

Lemma 4.5. *Let $\mathbf{M}$ be a De Morgan monoid. Then, (1) and (2) below are equivalent, and each of them implies (3).*

(1) $\mathbf{M}$ is subdirectly irreducible.
(2) There exists a greatest element in the set $\{a \in M : a < 1\}$.
(3) The element 1 is join prime.

Proof. Recall first that a commutative residuated lattice is subdirectly irreducible iff there exists a strictly negative element $c \in M$ such that for every strictly negative $d \in M$ there is a natural number k with $d^k \leq c$. As $\mathbf{M}$ is square-increasing, all negative elements are idempotent (see Lemma 4.4), so the characterisation above states that there is a $c < 1$ such that every $d < 1$ has $d \leq c$. The equivalence between (1) and (2) follows.

To show that (2) implies (3) notice first that if $a < 1$ and $b < 1$, we get $a \vee b \leq c < 1$, where c is the largest strictly negative element of M. Thus, (2) implies that 1 is join irreducible. Since $\mathbf{M}$ is distributive as a lattice, join irreducibility is equivalent to join primeness. $\square$

As an aside, let us observe that 1 can be join prime in a De Morgan monoid which is not subdirectly irreducible, a phenomenon well-known in the context of Heyting algebras. To obtain another equivalent characterisation of subdirect irreducibility, we would need a stronger concept of 1 being *strictly join irreducible*, i.e., such that $\bigvee X = 1$ implies $x = 1$ for some $x \in X$, where $X \subseteq M$. However, we will not go into further details here. Instead, we will look closer at De Morgan monoids satisfying $1 \not\leq 0$.

Since we will make frequent use of downward and upward closed subsets of De Morgan monoids, it is convenient to introduce shorthand notation for upward and downward closures. We write $\downarrow X$ for the downward closed set $\{y: y \leq x \text{ for some } x \in X\}$, and if $X = \{a\}$ we abbreviate $\downarrow\{a\}$ to $\downarrow a$. Similar notational devices apply to upward closed sets $\uparrow X$.

Lemma 4.6. *Let $\mathbf{M}$ be a De Morgan monoid. Then, $\mathbf{M}$ satisfies $1 \not\leq 0$ iff for any $a \in M$ at most one of $a, \sim a$ belongs to $\uparrow 1$.*

Proof. We have $a, \sim a \in \uparrow 1$ iff $1 \leq a$ and $1 \leq a \rightarrow 0$ iff $1 \leq a \leq 0$. □

Lemma 4.7. *Let $\mathbf{M}$ be a De Morgan monoid satisfying $1 \not\leq 0$. The following are equivalent:*

(1) $\mathbf{M}$ is subdirectly irreducible.
(2) 0 is the greatest element in the set $\{a \in M: a \not\geq 1\}$.
(3) The pair $(1, 0)$ splits the lattice reduct of $\mathbf{M}$.

Proof. To show that (1) implies (2), we employ Lemma 4.5. Let c be the largest strictly negative element of M. Consider $\uparrow 1 \cup \downarrow 0$. We will show that $\uparrow 1 \cup \downarrow 0 = M$. To get a contradiction, suppose $d \notin \uparrow 1 \cup \downarrow 0$. Since $d \vee \sim d \geq 1$, we have $1 = (d \vee \sim d) \wedge 1 = (d \wedge 1) \vee (\sim d \wedge 1)$. As $d \not\geq 1$, we get $d \wedge 1 \leq c$. Thus, $\sim d \wedge 1 = 1$ and so $\sim d \geq 1$, from which it follows that $d \leq 0$, contradicting the assumption. Since $\downarrow 0$ is the image of $\uparrow 1$ under $\sim$, these two sets are dually isomorphic as lattices. Now, as $0 \not\geq 1$ it follows that M is in fact a disjoint union of $\uparrow 1$ and $\downarrow 0$. Therefore $0 = \bigvee\{a \in M: a \not\geq 1\}$, proving the claim.

To prove that (2) implies (1), it suffices to observe that if $0 = \bigvee\{a \in M: a \not\geq 1\}$, then $c = 0 \wedge 1$ is the greatest strictly negative element. To see that, let $a < 1$. Then, $a \vee (0 \wedge 1) = (a \vee 0) \wedge (a \wedge 1) = (a \vee 0) \wedge a = a$, so $a \leq 0 \wedge 1$ as required.

We have thus shown that (1) and (2) are equivalent. That (3) is equivalent to (2) follows immediately by definition of splitting. □

A De Morgan monoid $\mathbf{M}$ is *normal* if for every $a \in M$ exactly one of $a, \sim a$ belongs to $\uparrow 1$. The next three lemmas gather some facts about them.

Lemma 4.8. *Let $\mathbf{M}$ be a De Morgan monoid. Then*

(1) If $\mathbf{M}$ is normal, then $\mathbf{M}$ is subdirectly irreducible.
(2) If $\mathbf{M}$ is subdirectly irreducible and $1 \not\leq 0$ in $\mathbf{M}$, then $\mathbf{M}$ is normal.
(3) If $\mathbf{M}$ is subdirectly irreducible and $1 \leq 0$ in $\mathbf{M}$, then there exists a normal De Morgan monoid $\mathbf{N}$ such that $\mathbf{M} \in H(\mathbf{N})$.

Proof. We begin the proof of (1) by showing that if **M** is normal, then $M = \uparrow 1 \cup \downarrow 0$. Suppose the contrary, then for some $d \in M$ we have $d \notin \uparrow 1 \cup \downarrow 0$, in particular, $d \not\geq 1$. By normality then, $\sim d \geq 1$, so $d \leq 0$, which means $d \in \downarrow 0$, contradicting the assumption.

Further, since $1 \not\leq 0$, the union $\uparrow 1 \cup \downarrow 0$ is disjoint, so $M = \uparrow 1 \uplus \downarrow 0$. We will show that $1 \wedge 0$. is the largest strictly negative element. Clearly $1 \wedge 0 < 1$. Take any $a < 1$. Then, by normality, $\sim a \geq 1$ and so $a \leq 0$. Therefore, $a \leq 1 \wedge 0$ as claimed. By Lemma 4.5(2), we get that **M** is subdirectly irreducible, proving (1).

To prove (2), suppose **M** is subdirectly irreducible and has $1 \not\leq 0$. Take any $a \not\geq 1$. Then, by Lemma 4.7(2), we get that $a \leq 0$, and thus $\sim a \geq 1$, proving that **M** is normal.

Finally, for (3) let **M** be subdirectly irreducible with $1 \leq 0$. Let **2** stand for a two-element Boolean algebra. $\mathbf{M} \times \mathbf{2}$. It is not difficult to verify that the set $N = \{(a, 0) \colon a \leq 0\} \cup \{(a, 1) \colon a \geq 1\}$ is a sublattice of $\mathbf{M} \times \mathbf{2}$. It is also immediate that N is closed under negation. We define a multiplication operation on N as follows

$$(a, c) \cdot (b, d) = \begin{cases} (ab, 0) & \text{if } cd = 0 \text{ and } ab \leq 0 \\ (ab, 1) & \text{otherwise.} \end{cases}$$

So defined, multiplication is obviously commutative. To show that it is associative, notice that if $((a, f)(b, g))(c, f) \neq (a, c)((b, d)(f, g))$, then without loss of generality we can assume $((a, f)(b, g))(c, f) = (abc, 0)$ and $(a, c)((b, d)(f, g)) = (abc, 1)$. It is not difficult to show that the first of these equalities holds if and only if $fgh = 0$ and $abc \leq 0$. But if $abc \leq 0$, the second equality can only hold if $f = 1$ and $(b, g)(c, h) = (bc, 1)$. Thus, $gh = 1$ or $bc \not\leq 0$. Now, $gh = 1$ is impossible, for it contradicts $fgh = 0$, so $bc \not\leq 0$. Thus, $\sim(bc) \not\geq 1$ and by subdirect irreducibility of **M** we get $bc \geq 1$. It follows that $a \leq abc \leq 0$. But if $a \leq 0$ and $f = 1$, then $a \geq 1$, so $bc \leq abc \leq 0$, which is a contradiction with $bc \leq 0$.

Further, $(1, 1)(a, b) = (a, 1)$ iff $b = 1$, so $(1, 1)(a, b) = (a, b)$ and thus $(1, 1)$ is a unit. It is also clear that multiplication is square-increasing. It remains to show that multiplication is residuated. We claim that the operation

$$(a, c) \rightarrow (b, d) = \begin{cases} (a \rightarrow b, 1) & \text{if } a \rightarrow b \geq 1 \\ (a \rightarrow b, 0) & \text{otherwise.} \end{cases}$$

provides a residual for multiplication. Verification of this claim is left to the reader. We have thus proved that $\mathbf{N} = \langle N, \vee, \wedge, \cdot, \rightarrow, (1, 1), (0, 0) \rangle$ is a De

Morgan monoid.

Now, observe that the element $(1,0)$ is by definition the largest element in S and thus $\mathbf{N}$ is subdirectly irreducible. Consider the congruence $\Theta = \mathbf{Cg}^{\mathbf{N}}\{(1,1),(1,0)\}$. It is clear from the construction that Θ is the restriction to $\mathbf{N}$ of the first projection on $\mathbf{M} \times \mathbf{2}$. Therefore, $\mathbf{N}/\Theta = \mathbf{M}$, proving (3). □

Lemma 4.9. *Let $\mathbf{M}$ be a subdirectly irreducible De Morgan monoid, and a,b,c,d be elements of M such that $ab \leq c+d$. Then $a \leq c$ or $b \leq d$.*

Proof. By Lemma 4.4(11) we have $ab \leq c+d$ iff $1 \leq \sim a + \sim b + c + d$. By (9) of the same lemma, and commutativity and associativity of addition, we obtain $1 \leq (\sim a + c) \vee (\sim b + d)$. Since $\mathbf{M}$ is subdirectly irreducible, Lemma 4.5(3) applies, and thus $1 \leq \sim a + c$ holds, or $1 \leq \sim b + d$ holds. It follows that $a \leq c$ or $b \leq d$, as claimed. □

Let us pause for a moment to comment on the quite extraordinary strength of Lemma 4.9. Namely, using commutativity of multiplication and addition, it is easy to see that $ab \leq c+d$ implies that least one of the following statements holds:

- $a \leq c \wedge d$,
- $b \leq c \wedge d$,
- $a \vee b \leq c$,
- $a \vee b \leq d$.

An analogy here would be that of the two-element subdirectly irreducible Boolean algebra, in which the property of Lemma 4.9 holds with $ab = a \wedge b$ and $c+d = c \vee d$. We will put Lemma 4.9 to good use in the next section.

The next result, due to Meyer, Dunn and Leblanc [21], is usually proved in a different way. Here it becomes an immediate corollary of Lemma 4.8.

Theorem 4.1. *Normal De Morgan monoids generate the variety of De Morgan monoids.*

The following well-known theorem, first proved in Dunn [16] (cf. Slaney [20] for a brief and clear statement of the first part) shows the importance of normal De Morgan monoids for relevant logic.

Theorem 4.2. *The following hold.*

(1) $\mathbf{R}$ *is sound and complete with respect to the class of bare normal De Morgan monoids.*

(2) $\mathbf{R}_t$ *is sound and complete with respect to the class of normal De Morgan monoids.*

5. Prime filters and ideals in De Morgan monoids

The embedding construction described in the next section makes essential use of prime filters and ideals on De Morgan monoids. Recall that a proper filter F of a lattice $\mathbf{L}$ is *prime* if $a \vee b \in F$ implies $a \in F$ or $b \in F$, for any $a, b \in L$. Dually, a proper ideal I on $\mathbf{L}$ is prime, if $a \wedge b \in I$ implies $a \in I$ or $b \in I$. If $\mathbf{L}$ is distributive, then the complement of a prime filter is a prime ideal.

Let $\mathbf{M}$ be a De Morgan monoid, and $X, Y \subseteq M$. We write $X \cdot Y$ or simply XY for the *complex product* of X and Y, that is, $\{xy \colon x \in X, \ y \in Y\}$. Similarly, $X + Y = \{x + y \colon x \in X, \ y \in Y\}$ is the *complex sum* of X and Y. Next, $\overline{X}$ is stands for $M \setminus X$. Finally, we let $\mathrm{Fil}(\mathbf{M})$ stand for the set of all proper filters of (the lattice reduct of) $\mathbf{M}$. Similarly, $\mathrm{Id}(\mathbf{M})$ will stand for the set of all proper ideals of $\mathbf{M}$.

Lemma 5.1. *Let* $\mathbf{M}$ *be a De Morgan monoid and* $F, G \in \mathrm{Fil}(\mathbf{M})$, $I, J \in \mathrm{Id}(\mathbf{M})$. *Then* $\uparrow(FG) \in \mathrm{Fil}(\mathbf{M})$ *and* $\downarrow(I + J) \in \mathrm{Id}(\mathbf{M})$.

Proof. Since $\uparrow(FG)$ is upward closed by definition, we only need to show that it is closed under meet. Take $a, b \in M$ such that for some $f_1, f_2 \in F$ and $g_1, g_2 \in G$ we have $f_1 g_1 \leq a$ and $f_2 g_2 \leq b$. Then, $f_1 g_1 \wedge f_2 g_2 \leq a \wedge b$ and by Lemma 4.4(8) we get $f_1 g_1 f_2 g_2 \leq a \wedge b$. Then, by commutativity of multiplication we conclude $f_1 f_2 g_1 g_2 \leq a \wedge b$, and so taking $f = f_1 f_2$ and $g = g_1 g_2$ we obtain $f \in F$, $g \in G$ with $fg \leq a \wedge b$, proving that $a \wedge b \in \uparrow(FG)$. The statement for ideals follows by duality. $\square$

It is well-known that in a distributive lattice every filter not containing an element can be extended to a prime one with this property. We will now establish two more extendability properties of prime filters of De Morgan monoids. The first two lemmas below set the stage with some preliminary observations; the third lemma states the properties we need.

Lemma 5.2. *Let* $\mathbf{M}$ *be a subdirectly irreducible De Morgan monoid and* $X, Y \subseteq M$ *be upward closed. Then* $\uparrow(XY) \cap \downarrow(\overline{X} + \overline{Y}) = \emptyset$

Proof. Suppose the contrary. Then there exist $x \in X$, $y \in Y$ and $a \notin X$, $b \notin Y$ such that $xy \leq a + b$. By Lemma 4.9 then $x \leq a$ or $y \leq b$, so $a \in X$ or $b \in Y$. Contradiction. $\square$

Lemma 5.3. *Let* **M** *be a De Morgan monoid,* $P \in \mathrm{Fil}(\mathbf{M})$ *be prime, and* $a, b \in M$ *with* $ab \in P$. *Then, there exist* $c, d \in M$ *such that* $c \not\geq a$, $d \not\geq b$, *and* $\downarrow c + \downarrow d \subseteq \overline{P}$.

Proof. Take any $u \notin P$ and put $c = d = a \wedge b \wedge u$. Then, $c \geq a$ implies $u \geq a$ implies $u \in P$: a contradiction. So, $c \not\geq a$ and similarly $d \not\geq b$. Since $c, d \in \overline{P}$ and $\overline{P}$ is an ideal, $c + d \in \overline{P}$. Thus, $\downarrow c + \downarrow d \subseteq \overline{P}$, as required. $\square$

Lemma 5.4. *Let* **M** *be a subdirectly irreducible De Morgan monoid and* $a, b \in M$. *Let* $F, G, H, P \in \mathrm{Fil}(\mathbf{M})$ *be prime. The following hold:*

(1) If $FGH \subseteq P$ *and* $\overline{F} + \overline{G} + \overline{H} \subseteq \overline{P}$, *then there exists a prime filter* R *such that* $GH \subseteq R$, $\overline{G} + \overline{H} \subseteq \overline{R}$, $FR \subseteq P$, *and* $\overline{F} + \overline{R} \subseteq \overline{P}$.
(2) If $FGH \subseteq P$ *and* $\overline{F} + \overline{G} + \overline{H} \subseteq \overline{P}$, *then there exists a prime filter* Q *such that* $FG \subseteq Q$, $\overline{F} + \overline{G} \subseteq \overline{Q}$, $QH \subseteq P$, *and* $\overline{Q} + \overline{H} \subseteq \overline{P}$.
(3) If $ab \in P$, *then there exist prime filters* Q *and* R *such that* $a \in Q$, $b \in R$, $QR \subseteq P$, *and* $\overline{Q} + \overline{R} \subseteq \overline{P}$.

Proof. For (1) suppose $FGH \subseteq P$ and $\overline{F} + \overline{G} + \overline{H} \subseteq \overline{P}$, and consider the set S of pairs $\langle Z, W \rangle \in \mathrm{Fil}(\mathbf{M}) \times \mathrm{Id}(\mathbf{M})$ satisfying the following conditions:

- $Z \cap W = \emptyset$
- $GH \subseteq Z$, $FZ \subseteq P$
- $\overline{G} + \overline{H} \subseteq W$, $\overline{F} + W \subseteq \overline{P}$

Observe that the pair $\langle \uparrow(GH), \downarrow(\overline{G} + \overline{H}) \rangle$ satisfies the conditions above. For we have $\uparrow(GH) \cap \downarrow(\overline{G} + \overline{H}) = \emptyset$ by Lemma 5.2, $GH \subseteq \uparrow(GH)$ trivially, and $F \cdot \uparrow(GH) \subseteq P$ follows from the assumption $FGH \subseteq P$. To see that, consider an arbitrary $fu \in F \cdot \uparrow(GH)$. Then, there are $g \in G, h \in H$ with $gh \leq u$ and so $fgh \leq fu$. Now, $fgh \in P$ by assumption, implying $fu \in P$, since P is a filter. Similarly, $\overline{G} + \overline{H} \subseteq \downarrow(\overline{G} + \overline{H})$, and $\overline{F} + \downarrow(\overline{G} + \overline{H}) \subseteq \overline{P}$ follows from $\overline{F} + \overline{G} + \overline{H} \subseteq \overline{P}$. Thus S is nonempty. It is not difficult to show that S is closed under unions of chains, so it contains a maximal disjoint pair $\langle R, R' \rangle$. Since M is a distributive lattice, R is a prime filter and $R' = \overline{R}$, proving (1). Then, (2) follows by symmetry.

For (3) we use a similar argument in two steps. First, pick elements c, d with $c \not\geq a$, $d \not\geq b$, and $\downarrow c + \downarrow d \subseteq \overline{P}$. Such elements exist by Lemma 5.3. Next, consider the set S_1 of pairs $\langle Z, W \rangle \in \mathrm{Fil}(\mathbf{M}) \times \mathrm{Id}(\mathbf{M})$ satisfying the following conditions:

- $Z \cap W = \emptyset$
- $a \in Z$, $Z \cdot \uparrow b \subseteq P$

- $c \in W$, $W + {\downarrow}d \subseteq \overline{P}$

By Lemma 5.3 again, S_1 is nonempty: the pair $\langle {\uparrow}a, {\downarrow}c \rangle$ satisfies the conditions. As before, it is straightforward to show that S_1 is closed under unions of chains. A maximal element of S_1, then, is of the form $\langle Q, \overline{Q} \rangle$ for some prime filter Q, such that $a \in Q$, $Q \cdot {\uparrow}b \subseteq P$, $c \in \overline{Q}$ and $\overline{Q} + {\downarrow}d \subseteq \overline{P}$. Now, consider the set S_2 of pairs $\langle Z, W \rangle \in \mathrm{Fil}(\mathbf{M}) \times \mathrm{Id}(\mathbf{M})$ satisfying the conditions:

- $Z \cap W = \emptyset$
- $QZ \subseteq P$
- $\overline{Q} + W \subseteq \overline{P}$

S_2 is nonempty, and closed under unions of chains. Clearly, $b \in Z$ for every pair $\langle Z, W \rangle \in S_2$. Any maximal element of S_2 is then a pair $\langle R, \overline{R} \rangle$ for some prime filter R, such that $a \in Q$, $b \in R$, $QR \subseteq P$ and $\overline{Q} + \overline{R} \subseteq \overline{P}$. Thus, Q and R satisfy all the requirements of (3). □

Lemma 5.5. *Let $\mathbf{M}$ be a normal De Morgan monoid. Then, ${\uparrow}1$ is a prime filter of $\mathbf{M}$ and ${\downarrow}0$ is a prime ideal of $\mathbf{M}$.*

Proof. Immediate from Lemma 4.5(3) and duality. □

Lemma 5.6. *Let $\mathbf{M}$ be a De Morgan monoid, F a prime filter on $\mathbf{M}$, and I a prime ideal in $\mathbf{M}$. Then, the following hold:*

(1) $\sim F = \{\sim a \in M : a \in F\}$ is a prime ideal.
(2) $\overline{F} = \{a \in F : a \notin F\}$ is a prime ideal.
(3) $\sim I = \{\sim a \in M : a \in I\}$ is a prime filter.
(4) $\overline{I} = \{a \in F : a \notin I\}$ is a prime filter.
(5) $\overline{\sim F} = \sim \overline{F}$ is a prime filter.
(6) $\overline{\sim I} = \sim \overline{I}$ is a prime ideal.

Proof. All these are well-known, so we only sketch the proofs. Thus, (1) follows by Lemmas 4.2 and 4.3. For (2), downward closedness of $\overline{F}$ is immediate, closure under join follows from primeness of F, and primeness follows from closure of F under meet. Further, (3) is dual to (1) and (4) is dual to (2). Now, (5) differs from the previous statements in that the equality in it is not definitional. To prove it holds, observe that $a \in \overline{\sim F}$ iff $a \notin \sim F$ iff $\sim a \notin F$ iff $\sim a \in \overline{F}$ iff $a \in \sim \overline{F}$. Then, combining (1) and (2) we have that both $\overline{\sim F}$ and $\sim \overline{F}$ are prime filters, proving (5). Finally, (6) follows from (5) by duality. □

6. Relation algebras from normal De Morgan monoids

Let $\mathbf{M}$ be a normal De Morgan monoid, and $\mathcal{F}$ be the set of all prime filters of $\mathbf{M}$. We will keep these fixed throughout this section. Consider $U = \wp(\mathcal{F})$ as (a universe of) a Boolean algebra under the standard set-theoretical operations $\cup$, $\cap$ and $-$. The atoms of U are of the form $\{F\}$, for $F \in \mathcal{F}$. This set will be denoted by $\mathrm{At}(U)$. For reasons that will become clear shortly, we call $\{\uparrow 1\}$ the *identity atom*, and all the others *diversity atoms*. We define a partial multiplication $\circ$ on U, putting

$$\{F\} \circ \{G\} = \{P \in \mathcal{F} : FG \subseteq P, \ \overline{F} + \overline{G} \subseteq \overline{P}\}.$$

It follows from Lemma 5.2 that $\{F\} \circ \{G\}$ is nonempty for any $F, G \in \mathcal{F}$. An easy calculation shows that $\{\uparrow 1\} \circ \{F\} = \{F\} = \{F\} \circ \{\uparrow 1\}$, so $\{\uparrow 1\}$ acts as an identity for $\circ$. Defining

$$u \circ w = \bigcup \{a \circ b : a, b \in \mathrm{At}(U), a \leq u, b \leq w\}$$

we extend multiplication onto the whole U. Observe that $u \circ \emptyset = \emptyset = \emptyset \circ u$ holds for all $u \in U$.

Lemma 6.1. *Multiplication on U defined as above is associative.*

Proof. It suffices to prove that $(a \circ b) \circ c = a \circ (b \circ c)$ holds for all atoms a, b, c. Let $a = \{F\}$, $b = \{G\}$ and $c = \{H\}$, for some $F, G, H \in \mathcal{F}$. We have

$$(\{F\} \circ \{G\}) \circ \{H\} = \{P \in \mathcal{F} : FG \subseteq P, \ \overline{F} + \overline{G} \subseteq \overline{P}\} \circ \{H\}$$
$$= \{R \in \mathcal{F} : PH \subseteq R, \ \overline{P} + \overline{H} \subseteq \overline{R}, \ FG \subseteq P, \ \overline{F} + \overline{G} \subseteq \overline{P}\}$$

It follows that $FGH \subseteq R$ and $\overline{F} + \overline{G} + \overline{H} \subseteq \overline{R}$ hold for any prime filter R with $R \in (\{F\} \circ \{G\}) \circ \{H\}$. Take any such prime filter R_0. By Lemma 5.4, there is a prime filter Q such that

$$GH \subseteq Q, \quad \overline{G} + \overline{H} \subseteq \overline{Q}, \quad FQ \subseteq R_0, \quad \overline{F} + \overline{Q} \subseteq \overline{R_0}$$

all hold. Therefore

$$R_0 \in \{R \in \mathcal{F} : FQ \subseteq R, \ \overline{F} + \overline{Q} \subseteq \overline{R}, \ GH \subseteq Q, \ \overline{G} + \overline{H} \subseteq \overline{Q}\}$$

But since we also have

$$\{F\} \circ (\{G\} \circ \{H\}) = \{F\} \circ \{Q \in \mathcal{F} : GH \subseteq Q, \ \overline{G} + \overline{H} \subseteq \overline{Q}\}$$
$$= \{R \in \mathcal{F} : FQ \subseteq R, \ \overline{F} + \overline{Q} \subseteq \overline{R}, \ GH \subseteq Q, \ \overline{G} + \overline{H} \subseteq \overline{Q}\}$$

it follows that $R_0 \in \{F\} \circ (\{G\} \circ \{H\})$, proving

$$(\{F\} \circ \{G\}) \circ \{H\} \subseteq \{F\} \circ (\{G\} \circ \{H\}).$$

The reverse inclusion is proved analogously. □

Next, we define a unary operation $\breve{\ }$ on U putting

$$\{F\}^{\breve{}} = \{\overline{\sim F}\}$$

for the atoms, and then setting

$$u^{\breve{}} = \bigcup \{a^{\breve{}} : a \in \mathrm{At}(U), \ a \leq u\}$$

to extend it to the whole U. Observe that $\emptyset^{\breve{}} = \emptyset$ holds.

Lemma 6.2. *For multiplication and converse defined as above, the following hold.*

(1) $a^{\breve{}\breve{}} = a$
(2) $(a \circ b)^{\breve{}} = b^{\breve{}} \circ a^{\breve{}}$
(3) $(a \cup b)^{\breve{}} = a^{\breve{}} \cup b^{\breve{}}$
(4) $a \circ (b \cup c) = a \circ b \cup a \circ c$
(5) $(a \cup b) \circ c = a \circ c \cup b \circ c$
(6) $(\neg a)^{\breve{}} = \neg(a^{\breve{}})$
(7) $a \circ b \cap c = 0$ *iff* $a^{\breve{}} \circ c \cap b = 0$ *iff* $b \circ c^{\breve{}} \cap a = 0$, *i.e., the triangle laws.*

Proof. Again, it suffices to verify these statements for atoms. Let $a = \{F\}$, $b = \{G\}$, $c = \{H\}$ for some $F, G, H \in \mathcal{F}$. Then (1) amounts to $\overline{\sim\overline{\sim F}} = \overline{\sim F}$, which is immediate by Lemma 5.6(5), and (2) holds by definition.

The proofs of both triangle laws are similar, so we only prove the first. Observe that if a, b, c are atoms, the contrapositive of the desired equivalence is

$$a \circ b \geq c \quad \text{iff} \quad a^{\breve{}} \circ c \geq b$$

which is the form we will use in the proof. Suppose $\{F\} \circ \{G\} \supseteq \{H\}$. Then, by definition of multiplication, we have $FG \subseteq H$ and $\overline{F} + \overline{G} \subseteq \overline{H}$. To get a contradiction, suppose $\{F\}^{\breve{}} \circ \{H\} \not\supseteq \{G\}$. This means that

$$\overline{\sim F} \cdot H \not\subseteq G \quad \text{or} \quad \sim F + \overline{H} \not\subseteq \overline{G}.$$

If the former, then for some $x \notin \sim F$ and some $y \in H$ we have $xy \notin G$. Now, $\sim x \notin F$, and by Lemma 4.4(13) we have $\sim x + xy \geq y$, so $\sim x + xy \in H$, contradicting $\overline{F} + \overline{G} \subseteq \overline{H}$. If the latter, then for some $x \in \sim F$ and some $y \notin H$ we have $x + y \in G$. Now, $\sim x \in F$ and by Lemma 4.4(14) we have $\sim x(x + y) \leq y$, so $\sim x(x + y) \notin H$, contradicting $FG \subseteq H$. □

Let $\mathbf{U}$ be the algebra $\langle U, \cup, \cap, \circ, \check{\ }, -, 1', 0, 1 \rangle$, where $\cup$, $\cap$, $-$ are the usual set-theoretical operations on $U = \wp(\mathcal{F})$, and $1' = \{\uparrow 1\}$, $0 = \emptyset$, $1 = \mathcal{F}$.

Theorem 6.1. *The algebra* $\mathbf{U}$ *is a square-increasing, commutative, integral relation algebra.*

Proof. Lemmas 6.1 and 6.2 together show that $\mathbf{U}$ is a relation algebra, and since the identity element of $\mathbf{U}$ is an atom by definition, $\mathbf{U}$ is integral. Commutativity follows from the fact that $FG = GF$ and $\overline{F} + \overline{G} = \overline{G} + \overline{F}$ hold for any prime filters F, G. Finally, to prove that $\mathbf{U}$ is square-increasing, it suffices to observe that for any prime filter F we have $FF \subseteq F$, and $\overline{F} + \overline{F} \subseteq \overline{F}$, which in turn follow easily by Lemma 4.4(8) and (9). $\qquad\square$

7. The embedding

For any normal subdirectly irreducible De Morgan monoid $\mathbf{M}$, we let $\mathbf{U_M}$ stand for the relation algebra defined in the previous section. Consider a map $\varepsilon\colon M \to U_{\mathbf{M}}$ defined by

$$\varepsilon(a) = \{F \in \mathcal{F} \colon a \in F\}$$

Lemma 7.1. *For any* $a \in M$, *the following hold:*

(1) $\varepsilon(\sim a) = -(\varepsilon(a))^{\check{\ }}$
(2) $\varepsilon(a \vee b) = \varepsilon(a) \vee \varepsilon(b)$
(3) $\varepsilon(a \wedge b) = \varepsilon(a) \wedge \varepsilon(b)$
(4) $\varepsilon(ab) = \varepsilon(a) \circ \varepsilon(b)$
(5) $\varepsilon(a \to b) = -\big(-\varepsilon(b) \circ (\varepsilon(a))^{\check{\ }}\big) = -\big((\varepsilon(a))^{\check{\ }} \circ -\varepsilon(b)\big)$

where ε *is the map defined above.*

Proof. For (1), take $a \in M$. We have

$$\begin{aligned}
\varepsilon(\sim a) &= \{F \in \mathcal{F} \colon \sim a \in F\} \\
&= -\{F \in \mathcal{F} \colon \sim a \notin F\} \\
&= -\{F \in \mathcal{F} \colon \sim a \in \overline{F}\} \\
&= -\{\overline{\sim F} \in \mathcal{F} \colon a \in F\} \\
&= -\{F \in \mathcal{F} \colon a \in F\}^{\check{\ }} \\
&= -(\varepsilon(a))^{\check{\ }}
\end{aligned}$$

as required. For (2) and (3) the claim follows from the usual representation of distributive lattices. For (4), we first calculate

$$\varepsilon(a) \circ \varepsilon(b) = \{F \in \mathcal{F} \colon a \in F\} \circ \{G \in \mathcal{F} \colon b \in G\}$$
$$= \{P \in \mathcal{F} \colon a \in F, \ b \in G, \ FG \subseteq P, \ \overline{F} + \overline{G} \subseteq P\}$$
$$\subseteq \{P \in \mathcal{F} \colon ab \in P\}$$
$$= \varepsilon(ab).$$

For the other inclusion, suppose $ab \in P_0$ for some $P_0 \in \mathcal{F}$. Then, by Lemma 5.4 there are prime filters F and G such that

$$a \in F, \quad b \in G, \quad FG \subseteq P_0, \quad \overline{F} + \overline{G} \subseteq \overline{P_0}$$

all hold. Therefore,

$$P_0 \in \{P \in \mathcal{F} \colon a \in F, \ b \in G, \ FG \subseteq P, \ \overline{F} + \overline{G} \subseteq P\}$$

and so P_0 belongs to $\varepsilon(a) \circ \varepsilon(b)$ as desired. Finally, (5) follows from (1) and (4) together with commutativity and the identity $x \to y = \sim(x \cdot \sim y)$, which, as Lemma 4.4 states, holds in all De Morgan monoids. $\square$

Theorem 7.1. *Every normal De Morgan monoid is embeddable as a bare De Morgan monoid into a square-increasing, commutative, integral relation algebra.*

Proof. Let $\mathbf{M}$ be a normal De Morgan monoid, $\mathbf{U_M}$ the relation algebra defined in Section 6, and ε the map defined in the present section. By definition of ε, it is an embedding of the lattice reduct of $\mathbf{M}$ into the lattice reduct of $\mathbf{U_M}$; in particular, ε is injective. Lemma 7.1 show that the multiplication, implication and De Morgan negation are preserved as well, so ε is an embedding of the constant-free reduct of $\mathbf{M}$ into the reduct of $\mathbf{U_M}$ to the type $\langle \cup, \cap, \circ, \to, \rangle$, where $\to$ is defined by $x \to y = -(a^{\smile} \circ -b)$. $\square$

8. Conclusions

From the embedding defined in the previous section the following result follows immediately.

Theorem 8.1. *The relevant logic $\mathbf{R}$ is sound and complete with respect to square-increasing, commutative, integral relation algebras.*

Observe however, that the *absence* of the Ackermann constant is crucial for the result. To see that, consider the algebraic counterpart of t in a normal De Morgan monoid $\mathbf{M}$, that is, the unit element $1 \in M$. Now, the embedding $\varepsilon \colon \mathbf{M} \to \mathbf{U_M}$ takes 1 to $\varepsilon(1) = \{F \in \mathcal{F} \colon 1 \in F\} \neq \{\uparrow 1\}$. Moreover, the element $\varepsilon(1)$ is not in general definable using only the operations of $\mathbf{U_M}$.

Several well-known results about $\mathbf{R}$ can be given new, simple proofs using the embedding. For instance, it immediately follows that adding Boolean negation to $\mathbf{R}$ results in a conservative extension, a fact well-known in the folklore of relevant logic. Moreover, we obtain an immediate proof of the celebrated γ admissibility. In the argot of relevant logicians, γ stands for the rule of inference

$$\frac{\alpha \quad \sim\alpha \vee \beta}{\beta}$$

which is not derivable in $\mathbf{R}$. One of major problems in the early stages of development of relevant logic was whether γ is *admissible* in $\mathbf{R}$. An inference rule I is said to be admissible in a logic $\mathbf{L}$ if the set of theorems of $\mathbf{L}$ is closed under I (intuitively, adding I to $\mathbf{L}$ does not produce new theorems). Admissibility of γ was first proved by Meyer and Dunn [22], and a number of other proofs have been published. Dunn and Restall [12] state that "at least four versions" exist. Here is a fifth proof.

Theorem 8.2. *The inference rule γ is admissible in $\mathbf{R}$.*

Proof. Let α and $\sim\alpha \vee \beta$ be theorems of $\mathbf{R}$. Suppose β is not a theorem of $\mathbf{R}$. Then, by Theorem 8.1 there is a square-increasing, commutative, integral relation algebra $\mathbf{A}$, and a valuation v such that $v(\alpha) \geq e$, $v(\sim\alpha \vee \beta) = \sim v(\alpha) \vee v(\beta) \geq e$ and $v(\beta) \not\geq e$ all hold in $\mathbf{A}$. Let $a = v(\alpha)$ and $b = v(\beta)$. Since $\sim a$ is interpreted in $\mathbf{A}$ as $\neg a^{\smile}$, we get $\neg a^{\smile} \vee b \geq e$ and $a \geq e$. Moreover, e is an atom, so $\neg a^{\smile} \vee b \geq e$ implies that $\neg a^{\smile} \geq e$ or $b \geq e$ holds. By monotonicity of $^{\smile}$ and the fact that $e^{\smile} = e$ we obtain $a^{\smile} \geq e$. Therefore, $\neg a^{\smile} \leq \neg e$, so $\neg a^{\smile} \not\geq e$. Thus, $e \leq b = v(\beta)$, a contradiction. $\square$

Quite apart from new proofs of old theorems, the embedding construction presented here seems to be applicable to all varieties of De Morgan monoids which are generated by their normal members. Exploring this further may lead to completeness results for extensions (and perhaps also weakenings) of $\mathbf{R}$. Moreover, a general possibility of transferring methods and results between relevant logics and relation algebras seems to be worth looking into.

It would also be of considerable interest to obtain a characterisation of these De Morgan monoids for which the embedding construction produces a *representable* relation algebra. Is this class the same as the class of De Morgan monoids embeddable (this way or another) into representable relation algebras?

References

1. R. D. Maddux, Relevance logic and the calculus of relations [abstract] (2007), International Conference on Order, Algebra and Logics, Vanderbilt University, June 2007. Available from http://www.math.vanderbilt.edu/~oal2007/viewabstracts.php.
2. S. Mikulás, Algebras of relations and relevance logic, *Journal of Logic and Computation* **19**, 305 (2009).
3. R. D. Maddux, Relevance logic and the calculus of relations, *Review of Symbolic Logic* **3**, 41 (2010).
4. K. Bimbó, J. M. Dunn and R. D. Maddux, Relevance logic and relation algebras, *Review of Symbolic Logic* **2**, 102 (2009).
5. R. Hirsch and S. Mikulás, Positive fragments of relevance logic and algebras of binary relations, *Review of Symbolic Logic* **4**, 81 (2011).
6. T. Kowalski, Weakly associative relation algebras hold the key to the universe, *Bulletin of the Section of Logic* **36**, 145 (2007).
7. R. K. Meyer, Ternary relations and relevant semantics, *Annals of Pure and Applied Logic* **127**, 195 (2004).
8. B. A. Davey and H. A. Priestley, *Introduction to Lattices and Order*, 2nd edn. (Cambridge University Press, 2002).
9. N. Galatos, P. Jipsen, T. Kowalski and H. Ono, *Residuated Lattices. An Algebraic Glimpse at Substructural Logics* (Elsevier, 2007).
10. R. D. Maddux, *Relation Algebras* (Elsevier, 2006).
11. R. Hirsch and I. Hodkinson, *Relation Algebras by Games* (Elsevier, 2002).
12. J. M. Dunn and G. Restall, *Relevance Logic*, in *The Handbook of Philosophical Logic*, eds. D. Gabbay and F. Guanther (Kluwer, 2002), pp. 1–136, 2nd edn.
13. A. R. Anderson and N. D. Belnap, *Entailment. The Logic of Relevance and Necessity, Vol. I* (Princeton University Press, 1975).
14. A. R. Anderson, N. D. Belnap and J. M. Dunn, *Entailment. The Logic of Relevance and Necessity, Vol. II* (Princeton University Press, 1992).
15. R. Routley and R. K. Meyer, The semantics of entailment (I), in *Truth, Syntax and Modality*, ed. H. Leblanc (North-Holland, 1973) pp. 199–243.
16. J. M. Dunn, The Algebra of Intensional Logics, PhD thesis, University of Pittsburgh, (Penn., USA, 1966).
17. R. K. Meyer and R. Routley, Algebraic analysis of entailment I, *Logique et Analyse* **15**, 407 (1972).
18. L. L. Maksimova, On models of the system E (in Russian), *Algebra i Logika* **6**, 5 (1967).
19. L. L. Maksimova, An interpolation and separation theorem for the logical systems E and R, *Algebra and Logic* **10**, 232 (1971).
20. J. K. Slaney, On the structure of De Morgan monoids with corollaries on

relevant logic and theories, *Notre Dame Journal of Formal Logic* **30**, 117 (1989).

21. R. K. Meyer, J. M. Dunn and H. Leblanc, Completeness of relevant quantification theories, *Notre Dame Journal of Formal Logic* **15**, 97 (1974).

22. R. K. Meyer and J. M. Dunn, E, R and γ, *Journal of Symbolic Logic* **34**, 460 (1969).

Van Lambalgen's theorem for uniformly relative Schnorr and computable randomness

K. MIYABE

Research Institute for Mathematical Sciences, Kyoto University,
Kyoto 606-8502 JAPAN
E-mail: kmiyabe@kurims.kyoto-u.ac.jp
www.kurims.kyoto-u.ac.jp/~kmiyabe/

J. RUTE

Department of Mathematical Sciences, Carnegie Mellon University,
Pittsburgh, PA 15217, USA
E-mail: jason.rute@andrew.cmu.edu
www.math.cmu.edu/~jrute

We correct Miyabe's proof of van Lambalgen's theorem for truth-table Schnorr randomness (which we will call uniformly relative Schnorr randomness). An immediate corollary is one direction of van Lambalgen's theorem for Schnorr randomness. It has been claimed in the literature that this corollary (and the analogous result for computable randomness) is a "straightforward modification of the proof of van Lambalgen's theorem." This is not so, and we point out why. We also point out an error in Miyabe's proof of van Lambalgen's theorem for truth-table reducible randomness (which we will call uniformly relative computable randomness). While we do not fix the error, we do prove a weaker version of van Lambalgen's theorem where each half is computably random uniformly relative to the other. We also argue that uniform relativization is the correct relativization for all randomness notions.

Keywords: van Lambalgen's theorem; uniformly relative Schnorr randomness; uniformly relative computable randomness; truth-table Schnorr randomness; truth-table reducible randomness.

1. Introduction

Recall van Lambalgen's theorem.

Theorem 1.1 (van Lambalgen [1]). $A \oplus B$ *is Martin-Löf random if and only if* A *is Martin-Löf random and* B *is Martin-Löf random relative to* A.

Merkle et al. [2] showed that the "⇒" direction of van Lambalgen's theorem does not hold for Schnorr or computable randomness. This has been extended by Yu [3], Kjos Hanssen [4, Remark 3.5.22], Franklin and Stephan [5], and Miyabe [6].

In [6] the first author claimed that van Lambalgen's theorem does in fact hold for Schnorr randomness if the usual notion of relativized Schnorr randomness is replaced with the weaker notion of uniformly relative Schnorr randomness—previously called truth-table Schnorr randomness in [7] and [6].

Theorem 1.2. $A \oplus B$ *is Schnorr random if and only if A is Schnorr random and B is Schnorr random uniformly relative to A.*

However, the proof given was incorrect and we provide a corrected proof. Our proof follows a standard proof of van Lambalgen's theorem using integral tests, except at the difficult point we apply a key lemma, which can be seen as an effective version of Lusin's theorem for a particular setting. Lusin's theorem, one of Littlewood's three basic principles of measure theory, is the basis behind the layerwise-computability framework that has been successively employed by Hoyrup, Rojas and others to relate algorithmic randomness and computable analysis.

The structure of the paper is as follows. In Section 2, we prove the key lemma and some corollaries.

In Section 3, as a warm-up, we show how our key lemma can be used to prove the "⇐" direction of van Lambalgen's theorem for Schnorr randomness. (A different proof was given recently by Franklin and Stephan [5].) Yu [3] had claimed that "the [⇐] direction of van Lambalgen's theorem is true for both Schnorr randomness and computable randomness. [...] The proof is just a straightforward modification of the proof of van Lambalgen's theorem." Downey and Hirschfelt [8] had made similar claims. Unfortunately, the proofs are not so straightforward, and we explain why in the case of Schnorr randomness.

In Section 4, we define uniformly relative Schnorr randomness, and prove van Lambalgen's theorem for this notion of randomness.

In Section 5, we discuss uniformly relative computable randomness—previously called truth-table reducible randomness. We remark that Miyabe's [6] proof of van Lambalgen's theorem for uniformly relative computable randomness is not correct in the "⇐" direction. While we do not provide a correction, we do prove the following weaker result.

Theorem 1.3. $A \oplus B$ *is computably random if and only if each of A and B are computably random uniformly relative to the other.*

This weakening of van Lambalgen's theorem is also known to hold for Kolmogorov-Loveland randomness. We leave as an open question the "$\Leftarrow$" direction of van Lambalgen's theorem for both computable randomness and uniformly relative computable randomness. We conjecture that it is false for both.

Finally, in Section 6 we prove that Franklin and Stephan's [7] truth-table Schnorr randomness is equivalent to our uniformly relative Schnorr randomness. The difference in terminology reflects the difference in definitions, and we discuss why the Franklin and Stephan definition, which uses truth-table reducibility, is very sensitive to the choice of test used.

We believe this paper gives a strong argument that uniform relativization (as in Definition 4.1) is the correct method to relativize a randomness notion (in contrast to the usual method of relativization). It is a natural definition that can be applied to all the standard randomness notions. Moreover, uniformly relative Martin-Löf randomness is equivalent to the usual relative Martin-Löf randomness (Section 5.1). Uniformly relative Schnorr randomness not only satisfies van Lambalgen's theorem, but also has well-behaved lowness properties [7]. Furthermore, it is not difficult to see that uniformly relative Demuth randomness is equivalent to Demuth$_{\mathrm{BLR}}$ randomness (see [9], [10]), which also satisfies van Lambalgen's theorem [10] and has natural lowness properties [9]. Indeed we suggest that if one wishes to explore either van Lambalgen's theorem or low-for-randomness with respect to other randomness notions, one should use uniform relativization.

2. The key lemma

For this paper, we will work in 2^ω with the fair-coin measure μ. Recall the following definitions. The reader may wish to consult the books [4,8] for further background.

Definition 2.1. A set $U \subseteq 2^\omega$ is *open* if it is a countable union of *basic open sets*, i.e. sets of the form $[\sigma] := \{X \in 2^\omega : X \succ \sigma\}$ for some $\sigma \in 2^{<\omega}$ as well as the empty subset $\varnothing \subset 2^\omega$. A *code* for an open set $U \subseteq 2^\omega$ is a sequence $\langle C_s \rangle_{s \in \mathbb{N}}$ of basic open sets such that $U = \bigcup_s C_s$. A set $U \subseteq 2^\omega$ is Σ_1^0, or *effectively open*, if it is open with a computable code.

Definition 2.2. A *Martin-Löf test* is a uniform sequence $\langle U_n \rangle$ of Σ_1^0 subsets of 2^ω such that $\mu(U_n) \leq 2^{-n}$. A *Schnorr test* is a Martin-Löf test

$\langle U_n \rangle$ such that $\mu(U_n)$ is uniformly computable in n. A set $X \in 2^\omega$ is said to be **covered by** a Martin-Löf (Schnorr) test $\langle U_n \rangle$ if $X \in \bigcap_n U_n$. The set $X \in 2^\omega$ is said to be **Martin-Löf** (resp. **Schnorr**) **random** if X is not covered by any Martin-Löf (resp. Schnorr) test.

Definition 2.3. A function $f \colon 2^\omega \to [0, \infty]$ is **lower semicomputable** if there is a uniform sequence of total computable functions $g_n \colon 2^\omega \to [0, \infty)$ such that $f = \sum_n g_n$.

A more standard definition of lower semicomputable is that f is lower semicomputable if $f(X)$ is uniformly lower semicomputable (left c.e.) from X. Our definition is easily seen to be equivalent. (This even remains true when 2^ω is replaced with the unit interval or another computable Polish space.)

Recall the following definitions.

Definition 2.4. A function $f \colon 2^\omega \to \mathbb{R}$ is **L^1-computable** if there is a uniformly computable sequence of bounded computable functions $\langle g_n \rangle$ such that $\|f - g_n\|_{L^1} \leq 2^{-n}$. (Since 2^ω is compact, all computable functions are bounded.)

The **distribution** of a function $f \colon 2^\omega \to \mathbb{R}$ is the probability measure ν on $\mathbb{R}$ defined by $\nu(A) = \mu(\{X \in 2^\omega : f(X) \in A\})$ for all Borel sets $A \subseteq \mathbb{R}$. This is also known as the **pushforward of the fair-coin measure along** f.

A distribution ν is **computable** if $\nu(U)$ is lower semicomputable (left c.e.) uniformly from any code for an open set $U \subseteq \mathbb{R}$.[a] (A code for an open set $U \subseteq \mathbb{R}$ is the same as in Definition 2.1, except that the basic open sets are open intervals with rational endpoints.)

We could not find a direct proof of this next fact, so we give one here.

Proposition 2.1. Let $f \colon 2^\omega \to \mathbb{R}$ be an L^1-computable function with distribution ν. Then ν is a computable distribution.

Proof. It is enough to prove that

$$\nu((a - r, a + r)) = \mu(\{X \in 2^\omega : |f(X) - a| < r\})$$

---

[a]This definition is equivalent to ν being a computable point in the Lévy-Prokhorov metric [11]. It is also equivalent to the map $f \mapsto \int f \, d\nu$ being a computable operator on bounded continuous functions f [12].

is lower semicomputable from a, r. (An open set $U \subseteq \mathbb{R}$ is encoded as a union of countably many rational intervals. To lower semicompute $\nu(U)$ it is enough to lower semicompute the measure of each finite subunion of intervals. A finite union of intervals can be made a disjoint union by joining overlapping intervals.)

If f is computable, then we are done, since $\{X \in 2^\omega : |f(X) - a| < r\}$ is Σ_1^0 relative to a and r.

Otherwise, we know

$$\mu(\{X \in 2^\omega : |f(X) - a| < r\})$$
$$= \sup_{\varepsilon > 0} \mu(\{X \in 2^\omega : |f(X) - a| < r - \varepsilon\}). \tag{1}$$

For any $\varepsilon > 0$ and $\delta > 0$ we can effectively approximate f by some computable function g such that $\|f - g\|_{L^1} < \varepsilon \cdot \delta$. Then by Chebeshev's inequality,

$$\mu(\{X \in 2^\omega : |f(X) - g(X)| \geq \varepsilon\}) \leq (\varepsilon \cdot \delta)/\varepsilon = \delta.$$

So outside a set of measure at most δ, we have $|f(X) - g(X)| < \varepsilon$, and therefore

$$|f(X) - a| < r - 2\varepsilon \quad \Rightarrow \quad |g(X) - a| < r - \varepsilon \quad \Rightarrow \quad |f(X) - a| < r.$$

Expressing this with measures gives us

$$\mu(\{X \in 2^\omega : |f(X) - a| < r\}) \geq \mu(\{X \in 2^\omega : |g(X) - a| < r - \varepsilon\}) - \delta$$
$$\geq \mu(\{X \in 2^\omega : |f(X) - a| < r - 2\varepsilon\}) - 2\delta.$$

Combining this with (1) we get

$$\mu(\{X \in 2^\omega : |f(X) - a| < r\})$$
$$= \sup_{\varepsilon > 0, \delta > 0} \mu(\{X \in 2^\omega : |g(X) - a| < r - \varepsilon\}) - \delta$$

where g depends on ε and δ. Finally, recall that $\mu(\{X \in 2^\omega : |g(X) - a| < r - \varepsilon\})$ is lower semicomputable from a, r. Using this we can approximate $\mu(\{X \in 2^\omega : |f(X) - a| < r\})$ from below. $\square$

The following lemma will be the key to this paper.

Lemma 2.1 (Key lemma). *Let t be a nonnegative lower semicomputable function with a computable integral $\int t \, d\mu$. There is a uniformly computable sequence $\langle h_n \rangle$ of total computable functions $h_n : 2^\omega \to [0, \infty)$ such that $h_n \leq t$ everywhere and if A is Schnorr random, there is some n such that $h_n(A) = t(A)$.*

Proof. Let $\langle g_k \rangle$ be a code for t, namely a sequence of total nonnegative computable functions such that $t = \sum_k g_k$. Find a sequence $\langle f_n \rangle$ of partial sums $f_n = \sum_{k < k_n} g_k$ (where $\langle k_n \rangle$ is increasing) such that $\int (t - f_n) \, d\mu < 2^{-2n}$. (This can be done since $\int g_n \, d\mu$ is uniformly computable from n.) By Chebeshev's inequality, for any $c > 0$,

$$\mu(\{X \in 2^\omega : t(X) - f_n(X) > c\}) \leq 2^{-2n}/c.$$

Moreover, we have this claim.

Claim 2.1. *There is a computable sequence $\langle c_n \rangle$ such that $2^{-n} < c_n < 2^{-(n-1)}$ for each n and $\mu(\{X \in 2^\omega : t(X) - f_n(X) > c_n\})$ is uniformly computable from n.*

Proof of claim. First, note that t is L^1-computable. (Use the sequence $\langle f_n \rangle$ from earlier in the proof.)

Now, fix n. Our goal is to find c_n. Since, t is L^1-computable, so is $t - f_n$. Let ν to be the distribution of $t - f_n$. By Proposition 2.1, ν is a computable distribution.

Hence for any real c,

$$\mu(\{X \in 2^\omega : t(X) - f_n(X) > c\}) = \nu((c, \infty))$$

is lower semicomputable uniformly from c and

$$\mu(\{X \in 2^\omega : t(X) - f_n(X) \geq c\}) = \nu([c, \infty)) = 1 - \nu((-\infty, c))$$

is upper semicomputable uniformly from c.

It is enough to find some c in the desired interval such that

$$\mu(\{X \in 2^\omega : t(X) - f_n(X) = c\}) = \nu(\{c\}) = 0.$$

Indeed, the set of all c such that $\nu(\{c\}) = 0$ is a computable intersection of dense Σ_1^0 sets. (To see this, note that $\nu(\{c\}) = 1 - \nu((-\infty, c) \cup (c, \infty))$ is upper semicomputable uniformly from c. So $\{c \in \mathbb{R} : \nu(\{c\}) < 2^{-n}\}$ is a Σ_1^0 set. This set is also dense since there are at most countable many c such that $\nu(\{c\}) > 0$.) Hence by the effective proof of the Baire category theorem (a basic diagonalization argument, see for example [13]), one can effectively find c_n in the desired interval such that $\mu(\{X \in 2^\omega : t(X) - f_n(X) > c_n\})$ is computable. This proves the claim. $\square$

To define h_m, first set

$$h_n^m = \min\{f_n, f_{n-1} + c_{n-1}, \ldots, f_m + c_m\} \quad \text{(for } n > m).$$

Define $h_m = \sup_{n>m} h_n^m$. For each X and $n > m$ we have $|h_m(X) - h_n^m(X)| \leq c_n$. So $h_m(X)$ is uniformly computable from X and m. Also, $h_m \leq t$ everywhere, and $t(X) > h_m(X)$ if and only if $t(X) > f_n(X) + c_n$ for some $n > m$. Then

$$\{X \in 2^\omega : t(X) > h_m(X)\} = \bigcup_{n>m} \underbrace{\{X \in 2^\omega : t(X) - f_n(X) > c_n\}}_{=:V_n^m} =: U_m.$$

Notice V_n is Σ_1^0 uniformly in n. By the claim and by inequality (2), $\mu(V_n)$ is uniformly computable from n and at most 2^{-n}. Therefore, $\mu(U_m) \leq \sum_{n>m} 2^{-n} = 2^{-m}$ and $\mu(U_m)$ is uniformly computable from m. Hence $\langle U_m \rangle$ is a Schnorr test.

Finally, for any Schnorr random A, there is some m such the $A \notin U_m$. Hence $h_m(A) = t(A)$. This completes the proof of the lemma. □

Remark 2.1. Notice, this proof gives a uniform algorithm for converting the codes for t and $\int t \, d\mu$ into a code for a Schnorr test $\langle U_m \rangle$ such that $t(X)$ is finite when $X \notin \bigcap_m U_m$. However, the uniformity is not (and cannot be) independent of the codes. Indeed even a different code for $\int t \, d\mu$ can change the sequences $\langle g_k \rangle$ and $\langle c_n \rangle$ in the proof leading to a different Schnorr test $\langle U_m \rangle$.

As a corollary, we get another proof of Miyabe's characterization of Schnorr randomness via Schnorr integral tests.

Definition 2.5.

(1) An *integral test* is a lower semicomputable function $t \colon 2^\omega \to [0, \infty]$ such that $\int t \, d\mu < \infty$.
(2) (Miyabe [14]) A *Schnorr integral test* is an integral test t such that $\int t \, d\mu$ is computable.

Proposition 2.2 (See for example [8]). *X is Martin-Löf random if and only if there is no integral test t such that $t(X) = \infty$.*

Corollary 2.1. *If t is a Schnorr integral test and A is Schnorr random, then $t(A)$ is finite and computable from A.*

Proof. By Lemma 2.1, there is some total computable h_n such the $t(A) = h_n(A)$. □

Corollary 2.2 (Miyabe [14]). *X is Schnorr random if and only if there is no Schnorr integral test t such that $t(X) = \infty$.*

Proof. ($\Rightarrow$) Assume A is Schnorr random and t is a Schnorr test. By Lemma 2.1, there is some total computable h_n such the $t(A) = h_n(A) < \infty$. ($\Leftarrow$) Assume A is not Schnorr random by the Schnorr test $\langle U_n \rangle$. Then let $t = \sum_n 1_{U_n}$ where 1_{U_n} is the characteristic function of U_n. □

3. Van Lambalgen's theorem for Schnorr randomness

As a warm up, we use the results from the previous section to give a simple proof of the "$\Leftarrow$" direction of van Lambalgen's theorem for Schnorr randomness. While our proof is simple, we argue that it is not a straightforward modification of van Lambalgen's theorem. (Franklin and Stephan [5] also recently gave a very different proof of this result.)

Theorem 3.1. *If A is Schnorr random and B is Schnorr random relative to A then $A \oplus B$ is Schnorr random.*

Proof. Assume $A \oplus B$ is not Schnorr random and that A is Schnorr random. There is some Schnorr integral test t such that $t(A \oplus B) = \infty$. We wish to show that B is not Schnorr random relative to A. For each X, define $t^X(Y) = t(X \oplus Y)$. Clearly, $t^A(B) = t(A \oplus B) = \infty$.

We will show that t^A is a Schnorr integral test relative to A. For each X, the function t^X is lower semicomputable with a code uniformly computable from X. Define, $u(X) = \int t^X(Y) \, d\mu(Y)$. Notice u is lower semicomputable and by Fubini's theorem

$$\int u(X) \, d\mu(X) = \iint t(X \oplus Y) \, d\mu(Y) \, d\mu(X) = \int t \, d\mu$$

which is computable. It follows that u is a Schnorr integral test. *Since A is Schnorr random, by Corollary 2.1, $u(A)$ is computable from A, and therefore so is $\int t^A(Y) \, d\mu(Y) = u(A)$.* Hence t^A is a Schnorr integral test relative to A, and B is not Schnorr random relative to A. □

Remark 3.1. Notice the emphasized line in the above proof. The key difficulty in adapting the standard (integral test) proof of van Lambalgen's theorem to Schnorr randomness is that for Martin-Löf randomness one need only show that $\int t^A(Y) \, d\mu(Y)$ is *finite* while here we must show it is *computable from A*. The same difficulties exist when trying to adapt a proof using Martin-Löf tests; given a Schnorr test $\langle U_n \rangle$ one must show the measure of $U_n^A := \{Y \in 2^\omega : A \oplus Y \in U_n\}$ is computable from A.

Consider this latter case. One may incorrectly think that $\mu(U_n^A)$ is uniformly computable from A, any code for U_n, and the measure $\mu(U_n)$.[b] This is false, as the following picture shows. The two open sets depicted (on $2^\omega \times 2^\omega$) are almost the same except one contains a small gap. Such a small gap could significantly change the value of $\mu(U_n^A)$. It is impossible to determine in a fixed number of steps whether such a small gap exists.

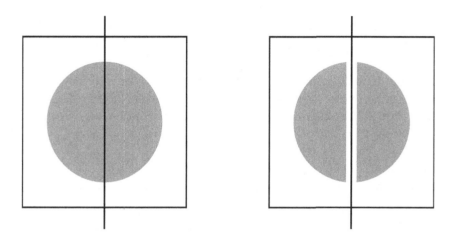

However, the key lemma does show that $\mu(U_n^A)$ is computable from A if A is Schnorr random. One could say such a function is "nearly computable," i.e. it is computable outside an arbitrary small set. This is an effectivization of Littlewood's "three principles" of measure theory, particularly Lusin's theorem which says that a measurable function is "nearly continuous". It is also the basis behind the layerwise-computability framework of Hoyrup and Rojas [15] and its extensions to Schnorr randomness. (See also [14,16,17].)

4. Uniformly relative Schnorr randomness

In this section we give the correction to Miyabe's proof of van Lambalgen's theorem for uniformly relative Schnorr randomness. But first, we define uniformly relative Schnorr randomness.

Recall, that a Schnorr test can be encoded by a function $f \in \mathbb{N}^{\mathbb{N}}$ which encodes a listing of basic open sets for each n which union to U_n (as in Definition 2.1) and also encodes a fast Cauchy sequence of rationals converging

---

[b]Indeed, this was the error in [6] when proving van Lambalgen's theorem for truth-table Schnorr randomness.

to each measure $\mu(U_n)$. (Recall, a sequence of rationals $\langle q_n \rangle$ is **fast-Cauchy** if for all $n \geq m$, we have $|q_n - q_m| \leq 2^{-m}$.)

Definition 4.1. A *uniform Schnorr test* is a total computable map $\Phi \colon 2^\omega \to \mathbb{N}^{\mathbb{N}}$ such that each $\Phi(X)$ encodes a Schnorr test. Also, call a collection $\langle U_n^X \rangle_{n \in \mathbb{N}, X \in 2^\omega}$ a *uniform Schnorr test* if it is given by a map $\Phi \colon 2^\omega \to \mathbb{N}^{\mathbb{N}}$ as above.

Definition 4.2 (Miyabe [6], following Franklin and Stephan [7]). Let $A, B \in 2^\omega$. Say A is **Schnorr random uniformly relative to** B *if there is no* uniform Schnorr test $\langle U_n^X \rangle_{n \in \mathbb{N}, X \in 2^\omega}$ such that $A \in \bigcap_n U_n^B$.

Remark 4.1. Uniformly relative Schnorr randomness has previously been called truth-table Schnorr randomness. This new name better reflects the exact nature of the tests. See Section 6 for more discussion on the differences between the two approaches to relativizing Schnorr randomness.

Also, it is possible to define uniform tests of other types similarly.

Definition 4.3. A *uniform Schnorr integral test* is a total computable map $\Phi \colon 2^\omega \to \mathbb{N}^{\mathbb{N}}$ such that each $\Phi(X)$ encodes a Schnorr integral test. Also, call a collection $\langle t^X \rangle_{X \in 2^\omega}$ a *uniform Schnorr integral test* if it is given by a map $\Phi \colon 2^\omega \to \mathbb{N}^{\mathbb{N}}$ as above.

Proposition 4.1. Let $A, B \in 2^\omega$. Then A is Schnorr random uniformly relative to B if and only if there is no uniform Schnorr integral test $\langle t^X \rangle_{X \in 2^\omega}$ such that $t^B(A) = \infty$.

Proof. Follow the proof of Corollary 2.2. The proof is uniform in that given a code for a test, some code for the other type of test is computable uniformly from the first code (see Remark 2.1). $\square$

Remark 4.2. While the reader should not confuse our definition of uniform test with that of Levin [18], the two ideas are closely related. The main difference is that in Levin's definition the uniform test is indexed by probability measures ξ and the test t^ξ was with respect to the measure ξ (i.e. $\int t^\xi \, d\xi < \infty$).

Now, we prove the main result. The proof will closely follow that of Theorem 3.1, but will use the full power of the key lemma (Lemma 2.1).

Theorem 4.1. $A \oplus B$ is Schnorr random if and only if A is Schnorr random and B is Schnorr random uniformly relative to A.

Proof. ($\Rightarrow$) This direction is correctly proved in [6].

($\Leftarrow$) Assume $A \oplus B$ is not Schnorr random and that A is Schnorr random. There is some Schnorr integral test t such that $t(A \oplus B) = \infty$. We wish to show that B is not Schnorr random uniformly relative to A. For each X, define $t^X(Y) = t(X \oplus Y)$. Clearly, $t^A(B) = t(A \oplus B) = \infty$.

While $\langle t^X \rangle_{X \in 2^\omega}$ may not be a uniform Schnorr integral test, we will construct a uniform Schnorr integral test $\langle \hat{t}^X \rangle_{X \in 2^\omega}$ such that $\hat{t}^A = t^A$. For each X, the function t^X is lower semicomputable with a code uniformly computable from X. Define, $u(X) = \int t^X(Y) \, d\mu(Y)$. Notice u is lower semicomputable and by Fubini's theorem

$$\int u(X) \, d\mu(X) = \iint t(X \oplus Y) \, d\mu(Y) \, d\mu(X) = \int t \, d\mu$$

which is computable. It follows that u is a Schnorr integral test. By the key lemma (Lemma 2.1), there is some total computable function $h \leq u$ such that $u(A) = h(A)$. Let $\hat{t}^X$ be t^X, but enumerated only so that $\int \hat{t}^X \, d\mu = h(X)$. More formally, for all X let $\langle g_n^X \rangle$ be some code for t, namely some sequence of total nonnegative computable functions such that $t^X = \sum_n g_n^X$. Label the integrals of the partial sums $c_n^X = \int \sum_{k<n} g_k^X \, d\mu$. Define

$$\hat{t}^X = \sum_n \begin{cases} g_n^X & \text{if } c_{n+1}^X < h(X) \\ \left(\frac{h(X) - c_n^X}{c_{n+1}^X - c_n^X} \right) \cdot g_n^X & \text{if } c_n^X \leq h(X) \leq c_{n+1}^X \\ 0 & \text{if } h(X) < c_n^X \end{cases}.$$

Then $\langle \hat{t}^X \rangle_{X \in 2^\omega}$ is a uniform Schnorr integral test where $\int \hat{t}^X \, d\mu = h(X)$ and if $h(X) = u(X)$ then $\hat{t}^X = t^X$. Since $h(A) = u(A)$, we have $\hat{t}^A(B) = t^A(B) = \infty$. Therefore B is not Schnorr random uniformly relative to A. □

5. Uniformly relative computable randomness

In this section we give a weaker version of van Lambalgen's theorem for uniformly relative computable randomness.

Recall that a *martingale* is a function $d \colon 2^{<\omega} \to [0, \infty)$ such that $d(\sigma 0) + d(\sigma 1) = 2d(\sigma)$ for all $\sigma \in 2^{<\omega}$. We will say a martingale d *succeeds* on a set $A \in 2^\omega$ if $\limsup_n d(A \upharpoonright n) = \infty$. A set A is *computably random* (or *recursively random*) if there is no computable martingale d which succeeds on A. A well-known alternate characterization is that A is computably random if and only if there is no computable martingale d such that $\lim_n d(A \upharpoonright n) = \infty$.

Definition 5.1. A *uniform martingale test* is a total computable map $\Phi: 2^\omega \to \mathbb{N}^\mathbb{N}$ such that each $\Phi(X)$ encodes a martingale. Also, call a collection $\langle d^X \rangle_{X \in 2^\omega}$ a *uniform martingale test* if it is given by a map $\Phi: 2^\omega \to \mathbb{N}^\mathbb{N}$ as above.

Definition 5.2 (Miyabe [6]). A **is computably random uniformly relative to** B *if there is no uniform martingale test* $\langle d^X \rangle_{X \in 2^\omega}$ *such that* d^B *succeeds on* A.

Uniformly relative computable randomness was previously called "truth-table reducible randomness" in [6]. In Section 6 we give an alternative definition using truth-table reducibility in the spirit of Franklin and Stephan.

The first author [6] claimed the following, but the proof of the "$\Leftarrow$" direction is incorrect.[c]

$A \oplus B$ is computably random if and only if A is computably random and B is computably random uniformly relative to A.

However, we can prove this weaker version of van Lambalgen's theorem for uniformly relative computable randomness.

Theorem 5.1. $A \oplus B$ *is computably random if and only if A is computably random uniformly relative to B and B is computably random uniformly relative to A.*

Before giving the formal proof, we give the main idea. Assume some computable martingale d satisfies $\lim_n d((A \oplus B) \upharpoonright n) = \infty$. Then split d into two martingales d_0^B and d_1^A. Here d_0^B uses B to bet on the nth bit of some set X with the same relative amount that d bets on the $2n$th bit of

---

[c]The logical error in [6, Theorem 5.10] is in the following formula.

$$\mu(\{Y \mid A \oplus Y \in W_n\} \cap [\tau]) \leq \frac{\nu(A \upharpoonright m \oplus \tau)}{\nu(A \upharpoonright m \oplus \lambda)} 2^{-n}$$

The first A should be $[A \upharpoonright m]$ and the $\in$ should be $\subseteq$. Then the next line about letting $m = 0$ does not hold.

The conceptual error is as follows. The proof as usual starts by assuming $A \oplus B$ is not computably random. The witnessing test is a bounded Martin-Löf test with bounding measure ν. The proof attempts to show that the bounding measure $\tau \mapsto \nu(\lambda \oplus \tau)$ (where λ is the empty string) witnesses that B is truth-table reducibly random relative to A. However, $\tau \mapsto \nu(\lambda \oplus \tau)$ (that is, the marginal distribution of the second coordinate) is computable (not just computable relative to A). Hence this argument would show that B is not computably random, which is too strong.

$X \oplus B$. The martingale d_1^A is defined similarly. We will show $d((A + B) \upharpoonright 2n) = d_0^B(A \upharpoonright n) \cdot d_1^A(B \upharpoonright n)$. So either d_0^B succeeds on A or d_1^A succeeds on B.

To prove the theorem, we introduce the following notation. Given a martingale d such that $d(\sigma) \neq 0$ for all $\sigma \in 2^{<\omega}$, define $\widetilde{d}(\sigma i) = \frac{d(\sigma i)}{d(\sigma)}$ for $i \in \{0, 1\}$. This codes the martingale by the relative changed at each step. The martingale can be recovered by $d(\sigma) = \prod_{m=1}^{|\sigma|} \widetilde{d}(\sigma \upharpoonright m)$ (assuming $d(\varnothing) = 1$). Also $\widetilde{d}$ codes a martingale if and only if $\widetilde{d}(\sigma 0) + \widetilde{d}(\sigma 1) = 2$ for all σ. Call $\widetilde{d}$ the *multiplicative representation* of the martingale d.

Another notion will be as follows. If $\sigma, \tau \in 2^{<\omega}$, then define $f = \sigma \oplus \tau$ as a partial function from $\mathbb{N}$ to $\{0, 1\}$ such that $f(2n) = \sigma(n)$ if $n < |\sigma|$ and $f(2n + 1) = \tau(n)$ if $n < |\tau|$.

Proof of Theorem 5.1. Assume $A \oplus B$ is not computably random. Then there is a computable martingale d such that $\lim_n d((A \oplus B) \upharpoonright n) = \infty$. (Without loss of generality, $d(\varnothing) = 1$ and $d(\sigma) > 0$ for all σ.) Define two uniform martingale tests $\langle d_0^X \rangle_{X \in 2^\omega}$ and $\langle d_1^X \rangle_{X \in 2^\omega}$ by their corresponding multiplicative representations as follows.

$$\widetilde{d_0^Y}(\sigma) = \widetilde{d}(\sigma \oplus (Y \upharpoonright |\sigma| - 1))$$
$$\widetilde{d_1^X}(\tau) = \widetilde{d}((X \upharpoonright |\tau|) \oplus \tau)$$

Then $\widetilde{d_0^Y}$ (and similarly $\widetilde{d_1^X}$) is a multiplicative martingale since

$$\widetilde{d_0^Y}(\sigma 0) + \widetilde{d_0^Y}(\sigma 1) = \widetilde{d}((\sigma \oplus (Y \upharpoonright |\sigma|))^\frown 0) + \widetilde{d}((\sigma \oplus (Y \upharpoonright |\sigma|))^\frown 1) = 2.$$

Now for $X = A$ and $Y = B$ we have that

$$d_0^B(A \upharpoonright n) \cdot d_1^A(B \upharpoonright n) = \left(\prod_{m=1}^{n} \widetilde{d_0^B}(A \upharpoonright m) \right) \cdot \left(\prod_{m=1}^{n} \widetilde{d_1^A}(B \upharpoonright m) \right)$$

$$= \prod_{m=1}^{2n} \widetilde{d}((A + B) \upharpoonright m)$$

$$= d((A + B) \upharpoonright 2n) \to \infty$$

Then either d_0^B succeeds on A or d_1^A succeeds on B. It follows that one of A, B is not computably random uniformly relative to the other. $\square$

We also obtain this easy corollary. Since it follows from the "$\Rightarrow$" direction of Theorem 5.1, it was provable from the results of [6].

Corollary 5.1. *There exist sets A and B such the A is computably random uniformly relative to B, but A is not computably random relative to B.*

Proof. Merkle et al. [2] showed that there is are two computable randoms A, B such that $A \oplus B$ is computably random, but A is not computably random relative to B. However, by Theorem 5.1, A is computably random uniformly relative to B. □

5.1. *Remark on Kolmogorov-Loveland and Martin-Löf randomness*

Recall that Kolmogorov-Loveland randomness is a notion of randomness similar to computable randomness, except the martingales do not need to bet on the bits in order. These are called **nonmonotonic martingales**. Also, the martingales may be **partial**, i.e. not defined on all inputs. (See [2,4,8] for formal definitions and more information on Kolmogorov-Loveland randomness.) The proof of Theorem 5.1 is similar to the proof of the following.

Theorem 5.2 (Merkle et al. [2]). *$A \oplus B$ is Kolmogorov-Loveland random if and only if A and B are Kolmogorov-Loveland random relative to each other.*

One may ask if there is a notion of "uniformly relative Kolmogorov-Loveland randomness" using uniform partial nonmonitonic martingale tests. The answer is that being Kolmogorov-Loveland random uniformly relative to B is the same as being Kolmogorov-Loveland random relative to B. Any nonmonotonic martingale computable from B, is easily extended to a uniform partial nonmonotonic martingale. This is easy to do because the nonmonotonic martingale needs only be partial computable from the oracle.

Similarly, "Martin-Löf random uniformly relative to B" is equivalent to Martin-Löf random relative to B. If $\langle U_n \rangle$ is a Martin-Löf test relative to B, then there is a uniform Martin-Löf test $\langle U_n^X \rangle_{n \in \mathbb{N}, X \in 2^\omega}$ given by using the same algorithm (as for $\langle U_n \rangle$) to enumerate the basic open sets of U_n^X, but we stop enumerating the basic open sets if doing so will make $\mu(U_n^X) > 2^{-n}$.

6. Truth-table Schnorr and truth-table reducible randomness

6.1. *Truth-table Schnorr randomness is equivalent to uniformly relative Schnorr randomness*

Truth-table Schnorr randomness was first defined by Franklin and Stephan [7] as follows using a truth-table relativized martingale test. Recall that a function $f \in \mathbb{N}^{\mathbb{N}}$ is **truth-table reducible** to $A \in 2^{\omega}$ if there is a total computable functional $\Phi \colon 2^{\omega} \to \mathbb{N}^{\mathbb{N}}$ such that $f = \Phi(A)$.

Definition 6.1 (Franklin and Stephan [7]). *A set A is* **truth-table Schnorr relative to** B *if there is no pair (d, f) consisting of a martingale d with code truth-table reducible to B and a function $f \colon \mathbb{N} \to \mathbb{N}$ truth-table reducible to B such that $\exists^{\infty} n \, [d(A \restriction f(n)) \geq n]$.*

Franklin and Stephan also remark that one may take f in the above definition to be computable (instead of truth-table reducible to B) with no loss.

The first author showed in [6] that truth-table Schnorr randomness is equivalent to (what we call) uniformly relative Schnorr randomness, but there was a small gap in the proof which we feel would be instructive to fill in here.

Lemma 6.1. *Let d be a martingale (with some code) truth-table reducible to A and let $f \colon \mathbb{N} \to \mathbb{N}$ be truth-table reducible to A. Then there is a uniform martingale test $\langle \hat{d}^{X} \rangle_{X \in 2^{\omega}}$ and a uniform function $\langle \hat{f}^{X} \rangle_{X \in 2^{\omega}}$ such that $\hat{d}^{A} = d$ and $\hat{f}^{A} = f$.*

Proof. Let d be a martingale truth-table reducible to A. And let $\Phi \colon 2^{\omega} \times 2^{<\omega} \to \mathbb{N}^{\mathbb{N}}$ be a total computable functional such that $\Phi(A, \sigma)$ encodes the fast-Cauchy code for $d(\sigma)$. Define $d^{X}(\sigma)$ to be the real coded by $\Phi(X, \sigma)$. Note, that there is no guarantee that, first, $\Phi(X, \sigma)$ is a Cauchy code for a nonnegative real for every X and σ, and that, second, d^{X} is a martingale for every X.

The first issue is easily fixed. We use a folklore trick to force $\Phi(X, \sigma)$ to be a fast Cauchy code for a nonnegative number. Let $\langle q_0, q_1, \ldots \rangle$ be the sequence of rationals given by $\Phi(X, \sigma)$. Firstly, replace each q_n with $\max\{q_n, 0\}$. Secondly, find the first n, if any, such that $|q_n - q_m| > 2^{-m}$ for $m \leq n$, then change the code to be $\langle q_0, q_1, \ldots, q_{n-1} \rangle ^\frown \langle q_{n-1}, q_{n-1} \ldots \rangle$. Notice this does not change the value of $d(A)$.

The second issue is also easily handled. Assuming, now that each $d^{X}(\sigma)$ is a nonnegative real, define $\hat{d}^{X}(\sigma)$ by recursion as follows.

$$\hat{d}^{X}(\varnothing) = d^{X}(\varnothing)$$
$$\hat{d}^{X}(\sigma 0) = \min\{d^{X}(\sigma 0), 2\hat{d}^{X}(\sigma)\}$$
$$\hat{d}^{X}(\sigma 1) = 2\hat{d}^{X}(\sigma) - \hat{d}^{X}(\sigma 0)$$

It is easy to check $\hat{d}^X$ is a nonnegative martingale whose code is uniformly computable from the code for d^X. Also if d^X is already a martingale, then $\hat{d}^X = d^X$. In particular, $\hat{d}^A = d^A$.

Last if f is truth-table reducible to A, then there is a total functional $\Psi \colon 2^\omega \to \mathbb{N}^{\mathbb{N}}$ such that $\Psi(A) = f$. Define $\hat{f}^X = \Psi(X)$. $\qquad\square$

Now we can show that the definitions are equivalent.

Proposition 6.1. *A set A is Schnorr uniformly relative to B if and only if A is truth-table Schnorr relative to B.*

Proof. By Lemma 6.1 (and its trivial converse) truth-table Schnorr randomness is equivalent to that obtained from uniform martingale tests of the above type. Now similar to Proposition 4.1, it is enough to show a uniform martingale test (of the above type) can effectively be converted to a Schnorr test which covers the same points that the martingale succeeds on, and vice versa. Indeed, the proofs in the literature are effective in this regard (see [4,7,8]). $\qquad\square$

Remark 6.1. There is an important subtlety in the last proof similar to Remark 2.1. We showed it is possible to compute *a code for* one test uniformly from *a code for* another test. However, it is not necessarily possible to do so in a way that is independent of the choice of codes. For example, it is known that one may effectively replace a real-valued martingale d with a rational-valued martingale $\hat{d}$ that succeeds on the same points. However, there is some d such that different codes for d lead to different rational approximations. This will become an issue if instead of relativizing with respect to a set $X \in 2^\omega$, one relativized with respect to a real $x \in [0, 1]$.

6.2. Truth-table reducible randomness is equivalent to uniformly relative computable randomness

We also have a similar result for computable randomness. For this reason, uniformly relative computable randomness is also known as **truth-table reducible randomness** [6].

Proposition 6.2. *A set A is computably random uniformly relative to B if and only if for all martingales d with a code truth-table reducible to B, we have that $\limsup_n d(A \restriction n) < \infty$.*

Proof. By the proof of Lemma 6.1 it is possible to pass from a martingale truth-table reducible to B to a uniform martingale test. The converse is trivial. □

6.3. Characterizations of "truth-table Schnorr randomness" by other tests

Notice that the motivation behind Definition 6.1 is to say that A is "truth-table Schnorr random" relative to B if there is no test for Schnorr randomness (e.g. Schnorr test, integral test, martingale test, etc.) which is truth-table reducible to B and witnesses that A is not random for that test. Unfortunately, this method is very sensitive to the choice of test. Lemma 6.1 does not hold for most other characterizations of Schnorr randomness, including the usual martingale and Schnorr test characterizations. More specifically we will show that the following two natural-looking definitions of "truth-table Schnorr randomness" are in fact strictly stronger than uniformly relative Schnorr randomness.

Definition 6.2.

(1) A set A is **"truth-table (martingale) Schnorr random" relative to B** if there is no pair (d, f) consisting of a martingale d which a code truth-table reducible to B and an order function $f: \mathbb{N} \to \mathbb{N}$ truth-table reducible to B such that $\exists^\infty n \, [d(A \upharpoonright n) \geq f(n)]$. (Recall, an **order** is an unbounded increasing function.)

(2) The set A is **"truth-table (test) Schnorr random" relative to B** if there is no sequence $\langle U_n \rangle$ of open sets with codes uniformly truth-table reducible to B such that $\mu(U_n) = 2^{-n}$ for all n and $A \in U_n$ for all n.

Proposition 6.3 (Due to Stephan and Franklin [7]). *There exist A and B such that A is Schnorr random uniformly relative to B but not "truth-table (martingale) Schnorr random" relative to B.*

Proof. Downey, Griffiths, and LaForte [19] showed that there is a Turing complete Schnorr trivial B. Nies, Stephan, and Terwijn [20] showed there is a Schnorr random $A \equiv_T \emptyset'$ that is not computably random. Take this to be our A and B. Franklin and Stephan [7] showed that every Schnorr trivial is low for truth-table Schnorr random. This means that since A is Schnorr random, we have that A is truth-table Schnorr random relative to B, and hence Schnorr random uniformly relative to B.

Given such A, B as above, Franklin and Stephan [7, Theorem 2.2] construct a computable martingale d (hence truth-table reducible to B) and an order function f truth-table reducible to B such that $\exists^\infty n\,[d(A \upharpoonright n) \geq f(n)]$. Therefore A is not "truth-table (martingale) Schnorr random" relative to B. $\qquad\square$

Proposition 6.4. *"Truth-table (test) Schnorr randomness" is equivalent to relative Schnorr randomness. Hence, there exist A and B such that A is Schnorr random uniformly relative to B but not "truth-table (test) Schnorr random" relative to B.*

Proof. We will show that "truth-table (test) Schnorr randomness" implies relative Schnorr randomness. (The other direction is trivial.) Take a Schnorr test $\langle U_n \rangle$ computable relative to B. We may assume $\mu(U_n) = 2^{-n}$. It remains to show that a code for $\langle U_n \rangle$ is truth-table reducible to B. There is a partial computable function $\Phi \colon 2^\omega \times \mathbb{N} \times \mathbb{N} \to \mathbb{N}$ such that $\Phi(B, -, -)$ encodes $\langle U_n \rangle$. Namely, each $\Phi(B, n, m)$ encodes a basic open set $C_{n,m}^B$ where $U_n = \bigcup_m C_{n,m}^B$. (Recall, we allow $\varnothing \subset 2^\omega$ as a basic open set.) Then one can modify Φ to be total by repeatedly adding $\varnothing$ to the code. Namely, define $\Psi \colon 2^\omega \times \mathbb{N} \times \mathbb{N} \to \mathbb{N}$ as follows. If m encodes the pair $\langle m', s \rangle$, and if $\Phi(X, n, m')$ halts by stage s, then let $\Psi(X, n, m) = \Phi(X, n, m')$. Otherwise let $\Psi(X, n, m)$ encode $\varnothing$. Since Ψ is total computable and $\Psi(B, -, -)$ still encodes $\langle U_n \rangle$, we have that $\langle U_n \rangle$ has a code truth-table reducible to B as desired.

The rest of the proof is similar to the previous one. Let B be a Turing complete Schnorr trivial set. Let $A \equiv_T \emptyset'$ be some Schnorr random. Hence A is Schnorr random uniformly relative to B. However, since $B \geq_T A$, then A is not Schnorr random relative to B, and hence not "truth-table (test) Schnorr random" relative to B. $\qquad\square$

Remark 6.2. A little thought reveals what is missing in Definition 6.2. It is not in general possible to extend an order f truth-table reducible to B to a uniform order $\langle f^X \rangle_{X \in 2^\omega}$. To do this, one would also need that the rate of growth of f is truth-table reducible to B. A similar phenomenon happens with Schnorr tests. One needs not only that the measure of $\mu(U_n)$ is truth-table reducible to B, but that given some code $\langle C_s \rangle$ for each U_n which is truth-table reducible to B, the rate of convergence of $\mu\left(\bigcup_{s < n} C_s\right)$ must also be truth-table reducible to B. After making these changes, then both parts of Definition 6.2 would be equivalent to uniformly relative Schnorr randomness.

Acknowledgments

We would like to thank Rod Downey and Denis Hirschfelt for encouraging us to publish this finding, as well as Johanna Franklin and Frank Stephan for clarifying the significance of Theorem 2.2 in [7] which led to the proof of Proposition 6.3. Last, we would like to thank the anonymous referee for many helpful corrections and suggestions.

The first author was supported by GCOE, Kyoto University and JSPS KAKENHI (23740072).

References

1. M. van Lambalgen, *J. Symbolic Logic* **55**, 1143 (1990).
2. W. Merkle, J. S. Miller, A. Nies, J. Reimann and F. Stephan, *Ann. Pure Appl. Logic* **138**, 183 (2006).
3. L. Yu, *Proc. Amer. Math. Soc.* **135**, 861 (2007).
4. A. Nies, *Computability and randomness*, Oxford Logic Guides, Vol. 51 (Oxford University Press, Oxford, 2009).
5. J. N. Y. Franklin and F. Stephan, *Notre Dame J. Form. Log.* **52**, 173 (2011).
6. K. Miyabe, *MLQ Math. Log. Q.* **57**, 323 (2011).
7. J. N. Y. Franklin and F. Stephan, *J. Symbolic Logic* **75**, 501 (2010).
8. R. G. Downey and D. R. Hirschfeldt, *Algorithmic randomness and complexity*, Theory and Applications of Computability, Vol. 1 (Springer, New York, 2010).
9. L. Bienvenu, R. Downey, N. Greenberg, A. Nies and D. Turetsky, Characterizing lowness for Demuth randomness, Submitted.
10. D. Diamondstone, N. Greenberg and D. Turetsky, A van Lambalgen theorem for Demuth randomness, to appear in *Proceedings of the 12th Asian Logic Colloquium*, (201x).
11. M. Hoyrup and C. Rojas, *Inform. and Comput.* **207**, 830 (2009).
12. M. Schröder, *MLQ Math. Log. Q.* **53**, 431 (2007).
13. V. Brattka, Computable versions of Baire's category theorem, in *Mathematical Foundations of Computer Science, 2001*, eds. J. Sgall, A. Pultr and P. Kolman, Lecture Notes in Comput. Sci., Vol. 2136 (Springer, Berlin, 2001) pp. 224–235.
14. K. Miyabe, L^1-computability, layerwise computability and Solovay reducibility, Submitted.
15. M. Hoyrup and C. Rojas, An application of Martin-Löf randomness to effective probability theory, in *Mathematical theory and computational practice*, eds. S. Albers, A. Marchetti-Spaccamela, Y. Matias, S. Nikoletseas and W. Thomas, Lecture Notes in Comput. Sci., Vol. 5635 (Springer, Berlin, 2009) pp. 260–269.
16. N. Pathak, C. Rojas and S. G. Simpson, Schnorr randomness and the Lebesgue differentiation theorem, to appear in *Proceedings of the American Mathematical Society*, (201x).

17. J. Rute, Algorithmic randomness, martingales, and differentiability I, In preparation.
18. L. A. Levin, *Dokl. Akad. Nauk SSSR* **227**, 33 (1976).
19. R. G. Downey, E. Griffiths and G. LaForte, *Math. Logic Quart.* **50**, 613 (2004).
20. A. Nies, F. Stephan and S. A. Terwijn, *J. Symbolic Logic* **70**, 515 (2005).

271

Computational aspects of the hyperimmune-free degrees

KENG MENG NG

School of Physical & Mathematical Sciences,
Division of Mathematical Sciences,
Nanyang Technological University,
Singapore
Email: kmng@ntu.edu.sg

FRANK STEPHAN

Department of Mathematics,
National University of Singapore,
Singapore
E-mail: fstephan@comp.nus.edu.sg

YUE YANG

Department of Mathematics,
National University of Singapore,
Singapore
E-mail: matyangy@nus.edu.sg

LIANG YU

Institute of Mathematical Science,
Nanjing University,
Nanjing, Jiangsu Province 210093,
China
E-mail: yuliang.nju@gmail.com

We explore the computational strength of the hyperimmune-free Turing degrees. In particular we investigate how the property of being dominated by recursive functions interacts with classical computability notions such as the jump operator, relativization and effectively closed sets.

Keywords: Hyperimmune-free, Turing degrees, effectively closed sets.

1. Introduction

The relative computational power between sets of natural numbers has traditionally been measured by Turing reducibility: If $A \leq_T B$ we think of B as containing at least as much algorithmic information as A. There have been various other well-studied methods of calibrating computational power; for instance, by examining the effective enumerations of A, by looking at the algorithmic randomness content of A, and by investigating the rate of growth of functions computable from A. These studies have all yielded deep results relating Turing reducibility with different aspects of computation. This paper is concerned with the last of these: The rate of growth of functions computed by A. Hence a set A can be viewed as being computationally powerful if it is able to compute functions which grow fast enough to dominate a certain other class of functions.

Domination properties have been studied extensively in the literature, and many relationships between domination and Turing degrees have been found. For instance it is easy to see that $A \oplus B \geq_T B'$ iff $A \oplus B$ computes a function dominating every B-partial recursive function. In some cases a class of degrees is first defined in terms of a domination property and subsequently results are then obtained about its computational properties; the almost everywhere dominating degrees are an example of such a class. It is more common to go the other way, for a class of degrees to be first introduced without mentioning domination and then subsequently characterized in terms of a domination property. For example, Martin [10] characterized the high Turing degrees as the degrees which compute a function dominating every recursive function; the class of array non-recursive degrees introduced by Downey, Jockusch and Stob [4] was shown in [5] to be the same as the class of degrees $\mathbf{a}$ where every ω-r.e. function fails to dominate some $\mathbf{a}$-recursive function.

The class studied in this paper is the class of hyperimmune-free (HIF) degrees (i.e. the degrees which contain no hyperimmune set). We recall that a set A is hyperimmune iff A is infinite and there is no disjoint strong array of finite sets each of which has non-empty intersection with A. The study of hyperimmune sets can be traced back to Post and attempts to solve Post's problem. Post [13] introduced the notion of a simple and a hypersimple r.e. set (A is hypersimple if $\overline{A}$ is hyperimmune) and it turned out that each hypersimple set is wtt-incomplete, but not necessarily Turing incomplete. Indeed Dekker [2] showed that every non-recursive r.e. degree is hyperimmune (i.e. not HIF), while Miller and Martin [11] showed that every degree $\mathbf{b}$ satisfying $\mathbf{a} < \mathbf{b} < \mathbf{a}'$ for some $\mathbf{a}$ is hyperimmune. Dekker and Myhill [3] showed that every non-recursive degree contains an immune set, hence it was rather surprising that Miller and Martin [11] were able to construct a non-recursive HIF degree. Indeed they characterized the HIF degrees using a domination property: a is HIF iff every function recursive in a is dominated by some recursive function. This property asserts that a is "almost recursive", in that a computes no fast-growing function (relative to the class of recursive functions). However the fact that no non-recursive Δ_2^0 degree possesses this property indicates that this notion of computational feebleness is intrinsically hard to understand.

The main aim of this paper is to shed some light on this class by investigating how the domination related property of HIF degrees is related to the other more traditional methods of measuring computational strength. It is easy to see that each HIF degree is of array recursive degree and is generalized low$_2$, hence each HIF degree cannot be computationally strong in this sense. On the other hand a HIF degree can be PA-complete.

The main difficulty that we face is how to translate between the information contained in a fast-growing function and the ability to code into a given set. The proofs given in this paper give various ways of doing this.

In section 2 we study the distribution of HIF degrees in Π_1^0 classes. The so-called "HIF basis theorem" of Jockusch and Soare [6] asserts that every non-empty Π_1^0 class contains a member of HIF degree. We construct an uncountable Π_1^0 class in which every member is generalized low (GL$_1$) and of HIF degree. This Π_1^0 class we construct will necessarily have recursive members (in fact, isolated paths).

In section 3 we investigate when a degree can be HIF relative to another. We introduce the notion of being HIF relative to $\mathbf{0}^n$, for $n > 0$. We show that there are uncountably many sets which are simultaneously HIF and HIF relative to $\mathbf{0}''$, but surprisingly we discover that no non-recursive set is both HIF and HIF relative to $\emptyset'$. On the other hand, we construct a perfect closed set of reals which are simultaneously HIF and HIF relative to every low r.e. set. We also obtain another characterization of the K-trivial sets as the Δ_2^0 sets A where some HIF set is A-random.

In section 4 we study the degrees which are the jump of some HIF degree. From folklore it is known that each degree above $\mathbf{0}''$ is the double jump of a HIF degree. However the degrees which are the jump of a HIF degree is not at all well-understood. Kučera and Nies [9] showed that each degree r.e. in and strictly above $\mathbf{0}'$ computes $\mathbf{a}'$ for some non-recursive HIF degree $\mathbf{a}$, while it follows from Jockusch and Stephan [7] that the jump of each HIF degree cannot be PA-complete relative to $\mathbf{0}'$. We will show that for each 2-generic degree $\mathbf{c}$, there is a HIF degree $\mathbf{a}$ such that $\mathbf{a}' = \mathbf{c} \cup \mathbf{0}'$. We conjecture that this is in fact a characterization of the degrees which are the jump of a HIF degree.

2. HIF and closed sets

Nies and Miller (unpublished) observed that no real can simultaneously be GL_1, HIF and of diagonally non-recursive (DNR) degree, although any combination of two are possible. It is a natural question to ask to what extent can these properties be reflected in Π_1^0 classes. We first show that there is an uncountable Π_1^0 class where every non-isolated path is GL_1 and of DNR degree. Hence GL_1 and DNR can be simultaneously realized by every non-recursive path in an uncountable Π_1^0 class. We note that the isolated paths are necessary, since every perfect Π_1^0 class contains a path of high degree, and clearly no set if HIF degree can be high.

Lemma 2.1. *Given any tree $T \leq_T \emptyset'$ there exists a recursive tree Q such that every path of T is Turing equivalent to a non-isolated path of Q, and vice versa.*

Proof. Let $T = \lim_s T_s$ for a recursive sequence $\{T_s\}$ of recursive trees. Define the partial recursive function $f(\sigma)$ to be the first stage s such that $\sigma \upharpoonright i \in T_s$ for every $i \leq |\sigma|$. Now define the Turing functional Ψ^X to output $X(0)2^{f(X \upharpoonright 1)} X(1) 2^{f(X \upharpoonright 2)} \cdots$. Here 2^s is the symbol 2 repeated s many times. If $f(X \upharpoonright k)$ is partial for some k then the functional outputs a sequence with a tail of 2s, otherwise Ψ^X is a ternary sequence with X coded. If A is on T then $f(A \upharpoonright n)$ is convergent for all n, so clearly $A \equiv_T \Psi^A$.

Now let $Q = \Psi$ applied to 2^ω. If A is on T then clearly Ψ^A is a non-isolated path of Q. On the other hand if A is not on T then let $\sigma \subset A$ be minimal such that $\sigma \notin T$. For all large enough s and every $\eta \supseteq \sigma$ of length s, $f(\eta) \uparrow$. Thus every infinite branch of Q extending Ψ^σ is isolated. □

Theorem 2.1. *There is an uncountable Π_1^0 class P such that every non-recursive path of P is GL_1 and computes a DNR function.*

Proof. Let $T \leq_T \emptyset'$ be a tree containing only 2-random reals. By Lemma 2.1 there exists a recursive tree P such that every path of P is either isolated or Turing equivalent to a 2-random. P is clearly uncountable, and every 2-random is DNR and GL_1.

Comment. The idea to construct this is the following: Assign to every σ the weight $w(\sigma)$ being 2^{-k} where k is the number of branching nodes below σ.

Make a branch at σ (that is, put $\sigma * 0$ and $\sigma * 1$ both into T) whenever all $e < |\sigma|$ satisfy that

(1) $w(\sigma) > 2^{-e-5}$ or $\varphi_e^\sigma(e)$ converges and is among the last share of 2^{-e-5} (according to weight) where $\varphi_e^\tau(e)$ converges and τ on T and $|\tau| = |\sigma|$,

(2) the string $a_0 a_1 \cdots a_e$ with $a_0, a_1, \cdots, a_e$ being the bits of σ at branching points (so called branching decisions) in ascending order up to level e satisfies that $H_{|\sigma|}(a_0 a_1 \cdots a_e) < e - 5$.

Otherwise only $\sigma * 0$ is put into T and $\sigma * 1$ is omitted.

Note that if a set A is an isolated branch in the tree T then it has positive weight being 2^{-k} for the number of branching nodes on A and it has stopped branching because one of the above two rules got violated permanently with some index e. The weight of all A for which this happens according to 1 adds up to 2^{-4} and similarly for 2, so that the overall weight of isolated branches is $1/8$. Then the non- isolated branches have weight $7/8$ or more and as each has measure 0, there are uncountably many of them. So far the proof idea.

The GL_1-ness follows from the fact that one can relative to K compute a stage s such that the weight of all A on which $\varphi_e^A(e)$ converges but needs at least s is at most 2^{-e-5}. Then one knows that for non-isolated infinite branches A of T, either $\varphi_e^A(e)$ does not converge or has already converged before stage s. Hence one can decide $\varphi_e^A(e)$ using K and A up to that stage s for all nonrecursive infinite branches of T. □

Next we argue that there can be no Π_1^0 class where every non-recursive path is HIF and DNR.

Theorem 2.2. *Suppose P is a Π_1^0 class where every path is HIF. Then P cannot contain a path of DNR degree.*

Sketch of proof. Suppose $A \in P$ and A is of DNR degree. By Kjos-Hanssen, Merkle and Stephan [8], there exists a function $f \leq_T A$ such that $C(f(n)) \geq n$ for every n. Since A is HIF, $f \leq_{tt} A$. The set

$$Q = \{X \in 2^\omega \mid \exists n C(f^X(n)) < n\}$$

is an open set. Observe that $P - Q$ is a non-empty Π_1^0 class as it contains A. Every path of $P - Q$ is of HIF degree and non-recursive. Applying the Low Basis Theorem gives a contradiction. □

Finally we turn to the apparently most difficult combination. Recall that a tree T has rank 1 if the Cantor-Bendixon derivative of T contains no isolated path. We show that there is a rank 1 uncountable Π_1^0 class such that every path is GL_1 and HIF. Again rank 1 is the best possible, since the isolated paths are necessary. We also note that every path of P has a strong minimal cover.

Theorem 2.3. *There is an uncountable rank 1 Π_1^0 class P such that every member of P is GL_1 and of HIF degree.*

Proof. As the strategies are rather complicated, we will sketch the proof briefly, and only discuss how to ensure that each path is of HIF degree. The complete details will appear in a later journal version of this paper. The requirements we must meet are R_e, for each $e \in \mathbb{N}$:

R_e: For all X in P, if Φ_e^X is total then there is a recursive function h such that Φ_e^X is dominated by h.

We first describe the strategy for a single requirement, and then consider the interaction between two requirements.

Strategy for a single requirement. We use R_0 as an example to illustrate the strategy in isolation. We start with $T_{-1} = 2^{<\omega}$, the full binary tree. We will describe how to compute $h(k)$ for each k and indicate the modifications to the tree needed to ensure that the bound $h(k)$ works. We will process each level at a time, as follows.

Level 0: At stage s, check whether there exists a node $\sigma \in T_s$ such that $|\sigma| \le s$ and $\Phi_0^\sigma(0)\downarrow$. If not, go to next stage; otherwise, let σ be the leftmost one found. Define $h(0) = \Phi_0^\sigma(0)+1$ and let $T_{s+1} = \{\tau \in T_s : \tau$ is compatible with $\sigma\}$. Declare all other nodes dead, i.e., all nodes α on $T_s - T_{s+1}$ which have length s are now dead ends.

Level 1: After processing level 0, we have a node σ. We look for two nodes $\tau_0 \supseteq \sigma^\smallfrown\langle 0\rangle$ and $\tau_1 \supseteq \sigma^\smallfrown\langle 1\rangle$ such that $\Phi_0^{\tau_0}(1)\downarrow$ and $\Phi_0^{\tau_1}(1)\downarrow$. We search for them one at a time. Each search is called a cycle, i.e. we will have cycle 0 and cycle 1.

While running cycle 0 we temporarily *isolate* $\sigma^\smallfrown\langle 1\rangle$. (To isolate a node α means that for each stage t, we ensure that $\alpha^\smallfrown\langle 0^{t-|\alpha|}\rangle$ is the only node extending α of length t on T_t.) For cycle 0 we look at the basic open set $[\sigma^\smallfrown\langle 0\rangle]$ (we refer to this as the current active zone). If we never find any node $\tau_0 \supseteq \sigma^\smallfrown\langle 0\rangle$ such that $\Phi_0^{\tau_0}(1)\downarrow$, then Φ_0^X is partial for all X in that open set. Furthermore since we end up isolating $\sigma^\smallfrown\langle 1\rangle$ forever, we get a single isolated (hence recursive) path extending $\sigma^\smallfrown\langle 1\rangle$. This means that R_0 is satisfied (vacuously) for every path on the tree. We refer to this outcome as the Σ_2^0-outcome for R_0.

On the other hand if we find, at some stage t, a node $\tau_0 \supseteq \sigma^\smallfrown\langle 0\rangle$ such that $\Phi_0^{\tau_0}(1)\downarrow$, we begin isolating τ_0 and we shift the active zone to the open set generated by $\sigma' = \sigma\string^\langle 1\rangle\string^\langle 0^{t-|\sigma|-1}\rangle$, and stop isolating $\sigma\string^\langle 1\rangle$. Since the active zone is now shifted to $[\sigma']$ we now search for a node $\tau_1 \supseteq \sigma'$ such that $\Phi_0^{\tau_1}(1)\downarrow$. As argued above, if we never see such a convergence then we will meet R_0 in a Σ_2^0 way. Now finally suppose we find such τ_1 at $t' > t$. We then modify the tree by defining $T_t = \{\alpha : \alpha$ is compatible with either $\tau_0\string^\langle 0^{t'-|\tau_0|}\rangle$ or $\tau_1\}$, and define $h(1) = \max\{\Phi_0^{\tau_0}(1), \Phi_0^{\tau_1}(1)\} + 1$. Level 1 is now concluded.

Level k: In general, suppose we have completed cycles $0, \cdots, k-1$ and have defined $h(0), h(1), \cdots, h(k-1)$. We then need to look for $N = 2^k$ incompatible strings $\tau_0, \tau_1, \cdots, \tau_{N-1}$ such that $\Phi_0^{\tau_i}(k)\downarrow$. We use the same strategy as above, except we now have N cycles. In general if we get stuck at some cycle associated with a level k (i.e. some cycle has a Σ_2^0-outcome described above) then we never begin a level $j > k$.

Thus the eventual outcomes for R_0 are as follow:

- A Σ_2^0-outcome for R_0: We get stuck at the i-th cycle in some k-th level. Let us use (k, i) to indicate this outcome. The final tree consists of the open set $[\tau_i]$ for some τ_i, together with $2^k - 1$ many isolated paths. Since $\Phi_0^X(k)$ is not defined for all X in this open set, and all other paths of the final tree are isolated, we win R_0.
- The Π_2^0-outcome for R_0: We finish each level and all cycles successfully, resulting in a perfect subtree T_0 and a recursive function h which dominates all Φ_0^X for all $X \in [T_0]$.

Interaction between two strategies. Consider now two requirements R_0 and R_1. We will have different versions of R_1 which are guessing different outcomes of R_0.

If R_0 has the Σ_2^0-outcome (k, i), then eventually only the R_1 which guesses (k, i) is active. This R_1 will work in a perfect tree (the open neighborhood generated by a certain σ in which R_0 is eventually stuck in), and in this environment R_1 works in a similar fashion as R_0 above. This version of R_1 will receive no interference from R_0. In this situation we need to add the following feature for technical reasons. We need the final tree to have a perfect kernel, so we do not want the combined effort of R_0 and R_1 to eventually trim the tree into one with only finitely many branches (in the case where R_0 and R_1 both have the Σ_2^0-outcome). Therefore, under each (k, i) outcome for R_0 we force two further versions of R_1: $R_{1,0}$ and $R_{1,1}$. Here $R_{1,0}$ will work on the left subtree of T_0 (the tree produced by R_0) and work towards producing $h_{1,0}$, while $R_{1,1}$ works on the right subtree in a similar fashion. (Note that this will increase further the non-uniformity of the dominating function h). This ensures that the resulting Π_1^0-class is uncountable.

Now suppose that R_0 has Π_2^0-outcome. The version of R_1 believing this will be handed a perfect tree T_0 (piecewise, level by level) by R_0. The requirement R_1 will act as in the basic strategy in this tree, i.e. R_1 will search for Φ_1-convergent strings in T_0 and isolate paths of T_0.

An interesting feature of this construction is that the actions of R_1 will have a direct impact on the strategy for R_0. Nevertheless we do not consider R_0 to be "injured" by this feedback. For instance R_0 may hand to the version of R_1 believing in a Π_2^0-outcome for R_0 a finite perfect tree S^* of size 2^k at some stage of the construction. R_1 will then begin its basic strategy in S^*. This will involve isolating certain nodes of S^*, and declaring certain nodes to be dead ends. Since we are building a Π_1^0-class, any dead end we declare cannot be later resurrected. This causes the finite tree S^* to be permanently reduced in size, and the R_1-modified tree $S^{**} \subseteq S^*$ will have fewer than 2^k leaves. At the next stage R_0 will have to work in this reduced tree S^{**} instead of the original S^* it had handed to R_1. R_0 will now have to run cycles which will look for convergent Φ_0-computations, where the active zones will now range over the leaves of S^{**}. This does not cause any injury to R_0, since the previously declared values for $h_0(i)$ will still dominate $\Phi_0^X(i)$ for $i < k$.

The full details of the construction will appear in the journal version of the paper. □

Since it takes $\emptyset''$ to pick out the perfect kernel of the final tree constructed above, we obtain the following corollary:

Corollary 2.1. *There exists a perfect tree $T \leq_T \emptyset''$ with no dead ends such that every path of $[T]$ is HIF and GL_1.*

3. HIF and relativization

In this section we study the properties obtained when relativizing the notion of HIF.

Definition 3.1. We say that X is HIF relative to A if every function recursive in $X \oplus A$ is dominated by an A-recursive function. For $n \geq 0$ we call A an $(n+1)$-HIF if every function recursive in $A^{(n)}$ is dominated by a $\emptyset^{(n)}$-recursive function.

Fact 3.1. If A is HIF relative to B and B is HIF then A is HIF.

Fact 3.2. A is $(n+1)$-HIF implies that A is HIF relative to $\emptyset^{(n)}$.

Fact 3.3. 2-HIF is equivalent to being GL_1 and HIF relative to $\emptyset'$.

Example 3.1. Every low_2 degree is $(n+3)$-HIF for every $n \geq 0$.

Proposition 3.1. *There exists uncountably many reals which are both HIF and 3-HIF.*

Proof. For every $C \geq_T \emptyset''$ there is a HIF A such that $A'' \equiv_T C$. Relativizing the construction of a HIF real to $\emptyset''$, we get uncountably many reals C which are HIF relative to $\emptyset''$. $\square$

We can show that there are sets which are HIF relative to every low r.e. set:

Theorem 3.4. *There exists uncountably many HIF sets which are HIF relative to every low r.e. set.*

The proof of Theorem 3.4 constructs an uncountable tree combined with the Robinson's technique for guessing Σ_1^0 facts about low sets. We refer the reader to the full paper for further details.

We now investigate the HIF sets which are HIF relative to $\emptyset'$. We first need the following lemmas.

Lemma 3.1 (Day, Reimann [1]). *Suppose that C is PA-complete and B is r.e so that $C \not\geq_T B$, then $C \oplus B \geq_T \emptyset'$.*

Lemma 3.2. *If a tree $T \leq_T \emptyset'$ contains a HIF path A then there is a recursive tree Q containing A such that $[Q] \subseteq [T]$.*

Proof. Let $T = \lim_s T_s$ for a uniformly recursive sequence of trees $\{T_s\}_{s \in \omega}$, and suppose $A \in [T]$ for some HIF set A. Define the A-recursive function f by $f(n) =$ the first stage $s > n$ such that $A \upharpoonright i \in T_s$ for every $i \leq n$. Let g be a recursive function majorizing f. Now let R be the set of all strings σ such that $\sigma \notin T_s$ for every $|\sigma| < s \leq g(|\sigma|)$. Now let Q be the recursive tree supporting the Π_1^0 class $2^\omega - [R]$.

Clearly $A \in [Q]$ because g majorizes f. Now let $X \notin [T]$. Then there is some least i such that $\sigma = X \upharpoonright i \notin T$. Let s_0 be such that $\sigma \notin T_s$ for every $s \geq s_0$. This means that every extension of σ of length s_0 is in R. Hence $X \in [R]$. $\square$

Theorem 3.5. *If A is HIF and HIF relative to some PA-complete set $B \leq_T \emptyset'$ then A is recursive.*

Proof. Assume that a non-recursive A and a B exist as above. There exists a uniformly B-recursive sequence $\{B_e\}_{e \in \omega}$ of reals such that for every e, either the eth Π_1^0 class is empty or B_e is a member of the eth Π_1^0 class. Let $f^{A \oplus B}(e)$ be the first x found such that $B_e \restriction x \neq A \restriction x$. Then $f^{A \oplus B}$ is total since $B \leq_T \emptyset'$ and so A cannot be recursive in B. This is majorized by some B-recursive function g^B. It is easy to see that there is a B-recursive and hence $\emptyset'$-recursive tree T containing exactly the paths X such that for every e, $X \restriction g(e) \neq B_e \restriction g(e)$. Clearly T contains the HIF path A and so by Lemma 3.2 there is a recursive tree Q such that $[Q] \subseteq [T]$. Since $[Q]$ is a non-empty Π_1^0 class, examining its index gives a contradiction. □

We obtain the following pleasing corollary, which says that every non-recursive HIF set cannot be HIF relative to $\emptyset'$:

Corollary 3.1. *If A is HIF and HIF relative to $\emptyset'$ then A is recursive.*

Corollary 3.2. *No PA-complete set can be both HIF and HIF relative to some non-recursive r.e. set.*

Proof. Let C be a PA-complete HIF and B be a non-recursive r.e. set. such that C is HIF relative to B. Since C forms a minimal pair with $\emptyset'$, by Lemma 3.1 we have $C \oplus B \geq_T \emptyset'$. Hence $C \oplus B$ computes the function $c_{\emptyset'}$ where $c_{\emptyset'}(n) = $ least stage s such that $\emptyset'_s \restriction n = \emptyset' \restriction n$. Since C is HIF relative to B, this is dominated by some function $g \leq_T B$. Hence $B \equiv_T \emptyset'$. By Corollary 3.1 we get that C is recursive, a contradiction. □

We now turn to investigating the interactions between HIF and randomness. By the HIF basis theorem, there are random sets of HIF degree. For which sets A are there A-random HIF sets? In the case for $A \leq_T \emptyset'$ we get exactly the class of K-trivial sets, yielding yet another characterization of K-triviality.

Theorem 3.6. *Let $A \leq_T \emptyset'$. Then A is K-trivial iff some HIF set is A-random.*

Proof. Left to right follows trivially from the existence of HIF random reals. Suppose that A is not low for Ω, and some HIF set B is A-random.

Then there exists a Π_1^0 class relative to A which contains B and only contains A-random reals. This class contains no left-r.e. path since A is not low for Ω. This contradicts Lemma 3.2. $\qquad\square$

Attempts to generalize this globally to obtain a characterization of low for Ω fails. Any non-recursive HIF set A cannot be low for Ω, yet by the relativized HIF basis theorem, there exists an A-random which is HIF relative to A and hence HIF.

We now study the situation when we replace "random" with "complex". Recall that a set B is complex if there exists a recursive function f such that $C(B \restriction m) > n$ whenever $m > f(n)$. B is A-complex if the same holds for an A-recursive f and C^A. A set B is autocomplex if there is a B-recursive function f such that $C(B \restriction m) > n$ whenever $m > f(n)$. B is A-autocomplex if the same holds for a $A \oplus B$-recursive f and C^A.

Theorem 3.7. *Let $A \leq_T \emptyset'$.*

(i) If A is K-trivial then some HIF set is A-complex.
(ii) If some HIF set is A-complex then A is low.
(iii) If A is a low r.e. set then some HIF set is A-autocomplex.

Proof. (i): Trivial.

(ii): Let B be a HIF A-complex set. By [8] Theorem 2.3 relativized to A, $B \oplus A$ computes a function $\Phi^{B \oplus A}$ which is DNR relative to A, where the functional $\Phi^{X \oplus A}$ converges for every X. The set of all X such that $\Phi^{X \oplus A}(n) = \Phi_n^A(n)$ for some n is $\Sigma_1^0(A)$, so there exists an A-recursive tree T containing B, where for every path X of T, $\Phi^{X \oplus A}$ is an A-DNR function. By Lemma 3.2 T must contain some left-r.e. path. Hence $\emptyset'$ computes an A-DNR function, and so by Rupprecht, Miller and Ng [12], A is low.

(iii): Suppose A is a low r.e. set. If A is recursive then we are done, so assume that A is non-recursive. Take B to be any HIF PA-complete set. By Lemma 3.1 we have $B \oplus A \geq_T \emptyset'$. $B \oplus A$ is able to compute for each n, a length $f(n)$ such that no string of length $f(n)$ or more has A-Kolmogorov complexity below n, since the oracle A' can search for the correct length. Hence B is A-autocomplex. $\qquad\square$

We remark that by [12], the class of sets $A \leq_T \emptyset'$ where $\emptyset'$ is A-autocomplex is exactly the low sets.

4. HIF and the jump operator

In this section we investigate the degrees which are the jump of a non-recursive HIF. Let $\mathcal{J}_\mathcal{H} = \{C \in 2^\omega \mid C >_T \emptyset'$ and there exists a non-recursive HIF A with $A' \equiv_T C\}$. Let $\mathcal{J}_\mathcal{R}$ be defined similarly with recursively traceable in place of HIF.

By Folklore, every degree computing $0''$ is the double jump of a HIF. However the situation for the single jump appears to be much more difficult. Several necessary conditions are known: By Jockusch and Stephan [6], no degree PA-complete relative to $0'$ is in $\mathcal{J}_H$. Also by Corollary 3.1 no set in $\mathcal{J}_H$ can be HIF relative to $0'$. A related result of Kučera and Nies states that each proper Σ_2^0 degree above $0'$ computes a member of $\mathcal{J}_H$:

Theorem 4.1 (Kučera, Nies [9]). *If $C >_T \emptyset'$ is Σ_2^0 then C computes a set in $\mathcal{J}_H$.*

It is easy to modify their construction to make C compute a set in $\mathcal{J}_R$ (this will also follow from Theorem 4.3 below). In contrast no member of $\mathcal{J}_H$ can have Σ_2^0 degree. In fact we show that no degree in $\mathcal{J}_H$ can compute a properly Σ_2^0 set:

Theorem 4.2. *Suppose A is HIF and $A' \geq_T C$ where C is a Σ_2^0 set. Then $C \leq_T \emptyset'$.*

Proof. Suppose that some Σ_2^0 set C is recursive in A. By the relativized limit lemma, let f be an A recursive function and R a recursive predicate such that for every x, $\lim_s f(x,s) = 1$ iff $(\exists s)(\forall t > s)R(x,t)$. Define $g(x,s)$ to be the first $t > s$ found such that $\neg R(x,t)$ or $f(x,t) = 1$. Then $g(x,s)$ is a total function recursive in A. Let $\tilde{g}$ be a recursive function majorizing g. Let $\tilde{R}(x,s) = \prod_{t=s}^{\tilde{g}(x,s)} R(x,t)$. For each x, $\lim_s \tilde{R}(x,s)$ exists. To see this, suppose that $\tilde{R}(x,s) = 0$ for infinitely many s. Then $x \notin C$ and hence $\lim_s f(x,s) = 0$. Hence for almost every s. there is some $s < t \leq g(x,s)$ for which $\neg R(x,t)$ holds. Hence $\tilde{R}(x,s) = 0$ for almost every s. Finally it is easy to check that $\lim_s f(x,s) = 1$ iff $\lim_s \tilde{R}(x,s) = 1$, and so $C \leq_T \emptyset'$. $\square$

Theorem 4.3. *Let C be 2-generic. Then there is a recursively traceable set A such that $A' \equiv_T A \oplus \emptyset' \equiv_T C \oplus \emptyset'$.*

Proof. We build a recursive sequence of total recursive functions $T_s : 2^{<\omega} \mapsto 2^{<\omega}$ such that for each s, T_s satisfies the usual definition of a tree and for every s and σ, there is some $\tau \supseteq \sigma$ such that $T_{s+1}(\sigma) = T_s(\tau)$. Provided that each σ is moved finitely often, we get that $T = \lim T_s$ exists and $[T]$ is a Π_1^0 class.

We start with T_0 the identity function. For each s and σ, we say that σ requires attention if there exists some $\tau \supset \sigma$ and $i, j < |\sigma|$ such that $\Phi_i^{T_s(\tau)}(j) \downarrow$ but $\Phi_i^{T_s(\sigma)}(j) \uparrow$. At s pick the lexicographically least σ requiring attention, and let $T_{s+1}(\sigma * \eta) = T_s(\tau * \eta)$ for every $\eta \in 2^{<\omega}$. If there is more than one pair (i, j) we move σ for the sake of the least pair in some fixed ordering of pairs of numbers. If no σ requires attention at s, set $T_{s+1} = T_s$.

Clearly each σ requires attention only finitely often. Hence $T \leq_T \emptyset'$. Let C be 2-generic, and $A = T(C)$. Clearly $A \oplus \emptyset' \equiv_T C \oplus \emptyset'$. It remains to verify that A is recursively traceable and $A' \leq_T C \oplus \emptyset'$. To see the former, fix e, and let $V = \{\sigma \mid |\sigma| > e$ and $\Phi_e(i)^{T(\sigma)} \uparrow$ for some $i \leq |\sigma|\} \leq_T \emptyset'$. Hence C must meet or strongly avoid V. If C meets V then by construction Φ_e^A is not total. Otherwise there exists $\eta \subset C$ such that $|\eta| > e$ and no extension of η is in C. This means that for every $\sigma \supseteq \eta$, $\Phi_e^{T(\sigma)}(|\sigma|) \downarrow$. Assume that η is never moved again. To compute a trace for $\Phi_e^A(i)$, $i > e$, we run the construction until a stage s is found such that $\Phi_e^{T_s(\sigma)}(i) \downarrow$ for every $\sigma \supseteq \eta$ of length i. There are at most 2^i many such values. Furthermore since T_s is an approximation to a Π_1^0 class, we have that $A \supset T_s(\sigma)$ for one such σ.

Finally to see that $A' \leq_T C \oplus \emptyset'$, note that $e \in A'$ if and only if $\Phi_e^{T(C\restriction e+1)}(e) \downarrow$. $\qquad\square$

We conclude with the following two questions.

Question 4.1. Do the degrees $\boldsymbol{a} \cup \boldsymbol{0}'$ where $\boldsymbol{a}$ is 2-generic characterize the class $\mathcal{J}_H$?

In the light of Theorems 4.1 and 4.2, the class of degrees $\boldsymbol{a} \cup \boldsymbol{0}'$ where $\boldsymbol{a}$ is 2-generic is a very natural candidate.

Question 4.2. Is the jump of each HIF degree also the jump of a recursively traceable degree?

Acknowledgments

Ng's research was partially supported by NTU reserach grant No. M4080797.110. Stephan was partially supported by NUS research grant WBS R 252-000-420-112. Yang was partially supported by NUS research grant WBS R 146-000-159-112 and NSFC (No. 11171031). Yu was partially supported by NSFC grant No. 11071114.

References

1. A. R. Day. Randomness and Computability. *Ph.D Thesis, Victoria University of Wellington.*
2. J. C. E. Dekker. A theorem on hypersimple sets. *Proceedings of the American Mathematical Society* **5**, 791–796, 1954.
3. J. C. E. Dekker and J. Myhill. Retraceable sets. *Canadian Journal of Mathematics* **10**, 357–373, 1958.
4. R. G. Downey, C. G. Jockusch Jr., and M. Stob. Array nonrecursive sets and multiple permitting arguments. In K. Ambos-Spies, G. H. Müller and G. E. Sacks, editors, *Recursion Theory Week. Proceedings of the Conference Held at the Mathematisches Forschungsinstitut, Oberwolfach,* 141–174, 1990.
5. R. G. Downey, C. G. Jockusch Jr., and M. Stob. Array nonrecursive degrees and genericity. In S. B. Cooper, T. A. Slaman, and S. S. Wainer, editors, *Computability, Enumerability, Unsolvability. Directions in Recursion Theory. London Mathematical Society Lecture Notes Series* **224**, 93–104, 1996.
6. C. G. Jockusch Jr. and R. I. Soare. Π_1^0 classes and degrees of theories. *Transactions of the American Mathematical Society* **173**, 33–56, 1972.
7. C. G. Jockusch Jr. and F. Stephan. A cohesive set which is not high. *Mathematical Logic Quarterly* **39**, 515–530, 1993.
8. B. Kjos-Hanssen, W. Merkle and F. Stephan. Kolmogorov complexity and the recursion theorem. *Transactions of the American Mathematical Society* **363(7)**, 5465–5480, 2011.
9. A. Kučera and A. Nies. Demuth randomness and computational complexity. *Annals of Pure and Applied Logic* **162(7)**, 504–513, 2011.
10. D. A. Martin. Classes of recursively enumerable sets and degrees of unsolvability. *Zeitschrift für Mathematische Logik und Grundlagen der Mathematik* **12**, 295–310, 1966.
11. W. Miller and D. A. Martin. The degrees of hyperimmune sets. *Zeitschrift für Mathematische Logik und Grundlagen der Mathematik* **14**, 159–166, 1968.
12. N. Rupprecht, J. Miller and K. M. Ng. Notions of effectively null and their covering properties. In preparation.
13. E. L. Post. Recursively enumerable sets of positive integers and their decision problems. *Bulletin of the American Mathematical Society* **50**, 284–316, 1944.

Calibrating the complexity of Δ_2^0 sets via their changes

André Nies

Dept. of Computer Science, University of Auckland

The computational complexity of a Δ_2^0 set will be calibrated by the amount of changes needed for any of its computable approximations. Firstly, we study Martin-Löf random sets, where we quantify the changes of initial segments. Secondly, we look at c.e. sets, where we quantify the overall amount of changes by obedience to cost functions. Finally, we combine the two settings. The discussions lead to three basic principles on how complexity and changes relate.

Keywords: Randomness; K-triviality; changes; cost functions; King Arthur

1. Introduction

In computability theory one studies the complexity of sets of natural numbers. A good arena for this is the class of Δ_2^0 sets, that is, the sets Turing below the Halting problem $\emptyset'$. For, by the Shoenfield Limit Lemma, they can be approximated in a computable way. More precisely, the lemma says that a set $Z \subseteq \mathbb{N}$ is Turing below the halting problem $\emptyset'$ if and only if there is a computable function $g \colon \mathbb{N} \times \mathbb{N} \to \{0, 1\}$ such that $Z(x) = \lim_s g(x, s)$ for each $x \in \mathbb{N}$. We will write Z_s for $\{x \colon g(x, s) = 1\}$. The sequence $\langle Z_s \rangle_{s \in \mathbb{N}}$ is called a *computable approximation* of Z.

The paper is set up as a play in three acts. The main topic of the play is to study the complexity of a Δ_2^0 set Z by quantifying the amount of changes that are needed in any computable approximation $(Z_s)_{s \in \mathbb{N}}$ of Z.

- In the first act, we will do this for random Δ_2^0 sets. They are played by knights living in a castle who do a lot of horseback riding.
- In the second act, we will do it mainly for computably enumerable (c.e.) sets. They are played by poor peasants living in a village who are trying to pay their taxes.
- In the final act, we will relate the two cases. The knights and the peasants meet.

The purpose of this work is to provide a unifying background for results in the papers [3,6–8,10,11,13]. It contains many new observations on the amount of changes of knights and peasants, and how they relate. However, it does not contain new technical results.

Martin-Löf randomness

Our central algorithmic randomness notion is the one due to Martin-Löf [18]. It has many equivalent definitions. We give one:

Definition 1.1. We say that a set $Z \subseteq \mathbb{N}$ is Martin-Löf random (ML-random) if for every computable sequence $(\sigma_i)_{i \in \mathbb{N}}$ of binary strings with $\sum_i 2^{-|\sigma_i|} < \infty$, there are only finitely many i such that σ_i is an initial segment of Z.

Note that $\lim_i 2^{-|\sigma_i|} = 0$, so this means that we cannot "Vitali cover" Z (viewed as the binary expansion of a real number) with the collection of dyadic intervals corresponding to $(\sigma_i)_{i \in \mathbb{N}}$. A sequence $(\sigma_i)_{i \in \mathbb{N}}$ as above is called a Solovay test (see e.g. [22, 3.2.2]).

Left-c.e. sets

We will often consider a special type of Δ_2^0 set. We say that $Z \subseteq \mathbb{N}$ is left-c.e. if it has a computable approximation $(Z_s)_{s \in \mathbb{N}}$ such that $Z_s \leq_L Z_{s+1}$, where $\leq_L$ denotes the lexicographical ordering. For instance, Ω, the halting probability of a universal prefix-free machine $\mathbb{U}$ (see, for instance, [22, Ch. 2]), is left-c.e. To see this, let Ω_s be the measure of $\mathbb{U}$-descriptions σ where the computation $\mathbb{U}(\sigma)$ has converged by stage s. This is a dyadic rational, which we identify with a binary string.

It is well known that Ω is ML-random. For general background on algorithmic randomness, see [4,22].

Quantifying changes

We introduce the terminology needed to quantify the changes of initial segments for a computable approximation of a Δ_2^0 set.

Definition 1.2. Let $g \colon \mathbb{N} \to \mathbb{N}$. We say that a Δ_2^0 set Z is a g-change set if it has a computable approximation $(Z_s)_{s \in \mathbb{N}}$ such that an initial segment $Z_s \upharpoonright n$ changes at most $g(n)$ times.

We also say that Z is g-computably approximable, or g-c.a. To be ω-c.a. means to be g-c.a. for some computable g.

We give an important example.

Proposition 1.1. *Every left-c.e. set is a g-change set for some $g = o(2^n)$.*

Proof. Fix a computable approximation $(Z_s)_{s \in \mathbb{N}}$ of Z such that $Z_s \leq_L$ Z_{s+1} for each s. It suffices to note that if $Z \restriction_k$ is stable by stage t, then for every $n > t$, $Z \restriction_n$ changes at most $t + 2^{n-k}$ times. □

If we say that a Δ_2^0 set *needs more than g changes*, we simply mean that it is not a g-change set. Figueira, Hirschfeldt, Miller, Ng and Nies [6] studied such lower bounds for the changes of random Δ_2^0 sets.

Proposition 1.2. *[6] Let Z be a random Δ_2^0 set. Let $q : \mathbb{N} \to \mathbb{R}^+$ be computable and nonincreasing. If Z is a $\lfloor q(n)2^n \rfloor$-change set then $\lim_n q(n) > 0$.*

For example, let $q(n) = 1/\log\log n$. Then $\lim_n q(n) = 0$. Thus, no Martin-Löf random set is a $\lfloor 2^n/\log\log n \rfloor$-change set. As a consequence, for the number of initial segment changes for Ω, the upper bound $o(2^n)$ is not far below 2^n.

Act 1: Martin-Löf random sets and initial segments

The players:

Ω, *the king.*

Z, *a raundon Δ_2^0-knight.*

More knights.

The scene: *The fields outside a castle.*

2.1. *Randomness enhancement*

The *randomness enhancement thesis* states that for a Martin-Löf raundon Z,

Z *gets more random* $\Leftrightarrow$ Z *is computationally less complex.*

The thesis was explicitly and in full generality first mentioned in Section 4 of Nies [20], and published in [23]. Particular instances were given in the literature much earlier on, possibly as far back as Kurtz [15].

The thesis was initially observed only for randomness notions not compatible with being Δ_2^0. Recall that a set Z is weakly 2-random if Z is not in any null Π_2^0 class; Z is 2-random if it is ML-random relative to $\emptyset'$;

Z is low for Ω if Ω is ML-random relative to Z. Lowness for Ω was first studied in Nies, Terwijn and Stephan [25].

Example 2.1. Let $Z \subseteq \mathbb{N}$ be ML-random. Then

Z and $\emptyset'$ form a minimal pair $\Leftrightarrow$ Z is weakly 2-random,
Z is low for Ω $\Leftrightarrow$ Z is 2-random.

The first example is due to Hirschfeldt and Miller; see [22, 5.3.16]. The second example follows from literature results: by Kurtz [15], 2-randomness is equivalent to randomness relative to $\emptyset'$. Chaitin realized that $\Omega \equiv_T \emptyset'$. Since Ω is ML-random, we can now invoke van Lambalgen's theorem to conclude that Ω is ML-random in Z iff Z is ML-random in Ω. The result was first explicitly mentioned in [25].

In contrast, the following, later result of Franklin and Ng [8] is also relevant for Δ_2^0 ML-random sets Z. A *difference test* consists of a sequence of uniformly given Σ_1^0 classes $\mathcal{A}_m$ and a further Σ_1^0 class $\mathcal{B}$ such that $\lambda(\mathcal{A}_m - \mathcal{B}) \le 2^{-m}$ for each m. To *pass* the test means to be out of $\mathcal{A}_m - \mathcal{B}$ for some m. A set Y is *difference random* if it passes all difference tests.

Example 2.2 ([8]). *Let Z be ML-random. Then*

Z *is Turing incomplete* $\Leftrightarrow$ Z *is difference random.*

Weak Demuth randomness is a property strictly in between weak 2-random and ML-random; see for instance [16,17]. Franklin and Ng have recently introduced a property of a c.e. set A called strong promptness, which strictly implies being promptly simple: there is a computable enumeration $(A_s)_{s \in \mathbb{N}}$ of A and an ω-c.a. bound g such that $|W_e| \ge g(e)$ implies that A promptly enumerates some element of W_e. They used this property to provide a further, related, example of randomness enhancement that is also relevant to Δ_2^0 sets: a ML-random Z does not compute a strongly prompt set if and only if it is weakly Demuth random.

2.2. *Malory's thesis*

All the quotes below are from Le Morte D'Arthur (1483) by Sir Thomas Malory[a].

Book III, Chapter IX: How Sir Tor rode after the knight with the brachet[b],

---

[a]Sir Thomas Malory was an English writer who died 1471. His major work, "Le Morte d'Arthur", is a prose translation of a collection of legends about King Arthur (OED). It was printed in 1483 by William Caxton, who also acted as a (somwehat sloppy) editor.
[b]A brachet is a small hunting dog.

and of his adventure by the way.

(...) And anon the knight yielded him to his mercy. But, sir, I have a fellow in yonder pavilion that will have ado with you anon. He shall be welcome, said Sir Tor. Then was he ware of another knight coming with great raundon[c], and each of them dressed to other, that marvel it was to see; but the knight smote Sir Tor a great stroke in midst of the shield that his spear all to-shivered. And Sir Tor smote him through the shield below of the shield that it went through the cost of the knight, but the stroke slew him not. (...)

From this quote one can derive what we will call Sir Thomas Malory's thesis.

Let Z be a Martin-Löf raundon Δ^0_2 set. Then

$$Z \text{ gets more raundon} \Leftrightarrow Z \text{ needs more changes.}$$

Combining the two theses

We combine the randomness enhancement thesis with Malory's thesis by "transitivity". This yields the main principle of this act: for a ML-random Δ^0_2 set Z,

$$\boxed{Z \text{ is computationally less complex} \Leftrightarrow Z \text{ needs more changes.}}$$

We will give multiple evidence for this principle. Firstly, we consider random Δ^0_2 sets that are complex. This should mean that they can be computably approximated with *few* changes. Thereafter, we consider random Δ^0_2 sets that are not complex. They should need *a lot* of changes.

Evidence for the main principle: Complex random Δ^0_2 sets.
1. Chaitin's halting probability Ω is Turing complete. By Fact 1.1, its rate of change is $o(2^n)$, which is at the bottom of the scale of possible changes for a random Δ^0_2 set.
2. Consider all the ML-random sets that are ω-c.a. (Def. 1.2). These sets change much less than a general Δ^0_2 set. By the already mentioned unpublished work of Hirschfeldt and Miller (see [22, 5.3.15]), it turns out that

---

[c]The Old French noun "randon", great speed, is derived from "randir", to gallop. It has been used in English since the 14th century. When used in a metaphorical way, "randon" meant "impetuosity" (OED). Malory's spelling "raundon" may have been an attempt to represent the French pronounciation.

they are "jointly" complex: there is an incomputable c.e. set Turing (even weak truth-table) below all of them. In contrast, by the low basis theorem with upper cone avoidance, for each incomputable c.e. set A, there is a ML-random Δ_2^0 set Z not Turing above A. The closer to computable A is, the more Z has to change; certainly Z is not ω-c.a. in general.

Evidence for the main principle: Non-complex random Δ_2^0 sets.
Recall that a set $Z \subseteq \mathbb{N}$ is *low* if $Z' \leq_T \emptyset'$, and *superlow* if $Z' \leq_{tt} \emptyset'$. To be superlow, a ML-random Δ_2^0 set needs to change considerably, by a result of Figueira, Hirschfeldt, Miller, Ng and Nies.

Theorem 2.1. *[6, Cor. 24] Suppose that a Martin-Löf random set Z is superlow. Then Z is not an $O(2^n)$ change set.*

In fact, in [6, Thm. 23] they showed the slightly stronger result that Z is not an $O(h(n)2^n)$ change set for some order function h.

In contrast, mere lowness can be achieved with fewer changes:

Theorem 2.2. *[6, Thm. 11] Some low Martin-Löf random set Z is an $o(2^n)$ change set.*

We note that the latter result also appears to give some contrary evidence to the main principle that Z is computationally less complex if and only if Z needs more changes: The set Z constructed in Theorem 2.2 has a rate of change similar to the one of Ω, but is low. This suggests that we would need a fine analysis of change bounds in $o(2^n)$ to differentiate between Ω and low random sets. In the the proof of Theorem 2.2, the function $m(k)$ quantifying the "o" in $o(2^n)$, that is, the minimal r such that for each $n \geq r$, $Z \restriction_n$ has at most 2^{n-k} changes, is an ω-c.a. function with $O(4^k)$ increases. In contrast, Ω only needs $O(2^k)$ increases of its analogous function.

Act 2: Computably enumerable sets and cost functions
The players:
Ω, *the King.*
A, *an abject Δ_2^0 peasant.*
The king's tax collector.
The scene: *A village.*

Book VIII, CHAPTER IV: How Sir Marhaus came out of Ireland for to ask truage of Cornwall, or else he would fight therefore.

(...) Then it befell that King Anguish of Ireland sent to Cornwall for his truage[d], that Cornwall had paid many winters. And all that time Cornwall was behind of the truage for seven years. And they gave unto the messenger of Ireland these words and answer, that they would none pay; and bade the messenger go unto his King Anguish, and tell him we will pay him no truage. (...)

Cost functions

Suppose the King issues a tax law. This is a computable function $\mathbf{c}\colon \mathbb{N} \times \mathbb{N} \to \mathbb{Q}^+$ that is nondecreasing in s, and nonincreasing in x. Consider a computable approximation $(A_s)_{s\in\mathbb{N}}$ of a Δ_2^0 peasant A. Suppose that on day s, the number x is least such that $A_s(x)$ changes. Then the tax the peasant pays is $\mathbf{c}(x,s)$. The established terminology for such a tax law is "cost function". Cost functions were used in an ad-hoc way in [5,14,21]. The general theory was developed in [22, Section 5.3], and in more depth in [11,24].

Definition 2.3 ([22]). *We say a Δ_2^0 set A obeys a cost function $\mathbf{c}$ if A has a computable approximation such that the total tax is finite.*

Let $\mathbf{c}^*(x) = \sup_s \mathbf{c}(x,s)$. We say that a cost function $\mathbf{c}$ has the *limit condition* if $\lim_x \mathbf{c}^*(x) = 0$. Informally, this is a fair tax law. We show that one can obey each fair tax law without being taxed to death (where death = computable). This result has roots in the work of Kučera and Terwijn [14] who built an incomputable low-for-random set. Downey et al. [5] gave a construction like this for the particular cost function

$$\mathbf{c}(x,s) = \sum_{w=x+1}^{s} 2^{-K_s(w)}$$

in order to build an incomputable K-trivial set (see below). In full generality, the construction was first stated in [22, Thm. 5.3.5].

Proposition 2.3. *Suppose a cost function $\mathbf{c}$ has the limit condition. Then there is a promptly simple set A obeying $\mathbf{c}$.*

Proof. We meet the usual prompt simplicity requirements

$$PS_e\colon\ |W_e| = \infty \ \Rightarrow\ \exists s\, \exists x\, [x \in W_{e,s} - W_{e,s-1} \ \wedge\ x \in A_s].$$

---

[d]tribute

We define a computable enumeration $\langle A_s \rangle_{s \in \mathbb{N}}$ as follows. Let $A_0 = \emptyset$. At stage $s > 0$, for each $e < s$, if PS_e has not been met so far and there is $x \geq 2e$ such that $x \in W_{e,s} - W_{e,s-1}$ and $\mathbf{c}(x, s) \leq 2^{-e}$, put x into A_s. Declare PS_e to be met.

Note that $\langle A_s \rangle_{s \in \mathbb{N}}$ obeys $\mathbf{c}$, since at most one number is put into A for the sake of each requirement. Thus the total tax the peasant A pays is bounded by $\sum_e 2^{-e} = 2$.

If W_e is infinite, there is an $x \geq 2e$ in W_e such that $\mathbf{c}(x, s) \leq 2^{-e}$ for all $s > x$, because $\mathbf{c}$ satisfies the limit condition. We enumerate such an x into A at the stage $s > x$ where x appears in W_e, if PS_e has not been met yet by stage s. Thus A is promptly simple. $\qquad \square$

In the traditional interpretation (such as [26]), being promptly simple would mean that the set changes quickly. So it seems the result says that a set can change quickly in that traditional sense, yet change little in the sense of the cost function. There is no contradiction because actually, A only has to change quickly *once* for each infinite c.e. set W_e. This is possible even if the global amount of changes is small.

We also note that the actual amount of tax paid is immaterial as long as it is finite: we can always modify the computable approximation so that the tax becomes arbitrarily small. Thus, a single cost function only distinguishes between sets that change little, and sets that change a lot. Later on, we will also consider classes of cost function. Jointly obeying each cost function in such a class yields a finer way to gauge the amount of changes.

When studying obedience to a single cost function, we can focus on the c.e. sets.

Proposition 2.4 ([22], Prop. 5.3.6). *Suppose a Δ_2^0 set A obeys a cost function $\mathbf{c}$. Then there is a computably enumerable set $D \geq_T A$ such that D also obeys $\mathbf{c}$. If A is ω-c.a., then we can in fact achieve that $D \geq_{tt} A$.*

Recall that $K(x)$ denotes the prefix-free complexity of a string x (see e.g. [22, Ch. 2], or [4]). The Levin-Schnorr theorem characterizes ML-randomness of Z via having an initial segment complexity $K(Z \upharpoonright n)$ of about n, which is near the upper bound (see e.g. [22, 3.2.9]). Recall that a set A is K-*trivial* if for some b, $\forall n\, K(A \upharpoonright n) \leq K(n) + b$. Since $K(n) \leq 2 \log n + O(1)$ is the lower bound, this means that A is far from random.

The following characterizes K-triviality among peasants by obedience to the King's tax law $\mathbf{c}^\Omega$, defined by $\mathbf{c}^\Omega(x, s) = \Omega_s - \Omega_x$. This is the amount

Ω increases from x to s. Note that $\mathbf{c}^\Omega$ actually depends on a particular computable approximation of Ω as a left-c.e. real.

Theorem 2.3 ([21], [24]). *A is K-trivial* $\Leftrightarrow$ *A obeys* $\mathbf{c}^\Omega$.

The implication '$\Leftarrow$' is not hard. The implication '$\Rightarrow$' is also not very hard for a c.e. set A, but needs the full power of the so-called golden run method of [21] in the case of a general Δ_2^0 set A. (The proof in [21] was for the cost function $\mathbf{c}_\mathcal{K}$.)

Corollary 2.1. *Every K-trivial set is Turing below a computably enumerable K-trivial set.*

Recall the main principle from Act 1: for a ML-random Δ_2^0 set Z.

Z is computationally less complex $\Leftrightarrow$ Z needs more changes.

For c.e. sets A, we propose a principle that is antipodal to the one for random Δ_2^0 sets:

> A is computationally less complex $\Leftrightarrow$ A obeys stricter cost functions.

Thus, for c.e. sets, being less complex means changing less. We give evidence for this principle, in fact also in the case of left-c.e. sets. Similar to Act 1, we proceed from sets of high complexity to sets of low complexity, and see that this complexity matches their changes in the predicted way. We first show that the King pays no taxes. Thereafter we see that peasants get poorer and poorer as they obey stricter and stricter tax laws.

Evidence 1. The left-c.e. set Ω is Turing complete. It obeys no cost function of any reasonable strength, by the following observation.

Proposition 2.5. *If* $\mathbf{c}$ *is a cost function with* $\mathbf{c}(x, s) \geq 2^{-x}$ *for all* x, s, *then no Martin-Löf random* Δ_2^0 *set* Z *obeys* $\mathbf{c}$.

Proof. We view Z_s as a binary string. At stage $s > 0$, if there is a least p such that $Z_s(p) \neq Z_{s-1}(p)$, we add the string $Z_s \restriction_{p+1}$ to an effective list of strings $(\sigma_i)_{i \in \mathbb{N}}$ as in Definition 1.1. If Z obeys $\mathbf{c}$ via $\langle Z_s \rangle_{s \in \mathbb{N}}$, then $\sum_i 2^{-|\sigma_i|} < \infty$. Since $\sigma_i \prec Z$ for infinitely many i, Z is not ML-random. $\square$

Evidence 2. Bickford and Mills [1] studied sets A such that $A' \leq_{\mathrm{tt}} \emptyset'$. They called these sets *abject*. They mainly studied this property for c.e. sets. Mohrherr [19] introduced the term "superlow" for this property and also provided results outside the c.e. sets.

The following was first proved using the so-called golden run method.

294

Theorem 2.4 ([21]). *Each K-trivial set is superlow. Thus, obeying $\mathbf{c}^\Omega$ implies superlowness.*

Evidence 3. Let J^A be a universal partial computable functional with oracle A. Strong jump traceability, introduced in [7], is a lowness property of a set A saying that the possible values of J^A are very limited: if $J^A(x)$ is defined at all, then it is contained in a tiny c.e. set T_x obtained uniformly from x. In [7] a c.e. but incomputable strongly jump traceable set was built. Cholak et al. [2] showed among other things that some c.e. K-trivial set is not strongly jump traceable. In later papers such as [3,10], strong jump traceability was studied in great depth. For general background, see Section 10.13 of the excellent book [4], and also Section 8.5 of [22].

A cost function $\mathbf{c}$ is called *benign* [11] if one can bound computably in k the number of pairwise disjoint intervals $[x, s)$ with increments $\mathbf{c}(x, s) \geq 2^{-k}$. For instance, the cost function $\mathbf{c}^\Omega$ used in Theorem 2.3 is benign via the bound $k \to 2^k$. Clearly, benignity implies the limit condition.

Theorem 2.5 ([11]). *Let A be c.e. Then*

A is strongly jump traceable $\Leftrightarrow$ A obeys each benign cost function.

Together with Greenberg et al. [10], this shows that a c.e. set is strongly jump traceable iff it is below each ω-c.a. ML-random set. This strengthens the result of Hirschfeldt and Miller from Act 1 that such a set can be incomputable.

Elaborating on Proposition 2.3, Franklin and Ng have shown that every benign cost function is obeyed by a strongly prompt set.

Act 3: Computably enumerable sets below random Δ_2^0 sets

The players:
Z, a raundon Δ_2^0-knight
A, a c.e. peasant.
Village people.

The scene:
A forest between village and castle.

Book VI, Chapter X: How Sir Launcelot rode with a damosel and slew a knight that distressed all ladies and also a villain that kept a bridge.

(...) And so Sir Launcelot and she departed. And then he rode in a deep forest two days and more, and had strait lodging. So on the third day he rode over a long bridge, and there stert upon him suddenly a passing foul churl[e], and he smote his horse on the nose that he turned about, and asked him why he rode over that bridge without his licence. Why should I not ride this way? said Sir Launcelot, I may not ride beside. Thou shalt not choose, said the churl, and lashed at him with a great club shod with iron. Then Sir Launcelot drew his sword and put the stroke aback, and clave his head unto the paps. At the end of the bridge was a fair village, and all the people, men and women, cried on Sir Launcelot, and said, A worse deed didst thou never for thyself (...)

We now consider the situation that $A \leq_T Z$, where Z is a raundon Δ_2^0 knight, and A is an incomputable c.e. peasant. We will see that

> the more Z is allowed to change, the less A can change.

The changes of Z are quantified in the sense of initial segments (Act 1). The changes of A are quantified by obeying cost functions (Act 2). This is in line with combining the main principles of these Acts: if Z changes more then Z is computationally less complex. So the set $A \leq_T Z$ is less complex as well, and hence can change less.

The situation above occurs by the following classical theorem of Kučera which says that every raundon Δ_2^0 knight has an incomputable c.e. peasant as a subject.

Theorem 2.6 ([13]). *Let Z be a random Δ_2^0 set. Then there is a c.e. incomputable set A such that $A \leq_T Z$.*

Greenberg and Nies [11] have given a cost function proof of Kučera's theorem: A is a set obeying a certain cost function c_Z associated with a computable approximation of Z.

Unless $Z \geq_T \emptyset'$, the peasant A in Kučera's theorem is quite obedient. That is, he is restricted in its amount of possible changes. This follows from a result of Hirschfeldt, Nies, and Stephan.

Theorem 2.7 ([12]). *If Z is Turing incomplete, then a set A as in Theorem 2.6 is necessarily K-trivial.*

Depending on the Δ_2^0 knight Z, a c.e. peasant subject to Z is can become arbitrarily obedient by a result of Greenberg et al. [10, Theorem 2.6].

---

[e]archaic: a person of low birth; a peasant (OED)

Theorem 2.8. *Let $\mathcal{P}$ be a non-empty Π_1^0 class consisting only of ML-random sets. Let $\mathbf{c}$ be a cost function with the limit condition. Then there is a Δ_2^0 set $Z \in \mathcal{P}$ such that every c.e. set $A \leq_T Z$ obeys $\mathbf{c}$.*

This result has a complicated history. It started with the main result of Greenberg [9].

Theorem 2.9. *There is a ML-random Δ_2^0 set Z such that every c.e. set A Turing below Z is strongly jump traceable.*

Greenberg built such a set Z directly in early 2009. Thereafter, Kučera and Nies [16] showed that any Demuth random set Z (see [22, Section 3.6]) does the job. (This is another instance of the main principle of this act: if Z is Δ_2^0, it needs to change a lot in order to be Demuth random. This means that A can only change little.) Greenberg et al. [10] defined a cost function $\mathbf{c}$ such that every c.e. set A obeying $\mathbf{c}$ is strongly jump traceable. They combined this with their Theorem 2.8 to obtain yet another proof of the result of Greenberg.

In fact, in Theorem 2.8, instead of ML-randomness of Z we can take membership in any non-empty Π_1^0 class by a result of Nies [24]. In that construction, the more restrictive $\mathbf{c}$, the more Z has to change. If $\mathbf{c}$ is benign as defined before Theorem 2.5, then it is not very restrictive. In this case, the construction makes the set Z ω-c.a. This is predicted by (the contrapositive of) the main principle of this act: if A is allowed more changes, then Z can change less. The extension to Π_1^0 classes shows that in Theorem 2.9, randomness of Z can for instance be replaced by PA completeness.

Exeunt omnes.

References

1. M. Bickford and C.F. Mills. Lowness properties of r.e. sets. Preprint, University of Madison, 1982. To appear in J. Symb. Logic.
2. P. Cholak, R. Downey, and N. Greenberg. Strongly jump-traceability I: the computably enumerable case. *Adv. in Math.*, 217:2045–2074, 2008.
3. R. Downey and N. Greenberg. Strong jump traceability II: K-triviality. *Israel J. Math.* 191: 647-667, 2012. In press.
4. R. Downey and D. Hirschfeldt. *Algorithmic randomness and complexity.* Springer-Verlag, Berlin, 2010. 855 pages.
5. R. Downey, D. Hirschfeldt, A. Nies, and F. Stephan. Trivial reals. In *Proceedings of the 7th and 8th Asian Logic Conferences*, pages 103–131, Singapore, 2003. Singapore University Press.

6. S. Figueira, D. Hirschfeldt, J. Miller, Selwyn Ng, and A Nies. Counting the changes of random Δ_2^0 sets. In *CiE 2010*, pages 1–10, 2010. Journal version to appear in J.Logic. Computation.

7. S. Figueira, A. Nies, and F. Stephan. Lowness properties and approximations of the jump. *Ann. Pure Appl. Logic*, 152:51–66, 2008.

8. J. Franklin and K. M. Ng. Difference randomness. *Proceedings of the American Mathematical Society.* To appear.

9. N. Greenberg. A random set which only computes strongly jump-traceable c.e. sets. *J. Symbolic Logic*, 76(2):700–718, 2011.

10. N. Greenberg, D. Hirschfeldt, and A. Nies. Characterizing the strongly jump traceable sets via randomness. Adv. Math. 231 (2012), no. 3-4, 2252 – 2293.

11. N. Greenberg and A. Nies. Benign cost functions and lowness properties. *J. Symbolic Logic*, 76:289–312, 2011.

12. D. Hirschfeldt, A. Nies, and F. Stephan. Using random sets as oracles. *J. Lond. Math. Soc. (2)*, 75(3):610–622, 2007.

13. A. Kučera. An alternative, priority-free, solution to Post's problem. In *Mathematical foundations of computer science, 1986 (Bratislava, 1986)*, volume 233 of *Lecture Notes in Comput. Sci.*, pages 493–500. Springer, Berlin, 1986.

14. A. Kučera and S. Terwijn. Lowness for the class of random sets. *J. Symbolic Logic*, 64:1396–1402, 1999.

15. S. Kurtz. *Randomness and genericity in the degrees of unsolvability.* Ph.D. Dissertation, University of Illinois, Urbana, 1981.

16. A. Kučera and A Nies. Demuth randomness and computational complexity. *Ann. Pure Appl. Logic*, 162:504–513, 2011.

17. Antonín Kučera and André Nies. Demuth's path to randomness. In *Proceedings of the 2012 international conference on Theoretical Computer Science: computation, physics and beyond*, WTCS'12, pages 159–173, Berlin, Heidelberg, 2012. Springer-Verlag.

18. P. Martin-Löf. The definition of random sequences. *Inform. and Control*, 9:602–619, 1966.

19. J. Mohrherr. A refinement of low_n and high_n for the r.e. degrees. *Z. Math. Logik Grundlag. Math.*, 32(1):5–12, 1986.

20. A. Nies. Applying randomness to computability. University of Auckland preprint based on a series of three lectures at the ASL summer meeting, Sofia, 2009. Available at http://hdl.handle.net/2292/19526.

21. A. Nies. Lowness properties and randomness. *Adv. in Math.*, 197:274–305, 2005.

22. A. Nies. *Computability and randomness*, volume 51 of *Oxford Logic Guides*. Oxford University Press, Oxford, 2009.

23. A. Nies. Studying randomness through computation. In *Randomness through computation*, pages 207–223. World Scientific, 2011.

24. A. Nies. Calculus of cost functions. In preparation, 2012.

25. A. Nies, F. Stephan, and S. Terwijn. Randomness, relativization and Turing degrees. *J. Symbolic Logic*, 70(2):515–535, 2005.

26. Robert I. Soare. *Recursively Enumerable Sets and Degrees.* Perspectives in Mathematical Logic, Omega Series. Springer–Verlag, Heidelberg, 1987.

TOPOLOGICAL FULL GROUPS OF MINIMAL SUBSHIFTS AND JUST-INFINITE GROUPS

Dedicated to the memory of Greg Hjorth

SIMON THOMAS

Mathematics Department, Rutgers University
New Brunswick, New Jersey 08854, USA
E-mail: sthomas@math.rutgers.edu

Using recent work on the algebraic structure of topological full groups of minimal subshifts, we prove that the isomorphism relation on the space of infinite finitely generated simple amenable groups is not smooth. As an application, we deduce that there does not exist an isomorphism-invariant Borel map which selects a just-infinite quotient of each infinite finitely generated group.

Keywords: Borel equivalence relation; Topological full group; Minimal subshift; Just-infinite group.

1. Introduction

In [24], confirming a conjecture of Hjorth-Kechris [16], Thomas-Velickovic proved that the isomorphism relation on the space $\mathcal{G}_{fg}$ of finitely generated groups is a universal countable Borel equivalence relation. (Here $\mathcal{G}_{fg}$ denotes the Polish space of finitely generated groups introduced by Grigorchuk [12]; i.e. the elements of $\mathcal{G}_{fg}$ are the isomorphism types of *marked groups* $(G, \bar{c})$, where G is a finitely generated group and $\bar{c}$ is a finite sequence of generators.) This result suggests the project of analyzing the Borel complexity of the isomorphism relation for various restricted classes of finitely generated groups; and the main result in this paper can be regarded as the first step in this analysis for both the class of infinite finitely generated simple groups and the class of infinite finitely generated amenable groups.

Theorem 1.1. *The isomorphism relation on the space of infinite finitely generated simple amenable groups is not smooth.*

The proof of Theorem 1.1 makes use of some recent work of Giordano-Putnam-Skau [11], Bezuglyi-Medynets [1], Matui [20] and Juschenko-

Monod [18] on the topological full groups of minimal subshifts. More precisely, if $X \subseteq n^{\mathbb{Z}}$ is a minimal subshift and $TF(X)$ is the topological full group, then the commutator subgroup $TF(X)'$ is an infinite finitely generated simple amenable group. Furthermore, if $Y \subseteq n^{\mathbb{Z}}$ is another minimal subshift, then $TF(X)' \cong TF(Y)'$ if and only if X and Y are flip conjugate. (A fuller discussion, including the relevant definitions, will be presented in Section 3.) Hence, in order to prove Theorem 1.1, it is enough to show that the flip conjugacy relation for minimal subshifts $X \subseteq n^{\mathbb{Z}}$ is not smooth. In Section 4, we will prove the stronger result that the flip conjugacy relation for Toeplitz subshifts is not smooth. In [3], Clemens showed that the topological conjugacy relation for *arbitrary* subshifts $X \subseteq n^{\mathbb{Z}}$ is a universal countable Borel equivalence relation. Unfortunately the subshifts constructed by Clemens are very far from minimal; and it is currently not known whether or not the topological conjugacy relation for minimal subshifts is strictly more complex than the Vitali equivalence relation E_0. However, it still seems reasonable to conjecture that the following strengthening of Theorem 1.1 should be true.

Conjecture 1.2. *The isomorphism relation on the space of infinite finitely generated simple amenable groups is countable universal.*

It should be pointed out that it is currently not known whether or not the isomorphism relation on the space $\mathcal{G}_{am}$ of infinite finitely generated amenable groups or on the space $\mathcal{G}_{sim}$ of infinite finitely generated simple groups is countable universal. Of course, it is also natural to consider the complexity of the isomorphism relation on the space $\mathcal{G}_{kaz}$ of finitely generated Kazhdan groups.

Conjecture 1.3. *The isomorphism relation on $\mathcal{G}_{kaz}$ is not smooth.*

Here it is worthwhile pointing out that a result of Ol'shanskii [21] implies that if G is any countable group, then there exists a finitely generated Kazhdan group K such that G embeds into K. More precisely, let H be an infinite hyperbolic Kazhdan group. (Such a group is necessarily finitely presented and hence finitely generated. For example, see Bridson-Haefliger [2, Proposition III.Γ.2.2].) Then, by Ol'shanskii [21], if G is any countable group, then G embeds into a quotient H/N of H; and, since the class of Kazhdan groups is closed under the taking of quotients, it follows that H/N is a finitely generated Kazhdan group. This suggests that the isomorphism relation on $\mathcal{G}_{kaz}$ is also a universal countable Borel equivalence relation.

In Section 5, we will present an application of Theorem 1.1 to the theory of just-infinite groups. Here an infinite group Γ is said to be *just-infinite* if every nontrivial normal subgroup of Γ has finite index. In [13], Grigorchuk observed that if G is an infinite finitely generated group, then G has a just-infinite homomorphic image. To see this, consider the poset

$$\mathbb{P} = \{\, N \trianglelefteq G \mid [\,G : N\,] = \infty \,\},$$

partially ordered by inclusion. If $\{\, N_i \mid i \in I \,\}$ is a chain in $\mathbb{P}$, then $N = \bigcup_{i \in I} N_i \in \mathbb{P}$, since otherwise $[\,G : N\,] < \infty$ and so N is finitely generated, which is a contradiction. Hence, by Zorn's Lemma, there exists a maximal element $N \in \mathbb{P}$ and clearly G/N is just-infinite. Of course, if G is an explicitly given infinite finitely generated group, then it is not necessary to use Zorn's Lemma in order to construct a just-infinite homomorphic image. More precisely, as we will explain in Section 2, there exists a Borel map $\theta : \mathcal{G}_{fg} \to \mathcal{G}_{fg}$ such that if $(\,G, \bar{c}\,) \in \mathcal{G}_{fg}$ is infinite, then $\theta(G, \bar{c})$ is a just-infinite homomorphic image of G. However, as our notation suggests, the definition of the just-infinite group $\theta(G, \bar{c})$ depends essentially upon the finite sequence of generators $\bar{c}$ and it is natural to ask whether there exists such a Borel map θ with the property that the isomorphism type of the just-infinite group $\varphi(G, \bar{c})$ only depends upon the isomorphism type of G. In Section 5, we will use Theorem 1.1 to show that no such map exists.

Theorem 1.4. *There does not exist a Borel map $\theta : \mathcal{G}_{fg} \to \mathcal{G}_{fg}$ such that for all infinite $(\,G, \bar{c}\,)$, $(\,H, \bar{d}\,) \in \mathcal{G}_{fg}$,*

(i) $\theta(G, \bar{c})$ is a just-infinite homomorphic image of G; and
(ii) if $G \cong H$, then $\theta(G, \bar{c}) \cong \theta(H, \bar{d})$.

The remainder of this paper is organized as follows. In Section 2, we will recall some basic notions and results from the theory of countable Borel equivalence relations, including the definition of the space $\mathcal{G}_{fg}$ of (marked) finitely generated groups. In Section 3, we will discuss some recent results concerning the structure of topological full groups of minimal subshifts; and in Section 4, we will prove that the flip conjugacy relation for Toeplitz subshifts is not smooth and hence also that the isomorphism relation on the space of infinite finitely generated simple amenable groups is not smooth. Finally, in Section 5, we will present the proof of Theorem 1.4.

2. Countable Borel equivalence relations

In this section, we will recall some basic notions and results from the theory of countable Borel equivalence relations, including the definition of the space $\mathcal{G}_{fg}$ of (marked) finitely generated groups.

Suppose that $(X, \mathcal{B})$ is a measurable space; i.e. that $\mathcal{B}$ is a σ-algebra of subsets of the set X. Then $(X, \mathcal{B})$ is said to be a *standard Borel space* if there exists a Polish topology $\mathcal{T}$ on X such that $\mathcal{B}$ is the σ-algebra of Borel subsets of $(X, \mathcal{T})$. If X, Y are standard Borel spaces, then a map $f : X \to Y$ is *Borel* if $f^{-1}(Z)$ is a Borel subset of X for each Borel subset $Z \subseteq Y$. Equivalently, $f : X \to Y$ is Borel if graph(f) is a Borel subset of $X \times Y$.

Let X be a standard Borel space. Then a *Borel equivalence relation* on X is an equivalence relation $E \subseteq X^2$ which is a Borel subset of X^2. If E, F are Borel equivalence relations on the standard Borel spaces X, Y respectively, then a Borel map $f : X \to Y$ is said to be a *homomorphism* from E to F if for all x, $y \in X$,

$$x \, E \, y \quad \Longrightarrow \quad f(x) \, F \, f(y).$$

If f satisfies the stronger property that for all x, $y \in X$,

$$x \, E \, y \quad \Longleftrightarrow \quad f(x) \, F \, f(y),$$

then f is said to be a *Borel reduction* and we write $E \leq_B F$. If both $E \leq_B F$ and $F \leq_B E$, then E and F are said to be *Borel bireducible* and we write $E \sim_B F$. Finally we write $E <_B F$ if both $E \leq_B F$ and $F \nleq_B E$.

In this paper, we will only be concerned with *countable Borel equivalence relations*; i.e. Borel equivalence relations E such that every E-equivalence class is countable. A detailed development of the general theory of countable Borel equivalence relations can be found in Jackson-Kechris-Louveau [17]. Here we will only recall some of the most basic results of the theory.

With respect to Borel reducibility, the least complex countable Borel equivalence relations are those which are *smooth*; i.e. those countable Borel equivalence relations E on a standard Borel space X such that E is Borel reducible to the identity relation Id_Y on some (equivalently every) uncountable standard Borel space Y. Next in complexity come those countable Borel equivalence relations E which are Borel bireducible with the *Vitali equivalence relation* E_0, which is defined on the space $2^{\mathbb{N}}$ of infinite binary sequences by

$$x \, E_0 \, y \quad \Longleftrightarrow \quad x(n) = y(n) \text{ for all but finitely many } n.$$

More precisely, by Harrington-Kechris-Louveau [15], if E is any (not necessarily countable) Borel equivalence relation, then E is nonsmooth if and only if $E_0 \leq_B E$. It turns out that there is also a most complex countable Borel equivalence relation E_∞, which is *universal* in the sense that

$F \leq_B E_\infty$ for every countable Borel equivalence relation F. (Clearly this universality property uniquely determines E_∞ up to Borel bireducibility.) Furthermore, E_∞ is strictly more complex than E_0. The universal countable Borel relation E_∞ has a number of natural realizations in many areas of mathematics, including algebra, topology and recursion theory. In particular, by Thomas-Velickovic [24], the isomorphism relation $\cong$ on the space $\mathcal{G}_{fg}$ of finitely generated groups is a universal countable Borel equivalence relation.

Most of our effort in this paper will be devoted to proving that various countable Borel equivalence relations are nonsmooth. Here we will make use of the following two observations.

Proposition 2.1. *If $E \subseteq F$ are countable Borel equivalence relations on the standard Borel space X and E is nonsmooth, then F is also nonsmooth.*

Proof. This is an easy consequence of the Feldman-Moore Theorem [8]. (For example, see Thomas [22, Lemma 2.1].) □

If E, E' are countable Borel equivalence relations on the standard Borel spaces Z, Z', then E is said to be *weakly Borel reducible* to E' if there exists a countable-to-one Borel homomorphism $f : X \to X'$ from E to E'. In this case, we write $E \leq_B^w E'$.

Proposition 2.2. *If E is a countable Borel equivalence relation on the standard Borel space X and $E_0 \leq_B^w E$, then E is nonsmooth.*

Proof. By Thomas [23, Theorem 4.4], if $E_0 \leq_B^w E$, then there exists a countable Borel equivalence relation $E' \subseteq E$ such that $E_0 \leq_B E'$. Hence the result follows from Proposition 2.1. □

In the remainder of this section, we will present a brief discussion of the Polish space $\mathcal{G}_{fg}$ of (marked) finitely generated groups, which is defined as follows. A marked group $(G, \bar{s})$ consists of a finitely generated group with a distinguished sequence $\bar{s} = (s_1, \cdots, s_m)$ of generators. (Here the sequence $\bar{s}$ is allowed to contain repetitions and we also allow the possibility that the sequence contains the identity element.) Two marked groups $(G, (s_1, \cdots, s_m))$ and $(H, (t_1, \cdots, t_n))$ are said to be *isomorphic* if $m = n$ and the map $s_i \mapsto t_i$ extends to a group isomorphism between G and H.

Definition 2.3. For each $m \geq 2$, let $\mathcal{G}_m$ be the set of *isomorphism types* of marked groups $(G, (s_1, \cdots, s_m))$ with m distinguished generators.

Let $\mathbb{F}_m$ be the free group on the generators $\{x_1, \cdots, x_m\}$. Then for each marked group $(G, (s_1, \cdots, s_m))$, we can define an associated epimorphism $\theta_{G,\bar{s}} : \mathbb{F}_m \to G$ by $\theta_{G,\bar{s}}(x_i) = s_i$. It is easily checked that two marked groups $(G, (s_1, \cdots, s_m))$ and $(H, (t_1, \cdots, t_m))$ are isomorphic if and only if $\ker \theta_{G,\bar{s}} = \ker \theta_{H,\bar{t}}$. Thus we can naturally identify $\mathcal{G}_m$ with the set $\mathcal{N}_m$ of normal subgroups of $\mathbb{F}_m$. Note that $\mathcal{N}_m$ is a closed subset of the compact space $2^{\mathbb{F}_m}$ of all subsets of $\mathbb{F}_m$ and so $\mathcal{N}_m$ is also a compact space. Hence, via the above identification, we can regard $\mathcal{G}_m$ as a compact space.

For each $m \geq 2$, there is a natural embedding of $\mathcal{N}_m$ into $\mathcal{N}_{m+1}$ defined by

$$N \mapsto \text{ the normal closure of } N \cup \{x_{m+1}\} \text{ in } \mathbb{F}_{m+1}.$$

Thus we can identify $\mathcal{N}_m$ with the clopen subset $\{ N \in \mathcal{N}_{m+1} \mid x_{m+1} \in N \}$ of $\mathcal{N}_{m+1}$ and form the locally compact Polish space $\mathcal{N}_{fg} = \bigcup \mathcal{N}_m$. Note that $\mathcal{N}_{fg}$ can be identified with the space of normal subgroups N of the free group $\mathbb{F}_\infty$ on countably many generators such that N contains all but finitely many elements of the basis $B = \{x_i \mid i \in \mathbb{N}^+\}$. Similarly, we can form the locally compact Polish space $\mathcal{G}_{fg} = \bigcup \mathcal{G}_m$ of finitely generated groups via the corresponding natural embedding

$$(G, (s_1, \cdots, s_m)) \mapsto (G, (s_1, \cdots, s_m, 1))$$

From now on, we will identify $\mathcal{G}_m$ and $\mathcal{N}_m$ with the corresponding clopen subsets of $\mathcal{G}_{fg}$ and $\mathcal{N}_{fg}$. If $\Gamma \in \mathcal{G}_{fg}$, then we will write $\Gamma = (G, (s_1, \cdots, s_m))$, where m is the least integer such that $\Gamma \in \mathcal{G}_m$. Following the usual convention, we will completely identify the Polish spaces $\mathcal{G}_{fg}$ and $\mathcal{N}_{fg}$; and we will work with whichever space is most convenient in any given context.

For example, to see that there exists a Borel map $\theta : \mathcal{G}_{fg} \to \mathcal{G}_{fg}$ such that if $(G, \bar{c}) \in \mathcal{G}_{fg}$ is infinite, then $\theta(G, \bar{c})$ is a just-infinite homomorphic image of G, it is convenient to work with $\mathcal{N}_{fg}$ as follows. Suppose that $(G, \bar{c}) \in \mathcal{G}_{fg}$ is infinite and that $N \in \mathcal{N}_{fg}$ is the corresponding normal subgroup of $\mathbb{F}_\infty$. Then, working with a fixed enumeration $\{ w_k \mid k \in \mathbb{N} \}$ of $\mathbb{F}_\infty$, we can define an increasing sequence of normal subgroups $N_k \in \mathcal{N}_{fg}$ by:

- $N_0 = N$.
- N_{k+1} is the normal closure M_k of $N_k \cup \{ w_k \}$ in $\mathbb{F}_\infty$ if $[\mathbb{F}_\infty : M_k] = \infty$. Otherwise, $N_{k+1} = N_k$.

Finally let $N_\omega = \bigcup_{k \in \mathbb{N}} N_k$. Then $\mathbb{F}_\infty / N_\omega$ is a just-infinite homomorphic image of G and the map $N \mapsto N_\omega$ is Borel.

In the remaining sections of this paper, the symbol $\cong$ will always denote the usual isomorphism relation on the space $\mathcal{G}_{fg}$ of finitely generated groups; i.e. two marked groups are $\cong$-equivalent if their underlying groups (obtained by forgetting about the distinguished sequences of generators) are isomorphic. Finally, we should mention that we will occasionally slightly abuse notation and denote the elements of $\mathcal{G}_{fg}$ by G, H, etc. instead of the more accurate $(G, \bar{c})$, $(H, \bar{d})$, etc.

3. Topological full groups of minimal subshifts

In this section, we will discuss some recent results concerning the structure of topological full groups of Cantor minimal systems. Let (X, T) be a *Cantor dynamical system*; i.e. X is a Cantor set and $T : X \to X$ is a homeomorphism. Then (X, T) is said to be a *Cantor minimal system* if X has no nonempty proper closed T-invariant subsets. It is well-known that a Cantor dynamical system (X, T) is minimal if and only if X is the closure of the orbit of an almost periodic point $x \in X$; and, in this case, every point $x \in X$ is almost periodic.

Definition 3.1. If (X, T) is a Cantor dynamical system, then the point $x \in X$ is *almost periodic* if for every open neighborhood U of x, the set

$$R = \{\, \ell \in \mathbb{Z} \mid T^\ell(x) \in U \,\}$$

of return times has bounded gaps; i.e. there exists a fixed $d \geq 1$ such that for all $z \in \mathbb{Z}$,

$$R \cap \{\, z, z+1, \cdots, z+d \,\} \neq \emptyset$$

Since Cantor minimal systems are infinite, it follows that they do not contain any genuinely periodic points; i.e. points with finite orbits.

Definition 3.2. If (X, T) is a Cantor minimal system, then the *topological full group* $[[\,T\,]]$ is the group of all homeomorphisms $\pi : X \to X$ such that there exists a partition $X = C_1 \sqcup \cdots \sqcup C_m$ into clopen subsets and $\ell_1, \cdots, \ell_m \in \mathbb{Z}$ such that $\pi \restriction C_i = T^{\ell_i} \restriction C_i$ for each $1 \leq i \leq m$.

The Cantor minimal systems (X, T) and (Y, S) are said to be *topologically conjugate* if there exists a homeomorphism $\pi : X \to Y$ such that $\pi \circ T = S \circ \pi$. If (X, T) is topologically conjugate to either (Y, S) or (Y, S^{-1}), then (X, T) and (Y, S) are said to be *flip conjugate*. The following theorem combines the work of Giordano-Putnam-Skau [11] and Bezuglyi-Medynets [1].

Theorem 3.3. *If* (X, T), (Y, S) *are Cantor minimal systems, then the following are equivalent.*

(i) (X, T), (Y, S) *are flip conjugate.*

(ii) *The topological full groups* $[[T]]$, $[[S]]$ *are isomorphic as abstract groups.*

(iii) *The commutator subgroups* $[[T]]'$, $[[S]]'$ *are isomorphic as abstract groups.*

If $n \geq 2$, then the shift transformation σ on the Cantor space $n^{\mathbb{Z}}$ is defined by $\sigma(x)_k = x_{k+1}$. An infinite subset $X \subseteq n^{\mathbb{Z}}$ is said to be a *subshift* if X is a closed σ-invariant subset. The subshift X is *minimal* if the corresponding Cantor dynamical system (X, σ) is minimal. In this case, we also say that (X, σ) is a minimal subshift. The following theorem is due to Matui [20].

Theorem 3.4. *Let* (X, T) *be a Cantor minimal system.*

(a) *The commutator subgroup* $[[T]]'$ *is an infinite simple group.*

(b) *The commutator subgroup* $[[T]]'$ *is finitely generated if and only if* (X, T) *is topologically conjugate to a minimal subshift over a finite alphabet.*

The following result, which confirms a conjecture of Grigorchuk-Medynets [14], was recently proved by Juschenko-Monod [18].

Theorem 3.5. *The topological full group of any Cantor minimal system is amenable.*

The following result is now an immediate consequence of Theorem 3.4 and Theorem 3.5.

Corollary 3.6. *If the Cantor minimal system* (X, T) *is topologically conjugate to a minimal subshift over a finite alphabet, then the commutator subgroup* $[[T]]'$ *is an infinite finitely generated simple amenable group.*

Throughout this paper, the collection $\mathcal{M}_n$ of minimal subshifts $X \subseteq n^{\mathbb{Z}}$ will be regarded as a subspace of the standard Borel space $K(n^{\mathbb{Z}})$ of closed subspaces of $n^{\mathbb{Z}}$. (This corresponds to identifying each minimal subshift X with the corresponding pruned tree $T_X \subseteq \bigcup_{k \in \mathbb{N}} n^{[-k,k]}$ such that X is the set $[T_X]$ of infinite branches through T_X.) Recall that a subshift $X \subseteq n^{\mathbb{Z}}$ is minimal if and only if some (equivalently every) point $x \in X$ is almost periodic. It follows easily that $\mathcal{M}_n$ is a Borel subset of $K(n^{\mathbb{Z}})$ and hence that $\mathcal{M}_n$ is a standard Borel space. We will make use of the following result in Section 4.

Proposition 3.7. *The topological conjugacy relation E_{tc} and the flip conjugacy relation E_{fp} are both countable Borel equivalence relations on $\mathcal{M}_n$.*

Proof. By Clemens [3, Lemma 9], E_{tc} is a countable Borel equivalence relation on $\mathcal{M}_n$; and, of course, this implies that E_{fp} is also a countable Borel equivalence relation. $\qquad\square$

From now on, if (X, σ) is a minimal subshift, then we will write $TF(X)$ for the corresponding topological full group.

Theorem 3.8. *For each $n \geq 2$, there exists a Borel map $X \mapsto G_X$ from $\mathcal{M}_n$ to $\mathcal{G}_{fg}$ such that $G_X \cong TF(X)'$.*

Let $X \in \mathcal{M}_n$ be a minimal subshift. In order to define the (marked) group $G_X \in \mathcal{G}_{fg}$, we first need to explicitly describe a finite generating set for $TF(X)'$. The following set of generators was originally extracted from Matui [20, Section 5] by Grigorchuk and Medynets in an early version of their paper [14].

Definition 3.9. Suppose that $A \subseteq X$ is a clopen subset such that the sets A, $\sigma(A)$ and $\sigma^2(A)$ are pairwise disjoint. Then the homeomorphism $\gamma_A \in TF(X)$ is defined by

$$
\gamma_A(x) = \begin{cases} \sigma(x) & \text{if } x \in A \cup \sigma(A); \\ \sigma^{-2}(x) & \text{if } x \in \sigma^2(A); \\ x & \text{otherwise.} \end{cases}
$$

By Matui [20, Section 5], each such homeomorphism γ_A is an element of $TF(X)'$; and, furthermore, $TF(X)'$ is generated by a suitably chosen finite subset of these homeomorphisms. In more detail, for each $m \geq 1$, let $\mathcal{B}_m(X)$ be the set of all *m-blocks* that occur in sequences $x \in X$; i.e. the words of the form

$$
x \restriction [k, k + m - 1] = x_k\, x_{k+1} \cdots x_{k+m-1}
$$

for some $x \in X$ and $k \in \mathbb{Z}$. And for each $w \in \mathcal{B}_m(X)$ and $k \in \mathbb{Z}$, let

$$
S_k(w) = \{\, x \in X \mid x \restriction [k, k + m - 1] = w \,\}.
$$

Then there exists an integer $m_0 \geq 1$ such that for each $w \in \mathcal{B}_{m_0}(X)$, $k \in \mathbb{Z}$ and $1 \leq i \leq 4$,

$$\sigma^i(S_k(w)) \cap S_k(w) = S_{k-i}(w) \cap S_k(w) = \emptyset.$$

(If not, then an easy compactness argument yields an element $x \in X$ such that $\sigma^i(x) = x$ for some $1 \leq i \leq 4$, which contradicts the fact that minimal subshifts contain no periodic points.) Finally, as pointed out by Grigorchuk and Medynets, the proof of Matui [20, Theorem 5.4] shows that the following result holds.

Proposition 3.10. *If (X, σ) is a minimal subshift, then $TF(X)'$ is generated by $D_{m_0} = \{\gamma_{S_0(w)} \mid w \in \mathcal{B}_{m_0+3}(X)\}$.*

We are now ready to present the proof of Theorem 3.8. Suppose that $X \in \mathcal{M}_n$ is a minimal subshift. Then, in a Borel manner, we can choose an integer $m_0 \geq 1$ such that $D_{m_0} = \{\gamma_{S_0(w)} \mid w \in \mathcal{B}_{m_0+3}(X)\}$ generates $T(X)'$, together with an ordering $\varphi_1, \cdots, \varphi_t$ of the elements of D_{m_0}. Also, again in a Borel manner, we can choose an element $p_X \in X$. (For example, see Kechris [19, Theorem 12.13].) Let $X \mapsto N_X \in \mathcal{N}_t \subseteq \mathcal{N}_{fg}$ be the Borel map defined by

$$w(x_1, \cdots, x_t) \in N_X \iff w(\varphi_1, \cdots, \varphi_t)(\sigma^n(p_X)) = \sigma^n(p_X) \text{ for all } n \in \mathbb{Z}.$$

Since $\{\sigma^n(p_X) \mid n \in \mathbb{Z}\}$ is dense in X and each φ_i is a homeomorphism, it follows that $\mathbb{F}_t/N_X \cong TF(X)'$. This completes the proof of Theorem 3.8.

4. Toeplitz subshifts

Combining Theorem 3.3, Corollary 3.6 and Theorem 3.8, we see that in order to prove that the isomorphism relation on the space of infinite finitely generated simple amenable groups is not smooth, it is enough to show that the flip conjugacy relation E_{fc} on the space $\mathcal{M}_n$ of minimal subshifts $X \subseteq n^{\mathbb{Z}}$ is not smooth. In [10, Section 9.3], Gao-Jackson-Seward proved that for each $n \geq 2$, the topological conjugacy relation E_{tc} on $\mathcal{M}_n$ is not smooth. Applying Proposition 2.1, since E_{tc}, E_{fc} are countable Borel equivalence relations and $E_{tc} \subseteq E_{fc}$, it follows that the flip conjugacy relation E_{fc} on $\mathcal{M}_n$ is also not smooth.

Since Gao-Jackson-Seward [10] work in the more general setting of minimal free G-subflows, where G is an arbitrary countably infinite group, their proofs are necessarily technically complex. (In fact, the existence of a single free G-flow for an arbitrary countably infinite group G has only recently been established by Gao-Jackson-Seward [9].) In this section, we will present an elementary proof that the topological conjugacy relation E_{tc} on the space $\mathcal{M}_n$ of minimal subshifts $X \subseteq n^{\mathbb{Z}}$ is not smooth.

Definition 4.1. An element $x \in n^{\mathbb{Z}}$ is said to be a *Toeplitz sequence* if for all $a \in \mathbb{Z}$, there exists $b \in \mathbb{N}^+$ such that $x(a + kb) = x(a)$ for all $k \in \mathbb{Z}$.

Clearly if $x \in n^{\mathbb{Z}}$ is a Toeplitz sequence, then x is almost periodic. Hence, if x is a nonperiodic Toeplitz sequence, then the closure X of its orbit is a minimal subshift of $n^{\mathbb{Z}}$; and, in this case, X is said to be a *Toeplitz flow*. By Downarowicz [6, Theorem 5.1], if X is a Toeplitz flow, then the set of all Toeplitz sequences $z \in X$ is a dense G_δ subset of X. It follows that the set $\mathcal{T}_n \subseteq \mathcal{M}_n$ of Toeplitz flows is a Borel subset of $\mathcal{M}_n$ and hence is a standard Borel space. (For example, see Kechris [19, Section 16.A].) The remainder of this section is devoted to the proof of the following result.

Theorem 4.2. *The topological conjugacy relation E_{tc} on the space $\mathcal{T}_n$ of Toeplitz flows is not smooth.*

Of course, applying Proposition 2.1, this immediately implies the following result.

Corollary 4.3. *The flip conjugacy relation E_{fc} on the space $\mathcal{T}_n$ of Toeplitz flows is not smooth.*

Clearly it is enough to prove Theorem 4.2 in the special case when $n = 2$. For each $z \in 2^{\mathbb{N}}$, let $\tilde{z} \in 2^{\mathbb{Z}}$ be the corresponding Toeplitz sequence defined as follows.

- For each $m \geq 1$, let $B_m = [0, 2^m - 1]$ and suppose inductively that we have defined the value $\tilde{z}(\ell)$ for all integers $\ell \in B_m \smallsetminus \{a_m, b_m\}$, where $0 \leq a_m < b_m \leq 2^m - 1$. Let $c_m = a_m$ if m is odd and $c_m = b_m$ if m is even. Then we define

$$\tilde{z}(c_m + k\, 2^m) = z(m - 1) \quad \text{for all } k \in \mathbb{Z}.$$

For example, at the beginning of stage 3 of the construction, $\tilde{z} \restriction B_3$ is given by

$$z(0) * z(0)\, z(1)\, z(0) * z(0)\, z(1)$$

where the $*$ indicates that the value has not yet been defined. We then define $\tilde{z}(1 + 8k) = z(2)$ for all $k \in \mathbb{Z}$ and hence obtain that for all $k \in \mathbb{Z}$,

$$\tilde{z} \restriction [\, 8k, 8(k + 1)\,) = z(0)\, z(2)\, z(0)\, z(1)\, z(0) * z(0)\, z(1)$$

Notice that if $z \in 2^{\mathbb{N}}$ is an eventually constant sequence, then $\tilde{z}$ is a periodic sequence and so the corresponding orbit $\{ \sigma^m(\tilde{z}) \mid m \in \mathbb{Z} \}$ is a finite closed subset of $2^{\mathbb{N}}$. From now on, let $\mathrm{Ec}(2^{\mathbb{N}})$ be the set of eventually constant sequences $z \in 2^{\mathbb{N}}$ and let $\mathrm{Nec}(2^{\mathbb{N}}) = 2^{\mathbb{N}} \smallsetminus \mathrm{Ec}(2^{\mathbb{N}})$.

Definition 4.4. Suppose that $x \in n^{\mathbb{Z}}$ is a Toeplitz sequence.

(i) For each $a \in \mathbb{Z}$, the corresponding *minimal period* $\mathrm{per}_x(a)$ is the least integer $b \geq 1$ such that $x(a + kb) = x(a)$ for all $k \in \mathbb{Z}$.

(ii) The set of *essential periods* of x is defined to be $\{ \mathrm{per}_x(a) \mid a \in \mathbb{Z} \}$.

Lemma 4.5. *If $z \in \mathrm{Nec}(2^{\mathbb{N}})$, then $\{ 2^m \mid m \in \mathbb{N}^+ \}$ is the set of essential periods of $\tilde{z}$.*

Proof. With the above notation, it is enough to show that each $c_m \in B_m$ has minimal period 2^m. This is clear when $m = 1$. So suppose that $m > 1$. Then at the beginning of stage m of the construction, $\tilde{z} \restriction B_m$ has the form $\bar{a} * \bar{b} \, \bar{a} * \bar{b}$, where $\bar{a} * \bar{b}$ has length 2^{m-1}; and at the end of stage m, we know that $\tilde{z}$ has the form

$$\cdots \bar{a} * \bar{b} \, \bar{a} \, z(m-1) \, \bar{b} \, \bar{a} * \bar{b} \, \bar{a} \, z(m-1) \, \bar{b} \, \bar{a} * \bar{b} \, \bar{a} \, z(m-1) \, \bar{b} \cdots$$

Clearly 2^m is a period of c_m. Also, since z is not eventually constant, we must eventually replace some $*$ by a value $z(\ell) \neq z(m-1)$ and so 2^{m-1} is not a period of c_m. Thus c_m has minimal period 2^m. $\square$

Definition 4.6. For each $z \in \mathrm{Nec}(2^{\mathbb{N}})$, and $m \in \mathbb{N}^+$, let $W_m(\tilde{z})$ be the set of subsequences of $\tilde{z}$ of the form $\tilde{z} \restriction [\, k \, 2^m, (k+1)2^m \,)$ for some $k \in \mathbb{Z}$.

Lemma 4.7. *If $z \in \mathrm{Nec}(2^{\mathbb{N}})$, then $|W_m(\tilde{z})| = 2$ for all $m \in \mathbb{N}^+$.*

Proof. If at the end of stage m of the construction, $\tilde{z} \restriction B_m$ has the form $\bar{c} * \bar{d}$, then $W_m(\tilde{z}) = \{ \bar{c} \, 0 \, \bar{d}, \bar{c} \, 1 \, \bar{d} \}$. $\square$

For each element $z \in \mathrm{Nec}(2^{\mathbb{N}})$, let $X_z \in \mathcal{T}_2$ be the closure of the orbit $\{ \sigma^n(\tilde{z}) \mid n \in \mathbb{Z} \}$ in $2^{\mathbb{Z}}$. Then it is clear that the map $z \mapsto X_z$ from $\mathrm{Nec}(2^{\mathbb{N}})$ to $\mathcal{T}_2$ is Borel.

Proposition 4.8. *If $y, z \in \mathrm{Nec}(2^{\mathbb{N}})$ and $y \, E_0 \, z$, then the Toeplitz flows X_y and X_z are topologically conjugate.*

We will make use of the following result, which is a special case of Downarowicz-Kwiatkowski-Lacroix [7, Theorem 1].

Lemma 4.9. *If y, $z \in \mathrm{Nec}(2^{\mathbb{N}})$, then the following statements are equivalent:*

(i) There exists a topological conjugacy $\pi : X_y \to X_z$ such that $\pi(\tilde{y}) = \tilde{z}$.
(ii) For some $m \in \mathbb{N}^+$, there exists a bijection $\Pi : W_m(\tilde{y}) \to W_m(\tilde{z})$ such that

$$\tilde{z} \restriction [\, k\, 2^m, (k+1)2^m\,) = \Pi(\, \tilde{y} \restriction [\, k\, 2^m, (k+1)2^m\,)\,)$$

for all $k \in \mathbb{Z}$.

Proof of Proposition 4.8. Suppose that $y(m) = z(m)$ for all $m \geq m_0$; and suppose that at the end of stage m_0 of the constructions of $\tilde{y}$, $\tilde{z}$, we have that $\tilde{y} \restriction B_{m_0} = \bar{a} * \bar{b}$ and $\tilde{z} \restriction B_{m_0} = \bar{c} * \bar{d}$. Then $W_{m_0}(\tilde{y}) = \{\, \bar{a}\,0\,\bar{b}, \bar{a}\,1\,\bar{b}\,\}$ and $W_{m_0}(\tilde{z}) = \{\, \bar{c}\,0\,\bar{d}, \bar{c}\,1\,\bar{d}\,\}$. Furthermore, for any $k \in \mathbb{Z}$, the unique $*$ in the interval $[\, k\, 2^{m_0}, (k+1)2^{m_0}\,)$ is replaced at the same stage $m > m_0$ in the constructions of $\tilde{y}$ and $\tilde{z}$ with the value of $y(m-1) = z(m-1)$. Hence the map $\Pi : W_{m_0}(\tilde{y}) \to W_{m_0}(\tilde{z})$, defined by $\Pi(\bar{a}\,\varepsilon\,\bar{b}) = \bar{c}\,\varepsilon\,\bar{d}$ for $\varepsilon = 0, 1$, satisfies statement (ii) of Lemma 4.9. $\square$

From now on, let $B = \{\, 2^{m+1} - 1 \mid m \in \mathbb{N}^+ \,\}$ and let Z be the standard Borel subspace of $\mathrm{Nec}(2^{\mathbb{N}})$ defined by

$$Z = \{\, z \in \mathrm{Nec}(2^{\mathbb{N}}) \mid z(n) = 0 \text{ for all } n \in \mathbb{N} \smallsetminus B \,\}.$$

Clearly $E_0 \restriction Z$ is Borel bireducible with E_0.

Proposition 4.10. *The Borel map $\theta : Z \to \mathcal{T}_2$ defined by $z \mapsto X_z$ is injective.*

Combining Propositions 4.8 and 4.10, we see that the map $z \mapsto X_z$ is a weak Borel reduction from $E_0 \restriction Z$ to the topological conjugacy relation E_{tc} on $\mathcal{T}_2$. Hence, applying Proposition 2.2, it follows that E_{tc} is not smooth. This completes the proof of Theorem 4.2.

Proof of Proposition 4.10. Suppose that $y \neq z \in Z$. Then we can assume that $y(n) = 0$ and $z(n) = 1$, where n is the least integer such that $y(n) \neq z(n)$. Let $n = 2^{m+1} - 1$ and let $s = 2^m$. Suppose that at the end of stage s of the constructions of $\tilde{y}$, $\tilde{z}$, we have that

$$\tilde{y} \restriction B_s = \bar{a} * \bar{b} = \tilde{z} \restriction B_s.$$

Then $W_s(\tilde{y}) = W_s(\tilde{z}) = \{\, \bar{a}\,0\,\bar{b}, \bar{a}\,1\,\bar{b}\,\}$ and $\tilde{y}$, $\tilde{z}$ are concatenations of the 2^s-blocks $\bar{a}\,0\,\bar{b}$ and $\bar{a}\,1\,\bar{b}$. Let $t = 2^{m+1} = 2s$ and consider $\tilde{z} \restriction B_t$. Since

$z(t-1) = z(n) = 1$, it follows that the concatenation of 2^s-blocks in $\tilde{z}$ contains a subsequence of period $2^t/|B_s| = 2^s$ in which $\bar{a}\,1\,\bar{b}$ occurs. Similarly, since $z(s) = 0$, the concatenation of 2^s-blocks in $\tilde{z}$ contains a subsequence of period 2 in which $\bar{a}\,0\,\bar{b}$ occurs. Thus each occurrence of $\bar{a}\,1\,\bar{b}$ in the expression of $\tilde{z}$ as a concatenation of 2^s-blocks is preceded and followed by occurrences of $\bar{a}\,0\,\bar{b}$. We claim that the sequence

$$\underbrace{\bar{a}\,0\,\bar{b}\,\bar{a}\,0\,\bar{b}\cdots\bar{a}\,0\,\bar{b}}_{2^s+1\,times} \tag{4.10}$$

cannot occur as a subsequence of $\tilde{z}$. For suppose that the sequence (4.10) occurs as the subsequence $\bar{u} = z_k\cdots z_{k+(2^s+1)2^s-1}$ of $\tilde{z}$. Then $\bar{u}$ must contain 2^s consecutive 2^s-blocks in the expression of $\tilde{z}$ as a concatenation of 2^s-blocks, one of which must be $\bar{a}\,1\,\bar{b}$. However, the sequence (4.10) clearly repeats with period 2^s and so we must obtain two consecutive occurrences of the 2^s-block $\bar{a}\,1\,\bar{b}$, which is impossible. On the other hand, since $y(\ell) = 0$ for all $s \le \ell \le 4s - 2 = 2^{m+2} - 2$, it follows easily that the sequence (4.10) occurs as a sub-block of $\tilde{y}$. Clearly this means that $X_y \ne X_z$. □

5. The proof of Theorem 1.4

In this final section, we will present the proof of Theorem 1.4. Our argument involves the following variant of the Vitali equivalence relation E_0.

Definition 5.1. For each $x \in 2^{\mathbb{N}}$, let $\bar{x} \in 2^{\mathbb{N}}$ be the element defined by

$$\bar{x}(n) = 1 - x(n) \qquad \text{for all } n \in \mathbb{N}.$$

Then E_0^* is the countable Borel equivalence relation on $2^{\mathbb{N}}$ defined by

$$x\,E_0^*\,y \qquad \Longleftrightarrow \qquad x\,E_0\,y \text{ or } x\,E_0\,\bar{y}.$$

Thus each E_0^*-class consists of exactly two E_0-classes. The proof of Theorem 1.4 makes use of the fact that there does not exist a Borel selection of an E_0-class within each E_0^*-class.[a] For the sake of completeness, we have included a proof of this standard result.

Proposition 5.2. *There does not exist a Borel homomorphism $\theta : 2^{\mathbb{N}} \to 2^{\mathbb{N}}$ from E_0^* to E_0 such that $\theta(x)\,E_0^*\,x$ for all $x \in 2^{\mathbb{N}}$.*

---

[a]I first learned of this "standard measure-theoretic fact" from Coskey-Schneider [4].

Proof. Suppose that $\theta : 2^{\mathbb{N}} \to 2^{\mathbb{N}}$ is such a Borel homomorphism. Let μ be the usual product probability measure on $2^{\mathbb{N}}$ and let

$$X = \{\, x \in 2^{\mathbb{N}} \mid x \, E_0 \, y \text{ for some } y \in \operatorname{ran} \theta \,\}.$$

Then X is a Borel tail event; and hence, by Kolmogorov's Zero-One Law, we have that $\mu(X) = 0, 1$. However, since the map $x \mapsto \bar{x}$ is measure preserving, it follows that $\mu(2^{\mathbb{N}} \smallsetminus X) = \mu(X)$, which is impossible. $\quad\square$

We are now ready to present the proof of Theorem 1.4. Suppose that there exists a Borel map $\varphi : \mathcal{G}_{fg} \to \mathcal{G}_{fg}$ such that for all infinite G, $H \in \mathcal{G}_{fg}$,

(i) $\varphi(G)$ is a just-infinite homomorphic image of G; and
(ii) if $G \cong H$, then $\varphi(G) \cong \varphi(H)$.

Applying Theorem 1.1 and Harrington-Kechris-Louveau [15], there exists a Borel reduction $z \mapsto S_z$ from E_0 to the isomorphism relation $\cong$ on the space of infinite finitely generated simple amenable groups. (The fact that S_z is amenable will play no role in the proof of Theorem 1.4.) Consider the Borel map $\psi : 2^{\mathbb{N}} \to \mathcal{G}_{fg}$ defined by $z \mapsto G_z = S_z \times S_{\bar{z}}$. Since S_z and $S_{\bar{z}}$ are nonabelian simple groups, the only nontrivial proper normal subgroups of G_z are S_z and $S_{\bar{z}}$; and it follows that:

(iii) ψ is a Borel reduction from E_0^* to $\cong$.
(iv) Each $\varphi(G_z)$ is isomorphic to either S_z or $S_{\bar{z}}$.

Thus the Borel map $\theta : 2^{\mathbb{N}} \to 2^{\mathbb{N}}$ defined by

$$\theta(z) = y \quad \Longleftrightarrow \quad y \in \{\, z, \bar{z} \,\} \text{ and } \varphi(G_z) \cong S_y$$

is a homomorphism from E_0^* to E_0 such that $\theta(x) \, E_0^* \, x$ for all $x \in 2^{\mathbb{N}}$, which contradicts Proposition 5.2.

Acknowledgments

I would like to thank Alexander Kechris and the referee for some very helpful comments on earlier versions of this paper. The research in this paper was partially supported by NSF Grant DMS 1101597.

References

1. S. Bezuglyi and K. Medynets, *Full groups, flip conjugacy, and orbit equivalence of Cantor minimal systems*, Colloq. Math. **110** (2008), 409–429.
2. M. R. Bridson and A. Haefliger, *Metric Spaces of Non-Positive Curvature*, Grundlehren Math. Wiss. **319**, Springer, Berlin, 1999.

3. J. D. Clemens, *Isomorphism of subshifts is a universal countable Borel equivalence relation*, Israel J. Math. **170** (2009), 113–123.
4. S. Coskey and S. Schneider, *Borel Cardinal Invariant properties of countable Borel equivalence relations*, preprint (2011).
5. R. Dougherty, S. Jackson and A. S. Kechris, *The structure of hyperfinite Borel equivalence relations*, Trans. Amer. Math. Soc. **341** (1994), 193–225.
6. T. Downarowicz, *Survey of odometers and Toeplitz flows*, in: *Algebraic and topological dynamics*, Contemp. Math. **385**, Amer. Math. Soc., Providence, 2005, pp. 7–37.
7. T. Downarowicz, J. Kwiatkowski and Y. Lacroix, *A criterion for Toeplitz flows to be topologically isomorphic and applications*, Colloq. Math. **68** (1995), 219–228.
8. J. Feldman and C. C. Moore, *Ergodic equivalence relations, cohomology and von Neumann algebras I*, Trans. Amer. Math. Soc. **234** (1977), 289–324.
9. S. Gao, S. Jackson and B. Seward, *A coloring property for countable groups*, Math. Proc. Cambridge Philos. Soc. **147** (2009), 579–592.
10. S. Gao, S. Jackson and B. Seward, *Group Colorings and Bernoulli Subflows*, preprint (2011).
11. T. Giordano, I. F. Putnam and C. F. Skau, *Full groups of Cantor minimal systems*, Israel J. Math. **111** (1999), 285–320.
12. R. I. Grigorchuk, *Degrees of growth of finitely generated groups and the theory of invariant means*, Math. USSR-Izv. **25** (1985), 259–300.
13. R. I. Grigorchuk, *Just infinite branch groups*, in *New Horizons in Pro-p Groups*, Birkhäuser, Boston, 2000, pp. 121–179.
14. R. Grigorchuk and K. Medynets, Topological full groups are locally embeddable into finite groups, preprint 2012.
15. L. Harrington, A. S. Kechris and A. Louveau. A Glimm-Effros dichotomy for Borel equivalence relations. *J. Amer. Math. Soc.* **3** (1990), 903–927.
16. G. Hjorth and A. S. Kechris, *Borel equivalence relations and classification of countable models*, Annals of Pure and Applied Logic **82** (1996), 221–272.
17. S. Jackson, A.S. Kechris, and A. Louveau, *Countable Borel equivalence relations*, J. Math. Logic **2** (2002), 1–80.
18. K. Juschenko and N. Monod, *Cantor systems, piecewise translations and simple amenable groups*, preprint 2012.
19. A. S. Kechris *Classical Descriptive Set Theory*, Graduate Texts in Mathematics **156**, Springer-Verlag, 1995.
20. H. Matui, *Some remarks on topological full groups of Cantor minimal systems*, Internat. J. Math. **17** (2006), 231–251.
21. A. Yu. Ol'shanski, *SQ-universality of hyperbolic groups*(Russian), Mat. Sb. **186** (1995), 119–132; translation in Sb. Math. **186** (1995), 1199–1211.
22. S. Thomas, *Continuous versus Borel Reductions*, Arch. Math. Logic **48** (2009), 761–770.
23. S. Thomas, *Popa superrigidity and countable Borel equivalence relations*, Annals Pure Appl. Logic. **158** (2009), 175–189.
24. S. Thomas and B. Velickovic, *On the complexity of the isomorphism relation for finitely generated groups*, J. Algebra **217** (1999), 352–373.

TW-MODELS FOR LOGIC OF KNOWLEDGE-CUM-BELIEF

SYRAYA CHIN-MU YANG

Department of Philosophy, National Taiwan University,
1, Section 4, Roosevelt Road, Taipei 106, Taiwan
Email: cmyang@ntu.edu.tw

I propose a class of TW-models for epistemic logic, Kripke's models in character with reflexivity as the only constraint on the required accessibility relation amongst states, which will satisfy the main theses of Timothy Williamson's *knowledge first epistemology*. I introduce a function δ from the set S of states to the power set of formulae of the language in use, referred to as ipk-function to signify Williamson's notion of 'the agent's being in a position to know a proposition in a state'. The semantic rules for the modal operators $\mathcal{K}$ (for 'knowing') and $\mathcal{B}$(for 'believing') will be stipulated, respectively. The proposed semantic rules not only illuminate Williamson's thesis of understanding the justification of belief in terms of knowledge but also illustrate that to construct models for an epistemic logic with both knowledge-operator $\mathcal{K}$ and belief-operator $\mathcal{B}$, we need not posit two distinct accessibility relations for $\mathcal{K}$ and $\mathcal{B}$ respectively. Moreover, TW-models will invalidate a list of problematic formulae/rules of inference, which are taken as the characterization of the problem of logical omniscience. Finally, I add to the original language one extra operator $I_\mathcal{K}$(corresponding to the ipk-function in that $I_\mathcal{K}\varphi$, meaning 'the agent being *actually* in a position to know φ', is true in a state α if and only if $\varphi \in \delta(\alpha)$), so that the derivability-power of the problematic formulae/rules of inference can be retained in the desired epistemic logic system by corresponding modified formulae.

Keywords: Epistemic logic; Logic of knowledge and belief; Kripke models; Logical omniscience; Knowledge first epistemology; Timothy Williamson.

INTRODUCTION

Ever since G. H. von Wright (1951, 1957) proposed the original idea of treating epistemic concepts, such as *knowing* and *believing*, as a kind of modal concepts, epistemic logic has flourished into a large family of logical systems.[a] In particular, with the application of Kripke's possible worlds

---

[a]Recently Heathcote (2004: 287) argues that knowledge would not have alleged modal character. He suggests that we should take 'knows' as a predicate. But Williamson (2004: 320) points out that this treatment makes no obvious difference.

semantics, thanks to Jaakko Hintikka (1962), numerous logical properties of related epistemic concepts can be characterized and clarified in epistemic logics. It is also well-observed that sometimes, a well-established epistemic logic may provide a certain way of developing new philosophical views. Yet, so far two notorious problems remain, which have been taken as intrinsic threat to the legitimacy of epistemic logic. The first lies in the difficulty with the choice of a correct, or the most appropriate, logical system amongst a huge family of epistemic logics, as far as a philosophical conception of knowledge/belief is concerned. The second one is known as the problem of logical omniscience: The agent should know all logical consequences and all tautologies. But this seems beyond human agents' epistemic limitation.

The main burden of this paper is to propose a reversed approach to the construction of a desired epistemic logic. Instead of dwelling on the pros and cons for which one is the most appropriate, or correct, epistemic logic, I start with a well-established conception of knowledge, that is, Timothy Williamson's (2000) *knowledge first epistemology*. Then I construct a class of models for the desired epistemic logic, referred to as TW-models, which would satisfy the main theses that Williamson proposes. Semantic rules for the knowledge-operator $\mathcal{K}$ and the believing-operator $\mathcal{B}$ will be given. In particular, the semantic rule for $\mathcal{B}$ indicates that the truth of a formula of the form $\mathcal{B}\varphi$ can be determined by the truth of $\mathcal{K}\varphi$ in some related states. In addition, I show that the problem of logical omniscience can be avoided on TW-models.

1. EPISTEMIC LOGIC, KRIPKE MODELS AND TWO MAIN PROBLEMS

Epistemic logic intends to provide a logical treatment of reasoning based on agents' epistemic states, typically knowledge and belief. At present, the standard approach to constructing an epistemic logic is to start with a modal language with some epistemic operators, typically $\mathcal{K}$ and $\mathcal{B}$, each of which is used to signify a certain corresponding epistemic concept, say knowledge and belief. For example, we may have a modal language $\mathcal{L}_{\mathcal{K}}$ for a logic of knowledge: $p|\neg\varphi|\varphi \to \psi|\mathcal{K}\varphi$ (p: All propositional letters are formulae; so are $\neg\varphi$, $\varphi \to \psi$, and $\mathcal{K}\varphi$, where $\mathcal{K}\varphi$ means 'the agent knows that φ'). A family of epistemic logic can be constructed by taking as the underlying system CPC, the classical propositional calculus, and then adding to CPC a set of so-called *characteristic formulae* in $\mathcal{L}_{\mathcal{K}}$ as axioms together with some rules of inference, typically,

$(K_\mathcal{K})$ $\mathcal{K}(\varphi \to \psi) \to (\mathcal{K}\varphi \to \mathcal{K}\psi)$ (The Distributive law)

$(T_\mathcal{K})$ $\mathcal{K}\varphi \to \varphi$ (The Truth Axiom; factiveness)

$(4_\mathcal{K})$ $\mathcal{K}\varphi \to \mathcal{K}\mathcal{K}\varphi$ (Positive Introspective Axiom)

$(5_\mathcal{K})$ $\neg\mathcal{K}\varphi \to \mathcal{K}\neg\mathcal{K}\varphi$ (Negative Introspective Axiom)

$(D_\mathcal{K})$ $\mathcal{K}\varphi \to \neg\mathcal{K}\neg\varphi$ ($\neg\mathcal{K}(\varphi \wedge \neg\varphi)$ or $\neg\mathcal{K}\bot$)

$(C_\mathcal{K})$ $\neg\mathcal{K}\neg\mathcal{K}\varphi \to \mathcal{K}\neg\mathcal{K}\neg\varphi$ (also known as $(4.2_\mathcal{K})$)

$(N_\mathcal{K})$ From $\vdash \varphi$, one can get $\vdash \mathcal{K}\varphi$.

to mention a few. As a result, we have a family of epistemic logic for knowledge, typically, $K(= CPC + (N_\mathcal{K}) + (K_\mathcal{K}))$, $T(= K + (T_\mathcal{K}))$, $S4(= T + (4_\mathcal{K}))$, and $S5(= S4 + (5_\mathcal{K}))$.

Also, Kripke models for epistemic logic can be constructed in the standard way by putting forth distinct accessibility relations on models in accordance with the so-called characteristic formulae involved in distinct epistemic logic.[b] The semantic rules are also standard, except that we need one for $\mathcal{K}\varphi$, namely

$\mathcal{M}, s \models \mathcal{K}\varphi$ iff for all states t in S accessible from s (i.e. Rst), $\mathcal{M}, t \models \varphi$.

Equipped with appropriated interpretation, an epistemic logic thus constructed not only 'theorizes about the abstract structure of epistemic agents' theorizing', (Williamson 2009: 439) but also 'shows its use as a tool for clarifying philosophical notions and arguments'. (van Benthem 2006: 71).

Yet, given that we have a family of epistemic logic of knowledge, we would have a variety of conceptions of knowledge, each of which is characterized in terms of the axioms of the logic we adopt. Now, the question is: Amongst the family of epistemic logic, which is the correct, or the most appropriate one? Some (see van Ditmarsch, van der Hoek, & Kooi 2008, and Fagin et al. 1995) accept the system $S5$; some (Hintikka 1962) are in favor of $S4$; while some would settle for the one stronger than $S4$ but weaker than $S5$, such as $S4.2$ (e.g. Lenzen 1978) and $S4.4$.[c] Still no conclusive argument for any of them has been proposed as yet.

Alternatively, some epistemologists suggest that a logic of belief, or *doxastic* logic, would be much more suitable and practical for modeling the rea-

---

[b] For instance, the system T requires a reflexive accessibility on the desired models. See Stalnaker (2006: 194), for a more detailed list of the correspondence between some well-known epistemic logics and the semantic conditions imposed on the constraints of the required accessibility relations on Kripke models for epistemic logic.

[c] Here, the logic $S4.2$ is $S4 + (4.2_\mathcal{K})$ and $S4.4$ is $S4 + (4.4_\mathcal{K})$. Both $(4.2_\mathcal{K})$ and $(4.4_\mathcal{K})$ are weaker than $(5_\mathcal{K})$. Note that $(4.4_\mathcal{K}) = \varphi \to (\neg\mathcal{K}\varphi \to \mathcal{K}\neg\mathcal{K}\varphi)$.

soning and the structure of agents' epistemic states in ordinary discourse, because our actual reasoning and epistemic states are mainly concerned with beliefs, rather than knowledge. Note that a logic of belief can be constructed in a similar way by substituting the belief-operator '$\mathcal{B}$' for '$\mathcal{K}$'. The language $\mathcal{L}_{\mathcal{B}}$ will be $p|\neg\varphi|\varphi \to \psi|\mathcal{B}\varphi$ (where $\mathcal{B}\varphi$ means 'The agent believes that φ'), and the semantic rule for $\mathcal{B}$ will be:

$\mathcal{M}, s \models \mathcal{B}\varphi$ iff for all states u in S accessible from s (i.e. Rsu), $\mathcal{M}, u \models \varphi$.

Again a family of logic of belief can be constructed in the same way. Still, the question concerning which one is the correct, or most appropriate, system remains open to dispute. Some (e.g. Stalnaker 1993) accept $K4$, or $K45$ as appropriate logic for belief. But a majority of theorists prefer the so-called $KD45$(or *weak* $S5$, obtaining by adding ($D_{\mathcal{B}}$) to $K45$. Some take both $K45$ and $KD45$ as appropriate for a logic of belief, e.g., Moses & Shoham (1993).

From a philosophical point of view, if an epistemic logic is to theorize the reasoning and the structures of the agent's epistemic states in ordinary discourse, then it seems more appealing to take both knowledge and belief into account, in short, a logic of knowledge-cum-belief, e.g. Halpern & Moses (1985a, 1985b), Kraus & Lehmann (1988), Voorbraak (1992), van de Hoek (1993), Meyer (2001, 2003), and Moses (2008), to mention a few. (Of course, we need a language $\mathcal{L}_{\mathcal{KB}} :: p|\neg\varphi|\varphi \to \psi|\mathcal{K}\varphi|\mathcal{B}\varphi$.) After all, there is a very close kinship between belief and knowledge and our reasoning in ordinary discourse sometimes on knowledge and sometimes on belief. A logic of this kind can be obtained easily by combining a logic of knowledge and a logic of belief, and the desired models can be constructed by putting forth two distinct accessibility relations $R_{\mathcal{K}}$ and $R_{\mathcal{B}}$ for $\mathcal{K}$ and $\mathcal{B}$, respectively in the desired models. For instance, we may have the system $S4[\mathcal{K}] + KD45[\mathcal{B}]$, a model for which requires a reflexive and transitive relation $R_{\mathcal{K}}$, and a serial, transitive and Euclidean relation $R_{\mathcal{B}}$. (Notation note: Hereafter, I may add $[\mathcal{K}]$ or $[\mathcal{B}]$ to the name of a logical system, e.g. $S4[\mathcal{K}]$ to indicate the logic of knowledge $S4$, and $KD45[\mathcal{B}]$, the logic of belief $KD45$) It can be further stipulated that (i) $R_{\mathcal{B}} \subseteq R_{\mathcal{K}}$ and (ii) $\forall w, u, v \in S$, if $R_{\mathcal{K}}wu$ and $R_{\mathcal{B}}uv$, then $R_{\mathcal{B}}wv$, so that the proposed models satisfy $(\mathcal{KB})\mathcal{K}\varphi \to \mathcal{B}\varphi$, and $(\mathcal{BKB})\mathcal{B}\varphi \to \mathcal{KB}\varphi$. Notice that $(\mathcal{KB})$ and $(\mathcal{BKB})$ were already suggested by Hintikka (1962) to link knowledge and belief. Adding $(\mathcal{KB})$ and $(\mathcal{BKB})$ to $S4 + KD45$ (correspondingly, to $S5 + KD45$), we can get $KL(S4/KD45)$ (correspondingly, $KL(S5/KD45)$, for details, see, Kraus & Lehmann (1988); also Battigalli & Bonanno (1999)).

Alternatively, we may construct a theory of knowledge based on some established theory of belief, e.g. $KD45$. For instance, as Meyer (2001: 195) notes, we can, thanks to W. Lenzen's work in the 1980s, define knowledge in terms of 'true belief' (i.e. $\mathcal{K}\varphi =_{def} \mathcal{B}\varphi \wedge \varphi$) on the basis of $KD45[\mathcal{B}]$, so as to get a system of knowledge $S4.4[\mathcal{K}]$.

By contrast, we may define beliefs in terms of knowledge, say 'belief as defeasible knowledge', as Moses & Shoham (1993) suggested.[d] Voorbraak (1991, 1992) maintains that the system of knowledge $S5$ comprehensively signifies the epistemic notion of *objective knowledge*; while $KD45$, *rational (introspective) belief*. A notion of *justified true belief* can be then explained in terms of notions of objective knowledge and rational introspective belief. The logic of *justified true belief* is shown to be $S4.2[\mathcal{K}]$. Accordingly, the desired objective-knowledge-cum-rational-belief model ($OK\&RIB$-model, for short) requires an equivalence relation $R_{\mathcal{K}}$ for $\mathcal{K}$, and a serial, transitive and Euclidean relation $R_{\mathcal{B}}$ for $\mathcal{B}$. Hence, the epistemic logic $OK\&RIB$ is the system obtained by combining $S5[\mathcal{K}]$ and $KD45[\mathcal{B}]$ by means of $(\mathcal{B}\mathcal{K}\mathcal{B})$ together with $(\mathcal{B}\mathcal{B}\mathcal{K})$(i.e. $\mathcal{B}\varphi \rightarrow \mathcal{B}\mathcal{K}\varphi$). Voorbraak claims that this should be the best candidate for an epistemic logic which deals with human agents' epistemic states.

Recently, Stalnaker (2006: 179) proposes a weaker logic of knowledge and belief, by taking $S4[\mathcal{K}]$ as the underlying system, and adding to it $(\mathcal{B}\mathcal{B}\mathcal{K})$, $(\mathcal{B}\mathcal{K}\mathcal{B})$, $(D_{\mathcal{B}})$, $(5_{\mathcal{B}})$, and $(\mathcal{K}\mathcal{B})$ as the required axiom-schemata. Stalnaker notes that the resulting epistemic logic for knowledge and belief yields $KD45[\mathcal{B}]$ and $S4.2[\mathcal{K}]$. Also the system contains '$\mathcal{B}\varphi \leftrightarrow \neg\mathcal{K}\neg\mathcal{K}\varphi$' as an equivalence theorem, which indicates in turn the definability of belief in terms of knowledge. However, the acceptance of both $(\mathcal{B}\mathcal{K}\mathcal{B})$ and $(\mathcal{K}\mathcal{B})$ indicates that $\mathcal{K}\mathcal{B}\varphi$ is reducible to $\mathcal{B}\varphi$. It follows that any claim of the agent's knowledge of her belief would be superfluous, as $\mathcal{B}\varphi \leftrightarrow \mathcal{K}\mathcal{B}\varphi$, which is apparently not acceptable, unless belief is luminous.

It is noteworthy that Halpern et al. (2009) examines several logics of knowledge and belief which result from a combination of $KD45[\mathcal{B}]$ and some well-established epistemic logics of knowledge, say $S4[\mathcal{K}]$, or $S4.4[\mathcal{K}]$, or $S5[\mathcal{K}]$, respectively, together with $(\mathcal{K}\mathcal{B})$ and $(\mathcal{B}\mathcal{K}\mathcal{B})$.They then conclude that knowledge cannot be defined explicitly in terms of belief in any of

---

[d]The credit for this line of thought, according to Voorbraak (1992:225), should go to Wolfgang Lenzen. In a 1978 paper, Lenzen proposed an epistemic logic containing a definition of belief in terms of a notion of knowledge which is at least as strong as $S4.2$.

the logics in which we are interested, in particular the logic containing $KD45[\mathcal{B}] + S5[\mathcal{K}] + (\mathcal{KB}) + (\mathcal{BKB})$. Nor can belief be defined explicitly in terms of knowledge in such a system. Here by 'explicit definability of knowledge in a logic' it is meant that we can find a formula of the form $\mathcal{K}\varphi \leftrightarrow \sigma$, where σ is a formula free of the epistemic operator $\mathcal{K}$. Although knowledge can be defined in terms of true belief, i.e., $\mathcal{K}\varphi \leftrightarrow (\varphi \wedge \mathcal{B}\varphi)$, in the system containing $S4.4[\mathcal{K}]$, no epistemologist would take this as an acceptable definition of knowledge.

Moreover, it appears to me that to construct a logic of knowledge-cum-belief simply by combining two well-established logics, one for knowledge and the other for belief, would not be acceptable if the logic is to serve as a characterization of the philosophical concepts of knowledge and belief respectively. For one thing, although it is not clear what exactly it is to distinguish knowledge from belief, one thing is certain, that is, the difference between knowledge and belief cannot be explained simply by the appeal to distinct accessibility relations involved. According to the standard Kripke models for the logic of knowledge and that of belief, one knows p iff p is true in all accessible states; also one believes p iff p is true in all accessible states. The difference between knowledge and belief obviously lies in the difference between the posited accessibility relations $R_\mathcal{K}$ for knowledge and $R_\mathcal{B}$ for belief. The posited accessibility relation is used to specify the notion of epistemic possibility, or so-called 'nearby cases'. As is well-known, under appropriate interpretation, each characteristic formula (justified by a certain constraint on the posited accessibility relation) can signify a certain property of the corresponding epistemological concept, such as knowledge or belief. For example, the acceptance of $(T_\mathcal{K})$ shows the factiveness of knowledge, while a rejection of $(T_\mathcal{B})$ indicates that belief is not factive. And the luminosity of knowledge—knowledge is known itself, can be characterized by $(4_\mathcal{K})$. Now, given that there is a family of epistemic logic for knowledge/belief based on distinct models which are in turn characterized by different accessibility relation, the best we can say is that we would have a variety of conceptions of knowledge/belief, each which may be of interest for research in a certain field, such as computer science, AI, epistemology, etc. Thus, the appeal to two distinct posited accessibility relations in a model just shows that we have two distinct conceptions of our epistemic possibility, and a fortiori, epistemic state. It would show nothing about the relation between knowledge and belief. Of course, some might argue that we can add some axioms such as $(\mathcal{KB})$, $(\mathcal{BKB})$ and $(\mathcal{BBK})$ to indicate the relation between knowledge and belief. However, it seems awkward to justify alleged ($\mathcal{KB}$), ($\mathcal{BKB}$) and ($\mathcal{BBK}$) on models with two distinct accessibility relations. A justification of formulae of this kind including ($\mathcal{KB}$), ($\mathcal{BKB}$) and ($\mathcal{BBK}$) relation is acceptable only in a model with a unified accessibility relation for our epistemic possibilities. The question is: Can we construct a kind of models for a logic of knowledge-cum-belief based on a sole accessibility relation? So far, to my knowledge, no such a kind of models has been proposed as yet. More importantly, is there any epistemic logic thus constructed powerful enough to capture alleged characteristic properties of knowledge and belief and their relation?

Apart from this problem, there is another intrinsic problem with the standard approach to epistemic logic, that is, the problem of logical omniscience, to which we next turn our attention.

From a proof-theoretical point of view, a modal operator $\mathcal{M}$ in a modal system is called *normal* if the system contains *Necessitation*—$(N_\mathcal{M}) \vdash \varphi \Rightarrow \vdash \mathcal{M}\varphi$, as a rule of inference and (K)—$\mathcal{M}(\varphi \to \psi) \to (\mathcal{M}\varphi \to \mathcal{M}\psi)$, as an axiom. Correspondingly, a modal system is said to be *normal* if its primitive modal operators are normal, such as T, $S4$, $S5$ for knowledge, and $K45$, and $KD45$ for belief.

Now, it is widely observed that in a normal epistemic logic, unrestricted applications of (N) and (K) would render a truism that if the agent knows/believes a proposition φ, and φ logically implies ψ, then the agent knows/believes ψ as well. That is, the set of sentences considered to be known or believed is closed under logical consequence. Clearly, unrestricted application of (N) assumes that the agent must be able to know or believe all valid sentences. This seems to be far-fetched. A rational agent in ordinary discourse, as a resource-limited being, may know (or believe) φ and $\varphi \to \psi$, but may not know (or believe) ψ due to a failure of drawing any connection between φ and ψ. Meyer (2001:191; 2003: 18) shows that the problem can be characterized in terms of the following list of formulae that an epistemic logic system may have:

- LO1 $\models \Box(\varphi \to \psi) \to (\Box\varphi \to \Box\psi)$
- LO2 $\models \varphi \;\Rightarrow\; \models \Box\varphi$
- LO3 $\models (\varphi \to \psi) \Rightarrow\models \Box\varphi \to \Box\psi$
- LO4 $\models (\varphi \leftrightarrow \psi) \Rightarrow\models (\Box\varphi \leftrightarrow \Box\psi)$
- LO5 $\models (\Box\varphi \land \Box\psi) \to \Box(\varphi \land \psi)$
- LO6 $\models \Box\varphi \to \Box(\varphi \lor \psi)$
- LO7 $\models \neg(\Box\varphi \land \Box\neg\varphi)$

where $\Box$ stands for either $\mathcal{K}$ or $\mathcal{B}$.

So far, various attempts have been proposed, but none is successful.[e] Hintikka (1975/1989) remarked that the possible worlds semantics is unsuitable for modeling human reasoning because humans do not seem to be logically omniscient. Together with the aforementioned problem concerning the pursuit of a correct, or best epistemic logic, should we then give up possible worlds semantics, and *a fortiori*, epistemic logic?

We should not be surprised to observe that both theorists of epistemic logics and epistemologists have some misgivings over the relevance of bring epistemic logic and epistemology together along issues of shared interests. For example, Hocutt (1972) challenged the applicability of epistemic logic to any realistic account of knowledge. Lenzen (1978) also casts doubt on the significant impacts of a correct analysis of knowledge on the construction of the desired epistemic logic. Recent development of epistemic logic, especially those which can be applied to computational engineering, has been shown to be far removed from genuine philosophical (epistemic) concepts and issues in epistemology.

The way out lies in the construction of an epistemic logic free of the problem of logical omniscience and also, providing appropriate explanation of some noticeable properties of knowledge and belief and especially the relation between knowledge and belief. In doing so, we need some guidelines. Perhaps, Hendricks & Symons (2006: 161) are right when they remark: 'the interplay between epistemic logic and epistemology works dialectically'. At present we have seen that most of the existing literature on epistemic logic have focused on the syntactical aspect (i.e. the variety of its axiomatic-deduction forms), as Hintikka (1986/1989) already noticed. It strikes me that perhaps, a better way of dealing with this issue is to adopt a reverse strategy by focusing on the construction of a class of models which would satisfy a certain presupposed conception of knowledge. It is my intention here to show how a somewhat informal analysis of the nature of knowledge and belief can be adopted to shape a formal construction of the required

---

[e]These attempts can be roughly classified into four types: (i) the *syntactic* approach which allows some rules of inference to be incomplete. (Eberle (1974), Moore & Hendrix (1979), and Konolige (1986)), (ii) the *semantic* approach which identifies a proposition with a set of possible worlds (Montague (1968, 1970), Levesque (1984), Vardi (1986)), (iii) the *impossible worlds* approach which accepts impossible worlds as well (Rantala (1982a), (1982b); Wansing (1990)), and (iv) the *nonstandard propositional logic* approach which takes a certain non-classical logic as the underlying system (Fagin et al. (1990)). For the sake of the limitation of space, I can only assume that the reader is familiar with the problem of logical omniscience and for a general survey, and much more detailed discussion, the interested reader is referred to Sim (1997).

models for a logic of knowledge-cum-belief.

The conception of knowledge I have in mind is Timothy Williamson's *knowledge first epistemology* proposed in his *Knowledge and Its Limits*. In what follows, I show how the desired models, a kind of Kripke's models in character, referred to as TW-models, can be constructed on the basis of knowledge first epistemology. It can be further shown that the proposed TW-models are not only appropriate for a logic of knowledge-cum-belief, but also capable of dealing with the problem of logical omniscience.

2. WILLIAMSON'S KNOWLEDGE FIRST EPISTEMOLOGY

Before showing how the desired TW-models can be constructed, let us summarize Williamson's knowledge first epistemology by listing several main theses that Williamson proposes in his *Knowledge and Its Limits* in what follows:

(i) Knowing is a state of mind.
(ii) The broadness of knowing (Externalist approach)
(iii) Knowing is factive.
(iv) The primeness of knowing (Knowledge first!)
(v) Take knowledge as central to our understanding of belief.
(vi) Cognitive-homeless thesis
(vii) The knowledge account of evidence—One's knowledge is just one's evidence.
(viii) The knowledge account of assertion—Assert p only if one knows that p.[f]

According to Williamson, for an agent A, for a given proposition p, A **knows** p if and only if A is in a certain mental state (assuming A is in a certain case α at a given time t), which holds, in turn, if and only if A is in a certain condition C which obtains in the assumed case α. It is noteworthy that for Williamson,

A *case* is a possible total state of a system, the system consisting of an agent at a time paired with an external environment, which may of course contain other subjects. A case is like a possible world, but with a distinguished subject and time. (Williamson 2000: 52)

---

[f]This thesis is mainly concerned with the norm of assertion and its relation to knowledge, which deserves a lengthy discussion elsewhere, and will not be discussed here.

It is then beyond reasonable doubt to treat a 'case' as a certain epistemic state of the agent at a certain time, or a possible world in possible worlds semantics with a distinguished subject and time: a 'centred world' in the terminology of David Lewis. Moreover, conditions are specified by 'that' clauses, so Williamson suggested. Thus, the condition that one is happy obtains in a case α if and only if in α the agent of α is happy at the time of α. (p.52) We may then take it for granted that the very condition C is to be signified as the content of the given proposition p so that when the condition C obtains in the assumed case α, the proposition p is said to be true in α, and when A is furthermore in a position to know it, then A knows p.

The second thesis indicates an externalist conception of knowledge. According to Williamson, a case α is *internally like* a case β if and only if the *total internal physical state of the agent* in α is exactly the same as the total internal physical state of the agent in β. While a condition C is *narrow* if and only if for all cases α and β, if α is internally like β then C obtains in α if and only if C obtains in β. In other words, narrow conditions supervene on or are determined by internal physical state: there is no difference in whether they obtain without a difference in that state. We may then call a condition C *broad* if and only if it is not narrow. Of course, a state w is *narrow* if and only if the condition that one is in is narrow; otherwise w is broad. Now Williamson proposes that internalism is the claim that all purely mental states are narrow; externalism is the denial of internalism. This also justifies Williamson's view that the notion of *knows p* is not conceptually prior to p. (p.243) It seems obvious that the second thesis indicates an externalist approach to knowledge and a realistic view about the obtaining of conditions: a condition obtains or fails to obtain in each case.

In view of (i) and (ii) and the heuristic exposition of the notions of 'a condition', 'a case', 'the obtaining of a condition in a case', and 'the broadness of knowing' in this way, it seems appealing to claim that the framework of Kripke's models fits Williamson's conception of knowledge perfectly.

The third thesis suggests that '*knows*' is a factive mental state operator. (p.39) For Williamson, whether one *knows p* constitutively depends on the state of one's external environment whenever the proposition is about that environment. (pp.49-50) This thesis then clearly implies that we should accept (T), the Truth axiom $(\mathcal{K}p \to p)$, so that the required accessibility involved in the proposed models must be reflexive.

The fourth thesis shows that every standard analysis of *knows* is incorrect: '[N]o analysis of the concept *knows* of the standard kind is correct'. (p.30) For Williamson, 'knowing does not factorize as standard analysis require' (p.33); also, 'the working hypothesis should be that the concept *knows* cannot be analysed into more basic concepts'. (p.33) So he insists that the primeness of knowing offers a better causal explanation of the connection between action and knowledge/belief.

It follows straightforwardly that we should not analyse knowledge in terms of justified true belief; instead, as the thesis (v) suggests, we should take knowledge as central to our understanding of belief. Williamson notes that although our '[a]ction is often more highly correlated with belief or with true belief than with knowledge', it is 'not always'. (p.86) In brief, 'the concept *knows* is fundamental, the primary implement of epistemological inquiry'. (p.185)

A naive consequence of the theses (vi) and (v) is that the idea of constructing a logic of knowledge on the basis of some well-established logic of belief should be rejected. Nor would belief can be defined in terms of knowledge. Of course, Williamson's rejection of 'the programme of understanding knowledge in terms of the justification of belief' is in fact rather controversial, but it is not my intention here to defend his thesis. All that I want to do is to stick to his idea that knowledge should play a central role to our understanding of belief, and show how this can be displayed in our models.

Now our primary concern is this: Can we construct a kind of models wherein the semantic rule for the belief-sentences can be so stipulated such that the truth value of a belief-sentence in a given state is determined by virtue of truth-values of some corresponding knowledge-sentence in some related (accessible) states? A positive answer to the question would then not only pave a way to the construction of an epistemic logic of knowledge-cum-belief, but also shed light on Williamson's appeal of 'understanding the justification of belief in terms of knowledge'. (pp.185-6) In particular, if this can be done, then we can have a kind of models without positing two distinct accessibility relations on possible states.

In fact, Williamson has already suggested a key to a positive answer when he puts forth a quite close relation between belief and knowledge by claiming that *to believe p is to treat p as if one knows p—that is, to treat p in ways similar to the ways in which subjects treat propositions which they know.* (pp.46-7) It is clear that if we take evidence as what justifies belief, then if we also accept Williamson's identification thesis of knowledge and evidence, then 'knowledge is what justifies belief'. (p.207) That is, knowledge, and only knowledge, justifies belief. (p.185)[g] Now, the question is: How to formulate this in the proposed models?

Intuitively, a natural reading of the phrase '*to believe p is to treat p as if one knows p*' is this: 'for an agent, to believe a proposition p is simply to *assume* that she knows p in case p is true'. From a model-theoretical point of view, this is tantamount to the claim that the agent believes p in a given state on the ground that the agent *assumes* that she knows p simply because p is true. This line of thought satisfies Williamson's original idea that knowledge, and only knowledge, justifies belief. But notice that here the phrase 'the agent assumes that she knows p' indicates clearly that the agent may not actually know p in the given state. Of course, when the agent actually knows p, she would believe p. Now given that the agent does not actually know p, she can only assume that she knows p. From a model-theoretical point of view, the best we can say with regard to such a case is to maintain that the agent knows p in some nearby case, or more straightforwardly in some state accessible from the given one. Following this line of thought, some might suggest, as a semantic rule for belief-operator, that if the agent knows p in some state t accessible from the given one, say s, then the agent believes p in s. However, such a characterization of belief in terms of knowledge merely shows '*to believe p as if one knows p*'. But then the metaphoric part, or the implicature, of the phrase '*to treat p*' in Williamson's original idea that '*to believe p is to treat p as if one knows p*' will vanish into thin air. It is striking that the underlying though of the phrase '*to treat p*' should play a certain significant role in an appropriate reading of the claim '*to treat p as if one knows p*' so that it should be, and can be, characterized explicitly in the proposed models.

From a model-theoretical point of view, the very notion of '*to treat p as if one knows p*' can be realized in a model in which whenever p is true the agent can dogmatically claim that she knows p without reference to any other nearby cases, or relevant possible states. That is, for the agent, knowing p holds in such a state solely on the grounds that p is true in that state; no other states should be referred to. It is noteworthy that model-theoretically, for a given model, there can be some states, referred to as

---

[g]From a logical point of view, a natural consequence is the fact that 'knowledge entails belief'. (p.33) This indicates that we should have as a logical truth $\mathcal{K}p \to \mathcal{B}p$. Later I shall show that the notion of justified true belief can be defined in TW-models, but the proposed semantic treatment for alleged justified true belief shows that justified true belief can hardly be equivalent to knowledge.

dead-end states, which cannot access to any states. However, models with a reflexive accessibility relation automatically reject the existence of dead-end states. But there can be a kind of state which can only access to itself, and to no others. Let us call a state of this kind a *self-isolated* state, that is, for any self-isolated state u, $\forall t \in S(Rut \rightarrow u = t)$. We may then assume that in a self-isolated state the agent knows whatever is true. I hope this will meet Williamson's idea '*to treat p as if one knows p*'. If this line of thought is acceptable, then we may say that if p is true in some self-isolated state u, then the agent believes p in any state accessible to u on the grounds that the agent will treat p as if she knows p. A formal treatment will be shown later.

The sixth thesis—the cognitive-homeless thesis is in essence a rejection of the so called $\mathcal{KK}$-principle. Note that $\mathcal{KK}$-principle says that if one knows p then one knows that one knows p, i.e. $\mathcal{K}p \rightarrow \mathcal{KK}p$, which is in essence precisely what the characteristic formula ($4_{\mathcal{K}}$) intends to say. To reject $\mathcal{KK}$-principle and argue for the cognitive-homeless thesis, Williamson appeals to a rather commonplace notion in ordinary discourse, namely *being in a position to know*. For Williamson, the fact that the proposition p is true in all nearby cases (i.e. all states accessible from the given one) is not sufficient for the agent to know p. To know p, in a given state, the agent must *be in a position to know* p in that state as well. As he remarks,

> If one is in a position to know p, and one has done what one is in a position to do to decide whether p is true, then one does know p... (2000:95)

Now Williamson points out that given that the agent knows p, it does not follow that the agent is in a position to know that one knows p. (p.11) As a matter of fact, 'one can know something without being in a position to know that one knows it'. (p.114)

In fact, the cognitive-homeless thesis is a natural consequence of the well-known principle of margin for error, arguably the most important and disputable principle, in his knowledge first epistemology. According to Williamson, knowledge requires a margin for error:

> When knowing p requires a margin for error, the cases in which p is known are separated from the cases in which p is false by a buffer zone, a protective belt of cases in which p is true but unknown. (2000:18)

At this point, Williamson has to appeal to the notion of being in a position to know again.

> Thus the area of conceptual space in which one is in a position to know is separated from the surrounding area in which p is false by a border zone in which p is true but one is not in a position to know. (2000: 17)

In other words, given the fact that one knows p, one may not be in a position to know that one knows p. Hence $\mathcal{K}\mathcal{K}$-principle fails to hold.[h] This justifies that knowing is not luminous. Meanwhile, Williamson also rejects the legitimacy of $(5_{\mathcal{K}})$. Since $(5_{\mathcal{K}})$ and (T) entail (B) (known as the Brouwersche schema, $\varphi \to \mathcal{K}\neg\mathcal{K}\neg\varphi$), but (B) would fail to hold in some cases in any reasonable sense of knowledge. Since (T) certainly holds, it must be $(5_{\mathcal{K}})$ which should be blamed. The rejection of $(4_{\mathcal{K}})$ and $(5_{\mathcal{K}})$, shows that the proposed models for a desired epistemic logic should posit neither transitivity, nor equivalence as a constraint on the required accessibility relation.

It is surprising to notice that Williamson's notion of *being in a position to know* receives little attention from epistemic logicians. However, it strikes me that if we can characterize this notion in the proposed models to show its intrinsic properties and its relation to knowledge, we may be able to construct a much more appropriate epistemic logic. But, how this can be done?

Roughly speaking, Williamson explicitly maintains that at least three conditions should be met when an agent *knows* φ: (i) φ is true in all *nearby* cases (p.17) (or *relevantly similar* cases—semantically, let us treat the two phrases as synonymous); (ii) the agent is in a position to know φ; and (iii) the agent '*has done what one is in a position to do to decide whether p is true*' (p.95). However, he did not offer more detailed descriptions about the difference between the first two conditions, except for a brief remark that the agent is in a position to know φ only if φ is true in all relevantly similar cases. (Recall that Williamson notes that knowledge requires a margin for error such that one is in a position to know that a condition C obtains only if C obtains in all relevantly similar cases.) It is then tempting to introduce straightforwardly a function, say α, from the set of states in a model to the power-set of the formulae of the language in use to capture Williamson's

---

[h]Williamson also argues that one believes p is not a luminous condition. (p.192) Hence, it may not hold that $\mathcal{B}\varphi \to \mathcal{K}\mathcal{B}\varphi$; it follows that $(4_{\mathcal{B}})$ (i.e. $\mathcal{B}\varphi \to \mathcal{B}\mathcal{B}\varphi$) may not always hold.

original idea of *'being in a position to know'* so that $\varphi \in \alpha(s)$ intends to mean that the agent is in a position to know φ in s. Then a further condition for $\alpha(s)$ should be added, that is, for any given formula φ, $\varphi \in \alpha(s)$ implies $\forall t \in S(Rst \to \mathcal{M}, t \models \varphi)$. In addition, we need a further function, say β, to capture Williamson's idea of the third condition for the agent to know φ in that $\varphi \in \beta(s)$ means that the agent *'has done what one is in a position to do to decide whether p is true'*. In view of the factiveness of being in a position to know, $\varphi \in \beta(s)$ implies that φ is true in s. We can then put forth the required semantic rule for $\mathcal{K}$ as what follows:

$$\mathcal{M}, s \models \mathcal{K}\varphi \text{ iff } \varphi \in \alpha(s) \text{ and } \varphi \in \beta(s)$$

Since, $\varphi \in \alpha(s)$ implies that φ is true in all nearby case, that is, true in all states accessible from s, the notion of being in a position to know φ characterized in this way says nothing more than what the standard semantic rule for the knowledge-operator $\mathcal{K}$ says, that is, *true in all states accessible from the given state*. However, no one would treat the condition that the agent *is in a position to know* φ the same as the condition that φ *is true in all nearby cases*. It seems to me that Williamson should have some other thought in mind but did not specify it explicitly.

Perhaps, what Williamson has in mind is to set as the condition for *being in a position to know p* by stipulating that p is true in all relevantly similar cases, and the condition for *knowing p*, by stipulating that the agent is not only *to be in a position to know p* but also *to have done what one is in a position to do to decide whether p is true*. (p.95) It is then tempting to introduce a function α^* to signify the set of formulae in a state so that $\varphi \in \alpha^*(s)$ means that the agent is *in a position to know* φ, and also *has done what one is in a position to do to decide whether φ is true*. But it follows that the agent knows φ in a state s iff $\varphi \in \alpha^*(s)$. Then we are in no position to distinguish *knowing* φ from being in a position to know φ. This seems unacceptable.

It strikes me that to capture the difference between the first two conditions, and to show the third condition explicitly, a more appealing approach is to introduce a function δ from the set of states in a model to the power-set of formulae in the language in use, referred to as ipk-function for short, so that $\varphi \in \delta(s)$ can be construed as meaning that the agent *is in a position to know φ* and *has done what one is in a position to do to decide whether φ is true*. Since I prefer to keep the standard semantic rule for $\mathcal{K}$, the condition required for Williamson's original notion of being in a position to know —i.e. $\varphi \in \alpha(s)$ only if $\forall t \in S(Rst \to \mathcal{M}, t \models \varphi)$, can be thereby ignored insofar as the semantic condition for the ipk-function δ is concerned.

Accordingly, instead of the stipulation that if $\varphi \in \delta(s)$ then δ must be true in all nearby cases as Williamson seemingly suggested, let us propose that the ipk-function δ assigns to each state s a set of *true* formulae $\delta(s)$ such that if $\varphi \in \delta(s)$ then the agent is *in a position so that she has actually done what she is in a position to decide whether φ is true*. For the sake of simplicity, let us say that the agent *is actually in a position to know φ* in s if $\varphi \in \delta(s)$. Note that the condition that φ must be true in s if $\varphi \in \delta(s)$ is required simply because of the factiveness of being in a position to know as Williamson proposes—'Thus being in a position to know... is factive: if one is in a position to know p, then p is true'. (p.95). So when the agent is *actually* in a position to know φ in s and when φ is true in all states accessible from s, the agent can be said to know φ in s. (This also shows that sometimes, even a proposition p is true in all nearby cases, the agent may not know p, simply because the agent is not *actually* in a position to know p.)

Admittedly, the proposed ipk-function δ is in fact a bit stronger than Williamson's original notion of being in a position to know. The intended interpretation of $\varphi \in \delta(s)$ as *'the agent is actually in a position to know φ'* is used to highlight this difference. By contrast, we my think of Williamson's original view of 'being in a position to know φ' as 'being in a position to know φ *in principle*'.

I hope that the proposed ipk-function δ with such an intended interpretation can reflect Williamson's real thought concerning his notion of *being in a position to know* and its relation to *knowing* without costing any substantial loss of the literal explanation that Williamson made. As it can be shown later, we can have formal treatment of the very function in our proposed models. Surprisingly, with the aid of the ipk-function, the problem of logical omniscience can be dealt with in a somewhat tricky but simple way.

Let us now turn our attention to the construction of TW-models.

3. TW-MODELS

First, let us fix a language in use, say $\mathcal{L} :: p|\neg\varphi|\varphi \to \psi|\mathcal{K}\varphi|\mathcal{B}\varphi$. A TW-model, as a complex, can be then described in what follows:

$$\mathcal{M} = \langle S, \sigma, R, \delta \rangle$$

S: A non-empty set of states, or possible worlds;

$\sigma : (S \to (\mathcal{P} \to \{T, F\}))$, an assignment of a truth value of $\{T, F\}$ to the propositional letters of the language in use in every state.

$R \subseteq S \times S$: a partial ordering with reflexivity to serve as the required accessibility relation amongst all states.[i]

$\delta : S \to \wp(\mathcal{L})$, such that for any $s \in S$, $\delta(s) \subseteq \{\varphi \mid \mathcal{M}, s \models \varphi, \varphi \in \mathcal{L}\}$, in particular, for any self-isolated state $u \in S$, $\delta(u) = \{\varphi \mid \mathcal{M}, u \models \varphi, \varphi \in \mathcal{L}\}$, i.e. $\varphi \in \delta(u)$ iff $\mathcal{M}, u \models \varphi$.

Note that δ intends to signify the aforementioned ipk-function so that for a given state s, whenever a formula $\varphi \in \delta(s)$, the agent is *actually* in a position to know φ in s. The condition implies that in a given state, some true propositions appear to be true in all similar cases but the agent is not actually in a position to know them; hence, the agent may not know them. Also the special condition for a self-isolated state u, such that $\varphi \in \delta(u)$ iff $\mathcal{M}, u \models \varphi$ is to guarantee, as later we will show, that $\mathcal{M}, u \models \varphi$ iff $\varphi \in \delta(u)$ iff $\mathcal{M}, u \models \mathcal{K}\varphi$ (based on the proposed semantic rule for $\mathcal{K}$).

Next, the semantic rule for $\mathcal{K}$ can be stipulated in what follows:

$$(\mathcal{K}^S) \quad \mathcal{M}, s \models \mathcal{K}\varphi \quad \text{iff} \quad \forall t \in S(Rst \to \mathcal{M}, t \models \varphi) \wedge \varphi \in \delta(s).$$

---

[i]Note that as far as the constraint on the required accessibility relation is concerned, we should first of all give up transitivity, and *a fortiori*, the equivalence relation, in view of Williamson's rejection of $\mathcal{K}\mathcal{K}$-principle and $(5_\mathcal{K})$. Also the thesis concerning the factiveness of knowing indicates the acceptance of the truth axiom, i.e. $\mathcal{K}\varphi \to \varphi$; hence reflexivity is required. Assuming reflexivity, *serial* is redundant, as every state must be accessible from itself. Also, (D_K) $\mathcal{K}\varphi \to \neg\mathcal{K}\neg\varphi)$ $(\neg\mathcal{K}(\varphi \wedge \neg\varphi)$, or $\neg\mathcal{K}\bot)$ can be satisfied in the model with reflexivity. Of course, (D_B) $\mathcal{B}\varphi \to \neg\mathcal{B}\neg\varphi)$ may not hold. Some may feel a bit uneasy about the rejection of (D_B). After all, it appears rather awkward to claim that one may believe p and at the same time believe not p. But this would not show any irrationality, as in no case do we have $\mathcal{B}(\varphi \wedge \neg\varphi)$. This will justify that we need only reflexivity as the sole constraint on the required accessibility relation.

Interestingly, Williamson (2000:306) suggests that if we add symmetry, the model can satisfy the Brouwersche schema (B)—$\varphi \to \mathcal{K}\neg\mathcal{K}\neg\varphi$. We may then get the system KTB. He acknowledges:

> KTB is exactly the logic for knowledge determined by the simplest version of the margin for error considerations. The only logic features essential to the binary similarity relation are reflexivity and symmetry, accessibility in the model plays the role of similarity, and KTB is the logic determined by the constraints of reflexivity and symmetry on accessibility. (2000: 306)

It should be noted that Williamson's suggestion to accept symmetry is based on the assumption that the alleged nearby cases can be specified by similarity. Admittedly, similarity is substantially a symmetric relation. However, model-theoretically, the very notion of accessibility relation in models for an epistemic logic intends to show that whether the agent knows a proposition in a certain epistemic state will be determined by what happens in the accessible states. If this understanding of the notion of accessibility relation is right, then the requirement of symmetry is too strong to hold. After all, the agent may not have similar belief or knowledge in two distinct states although they are accessible from each other. Perhaps, all that we need is a reflexive accessibility relation on the intended TW-models.

The first condition, $\forall t \in S(Rst \to \mathcal{M}, t \models \varphi)$, simply follows Hintikka's proposal that knowledge can be viewed as '*truth throughout the logical space of possibilities that the agent consider relevant*', or in Williamson's words, 'we know p only if p is true in nearby cases'. (p.17) The second condition, $\varphi \in \delta(s)$, indicates the requirement that to know φ, the agent must be *actually in a position to know φ* in the given state. This indicates that *being true in all relevant states* is not sufficient for knowledge. In order to know what actually is, the agent must be *actually* in a certain condition so that when the case obtains, the agent can *actually* get what obtains.

Finally, let us set forth the following semantic rule for $\mathcal{B}$:

$\mathcal{M}, s \models \mathcal{B}\varphi$ iff (i) $\mathcal{M}, s \models \mathcal{K}\varphi$; or

(ii) $\exists u \in S((Rsu \wedge \forall t \in S(Rut \to u = t)) \wedge \mathcal{M}, u \models \mathcal{K}\varphi)$, that is, there is a self-isolated state u in S accessible from s, such that $\mathcal{K}\varphi$ is true in u (and *a fortiori*, φ is true in u).

The condition (i) shows the rationality of human agent: One must believe whatever one knows. This also indicates that TW-models satisfy $(\mathcal{KB})$ $\mathcal{K}\varphi \to \mathcal{B}\varphi$. Meanwhile, as we have already remarked, the condition (ii) indicates that the agent believes φ in s simply on the ground that φ is true in some self-isolated state accessible from s. Note that the specification of the ipk-function in self-isolated states—$\delta(u) = \{\varphi | \mathcal{M}, u \models \varphi\}$ together with the reflexivity accessibility relation already shows that $\mathcal{M}, u \models \varphi$ iff $\varphi \in \delta(u)$ iff $\mathcal{M}, u \models \mathcal{K}\varphi$. I hope this formulation can capture Williamson's heuristic account of belief in terms of knowledge: to believe p is *to treat p as if she knows p, that is, to treat p in ways similar to the ways in which subjects treat propositions which they know.* (pp.46-7)

This completes our construction of TW-models.

4. A SOLUTION TO THE PROBLEM OF LOGICAL OMNISCIENCE

Having constructed TW-models, we can now show how the problem of logical omniscience can be avoided in our proposed TW-models.

First of all, it can be easily checked that TW-models invalidate K-LO1–K-LO6. But notice that K-LO7 (i.e. $\models \neg(\mathcal{K}\varphi \wedge \mathcal{K}\neg\varphi)$) will hold in TW-models. It is noteworthy that K-LO7 should be compatible with the realist's conception of the rationality of human agents, which has been one of the underlying thoughts of Williamson's knowledge first epistemology. For instance, consider K-LO2, this will be invalidated in TW-models simply because in some state the agent may not be *actually* in a position to know

φ, though φ is true in all accessible states.

But then, should we reject K-LO1–K-LO6 as axiom schemata or rules of inference, or theorems in an epistemic logic of knowledge and belief? It seems more likely that getting rid of all these problematic formulae is just like throwing the baby out with the bath water. If so, that is, once we get rid of all the listed formulae from an epistemic logic, what remains would be too weak to deserve being called a logic of knowledge. The question is then this: can we retain all these important axioms in a desired epistemic logic?

Our aforementioned discussion shows that it is the introduction of ipk-function to TW-models which invalidates the involved problematic characteristic formulae. E.g., K-LO2 can be invalidated simply because in some state the agent may not be *actually* in a position to know φ, though φ is true in all states. Thus a simple way to retain the derivability capacity of these formulae as axioms in the desired logic is to formulate this function explicitly. Firstly, we need to add to the language in use an extra modal operator, '$I_{\mathcal{K}}$ ($I_{\mathcal{K}}\varphi$ should read as 'The agent is *actually* in a position to know φ). Accordingly, the semantic rule for $I_{\mathcal{K}}$ can be stipulated as what follows:

$$(I_{\mathcal{K}}^S) \quad \mathcal{M}, s \models I_{\mathcal{K}}\varphi \quad \text{iff} \quad \varphi \in \delta(s).$$

Accordingly, we may have an alternative semantic rule for $\mathcal{K}$:

$$(\mathcal{K}^{S^*}) \quad \mathcal{M}, s \models \mathcal{K}\varphi \quad \text{iff} \quad \forall t \in S((Rst \rightarrow \mathcal{M}, t \models \varphi) \wedge \mathcal{M}, s \models I_{\mathcal{K}}\varphi).$$

This says that the agent knows φ in a certain case if and only if (i) φ is true in all nearby cases and (ii) the agent is in a position to know φ, also (iii) the agent has done *what she is in a position to decide whether p is true*. Note that given that the required accessibility relation is reflexive, it follows automatically that if $\mathcal{M}, s \models I_{\mathcal{K}}\varphi$ then $\mathcal{M}, s \models \varphi$. This meets the factiveness of *being in a position to know*, as Williamson proposes. Of course, for any self-isolated state $u \in S$, $\mathcal{M}, u \models I_{\mathcal{K}}\varphi$ iff $\mathcal{M}, u \models \varphi$.

Now, to retain the derivability-capacity of the original problematic formulae, while invalidating them on TW-models, we may modify them in the following way:

K-LO1 $\vdash \mathcal{K}(\varphi \rightarrow \psi) \rightarrow (\mathcal{K}\varphi \rightarrow (I_{\mathcal{K}}\psi \rightarrow \mathcal{K}\psi))$

K-LO2 $\vdash \varphi \Rightarrow \vdash I_{\mathcal{K}}\varphi \rightarrow \mathcal{K}\varphi$

K-LO3 $\vdash \varphi \rightarrow \psi \Rightarrow \vdash \mathcal{K}\varphi \rightarrow (I_{\mathcal{K}}\psi \rightarrow \mathcal{K}\psi)$

K-LO4 $\vdash \varphi \leftrightarrow \psi \Rightarrow \vdash (I_{\mathcal{K}}\varphi \rightarrow \mathcal{K}\varphi) \leftrightarrow (I_{\mathcal{K}}\psi \rightarrow \mathcal{K}\psi)$

K-LO5 $\vdash (\mathcal{K}\varphi \wedge \mathcal{K}\psi) \to (I_{\mathcal{K}}(\varphi \wedge \psi) \to \mathcal{K}(\varphi \wedge \psi))$

K-LO6 $\vdash \mathcal{K}\varphi \to (I_{\mathcal{K}}(\varphi \vee \psi) \to \mathcal{K}(\varphi \vee \psi))$

Alternatively, we may introduce a convention by putting all related for-mulae of the form $I_{\mathcal{K}}\varphi$ in a given formula in question to form a conjunction Φ to stand for the conjunction $\bigwedge I_{\mathcal{K}}\varphi_i$. Then we may have the following schemata:

K-LO1 $\vdash \Phi \to (\mathcal{K}(\varphi \to \psi) \to (\mathcal{K}\varphi \to \mathcal{K}\psi))$

K-LO2 $\vdash \varphi \Rightarrow \vdash I_{\mathcal{K}}\varphi \to \mathcal{K}\varphi$

K-LO3 $\vdash \varphi \to \psi \Rightarrow \vdash \Phi \to (\mathcal{K}\varphi \to \mathcal{K}\psi)$

K-LO4 $\vdash \varphi \leftrightarrow \psi \Rightarrow \vdash \Phi \to (\mathcal{K}\varphi \leftrightarrow \mathcal{K}\psi)$

K-LO5 $\vdash \Phi \to ((\mathcal{K}\varphi \wedge \mathcal{K}\psi) \to \mathcal{K}(\varphi \wedge \psi))$

K-LO6 $\vdash \Phi \to (\mathcal{K}\varphi \to \mathcal{K}(\varphi \vee \psi))$

Hintikka (1986/1989: 23-24) argues that the problem of logical omni-science can be avoided by proposing two equivalent ways of delineating the subclass of logical consequence (i.e. of the form $p \to q$), such that $\mathcal{K}p \to \mathcal{K}q$ holds, given that $p \to q$ is in the delineated subclass. The first one is to put certain syntactical restriction on the deductive argument which leads from p to q. This can be done in several different but equivalent ways. One can see that my approach here can be seen as a variant of this approach. But Hintikka himself notes that so far 'no simple axiomatic-deductive system codifying it (the proposed restriction) has been presented in the literature.' (p.23) It seems promising that based on our TW-models, with the help of the proposed modification, we can have a relatively simple way of con-structing a desired axiomatic-deductive system.

5. FURTHER REMARKS

Some interesting results are noteworthy. For one thing, on TW-models, the semantic rule for the belief-operator $\mathcal{B}$ is so stipulated such that the truth value of $\mathcal{B}\varphi$ in a given epistemic state can be determined by virtue of the truth-value of $\mathcal{K}\varphi$ in some related states. This will illustrate that the concept of knowledge is central to our understanding of belief. More importantly, we can then have a kind of models, i.e. TW-models, which can deal with both $\mathcal{K}$ and $\mathcal{B}$ without positing two different accessibility relations.

Moreover, on TW-models we may further define a notion of justified belief such that it is not the case that knowledge implies justified true belief, nor the converse would hold. It would be beyond reasonable doubt to claim that an agent A's believing φ is justified if and only if A has good reason for believing φ. This in turn requires that A would believe φ in every relevantly similar state with regard to the given state (i.e., in every state accessible from the given one). Of course, to deal with the notion of 'justified belief', we need to add to the language in use an extra modal operator, say $\mathcal{B}^j$, and the desired semantic rule for $\mathcal{B}^j \varphi$ can be stipulated in what follows:

$$\mathcal{M}, s \models \mathcal{B}^j \varphi \quad \text{iff} \quad \forall t \in S(Rst \to \mathcal{M}, t \models \mathcal{B}\varphi).$$

It is then clear that on the proposed TW-models with the stipulated semantic rules for $\mathcal{K}$, $\mathcal{B}$, and $\mathcal{B}^j$, neither $\mathcal{K}\varphi \to (\mathcal{B}^j \varphi \wedge \varphi)$ nor $(\mathcal{B}^j \varphi \wedge \varphi) \to \mathcal{K}\varphi$. This would provide a justification not only for the rejection of the analysis of knowledge with justified true belief, but also for a rejection of identification of knowledge with justified true belief plus some further elements.

CONCLUSION

If my proposal is acceptable, then we may have a much more flexible choice amongst the family of modal system with regard to the question concerning which one is the most satisfactory epistemic logic for our philosophical conception of knowledge and belief. It looks promising to claim that with the aid of ipk-function and the modal operator '$I_\mathcal{K}$', some expansion of $KT[\mathcal{K}]$ together with some formulae specifying the relation between knowledge and belief such as $\mathcal{K}\varphi \to \mathcal{B}\varphi$ and $\mathcal{K}\neg\varphi \to \neg\mathcal{B}\varphi$ would be a better choice for the logic of knowledge-cum-belief. Since the primary concern of this paper is with the model-theoretic perspective of a logic of knowledge-cum-belief, I shall leave the construction of a certain suitable deductive-system somewhere else.

ACKNOWLEDGMENTS

Some parts of previous versions of this paper were presented at Asian Workshop on Philosophical Logic (February 15-17, 2012, Japan Advanced Institute of Science and Technology (JAIST), Ishikawa, Japan), and The 2012 Taiwan Philosophical Logic Colloquium (December 7, 2012, The Department of Philosophy, National Taiwan University, Taiwan). I am extremely grateful to all participants on these occasions for their helpful remarks and comments. I am also deeply indebted to Rob Goldblatt and an anonymous referee for their valuable comments and suggestions which helped me to clarify some of my ideas and to rectify some mistakes in previous versions. I should also thank Kok Yong Lee for his critical remarks on an early version. The research for this paper is partly supported by National Science Councils, Taiwan (NSC No.:100-2410-H-002-112).

References

Battigalli, P. and Bonanno, G. (1999), 'Recent results on belief, knowledge and the epistemic foundations of game theory', *Research in Economics* 53: 149-225.

Eberle, R. A. (1974), 'A logic of believing, knowing and inferring', *Synthese* 26: 356-382.

Fagin, R., Halpern, J. Y. and Vardi, M. (1990), 'A non-standard approach to the logical omniscience problem', in *TARK'90 Proceedings of the Third Conference on Theoretical Aspects of Reasoning about Knowledge*, San Francisco, CA: Morgan Kaufman, pp. 41-55.

Fagin, R., Halpern, J., Moses, Y. and Vardi, M. (1995), *Reasoning about Knowledge*, Cambridge, Mass.: MIT Press.

Halpern, J. Y. and Moses, Y. (1985a), 'Toward a theory of knowledge and ignorance', in *Logics and models of concurrent systems*, Krzysztof R. Apt, ed., Berlin & New York: Springer-Verlag, pp. 459-476; a full version of an original, preliminary report—Techn. Rep. IBM, RJ 4448, 1984.

Halpern, J. Y. and Moses, Y. (1985b), 'A guide to completeness and complexity for modal logics of knowledge and beliefs', *Artificial Intelligence* 54: 319-379.

Halpern, J. Y., Samet, D. and Segev, E. (2009), 'Defining knowledge in terms of belief: The modal logical perspective', *The Review of Symbolic Logic* 2(3): 469-487.

Heathcote, A. (2004), 'KT and the diamond of knowledge', *Philosophical Books* 45(4): 286-295.

Hendricks, V. F. and Symons, J. (2006), 'Where's the bridge? Epistemology and epistemic logic', *Philosophical Studies* 128 (1)–A special issue with subtitle: *Bridges between Mainstream and Formal Epistemology*, pp. 137-167.

Hintikka, J. (1962), *Knowledge and Belief*, Ithaca, NY: Cornell University Press.

Hintikka, J. (1975/1989), 'Impossible possible worlds vindicated', *Journal of Philosophical Logic* 4, 475-484; reprinted in *The Logic of Epistemology and The Epistemology of Logic: Selected Essays*, by J. Hintikka and M. B. Hintikka, Dordrecht: Kluwer Academic Publishers, 1989, pp. 63-72.

Hintikka, J. (1986/1989), 'Reasoning about knowledge in philosophy: The paradigm of epistemic logic', in *TARK'86 Proceedings of the 1986 Conference on Theoretical Aspects of Reasoning about Knowledge*, CA: Morgan Kaufmann, 1986, pp. 63-80; reprinted in *The Logic of Epistemology and The Epistemology of Logic, selected Essays*, by J. Hintikka and M. B. Hintikka, Dordrecht: Kluwer Academic Publishers, 1989, pp. 17-35.

Hocutt, M. O. (1972), 'Is epistemic logic possible?' *Notre Dame Journal of Formal Logic* 13: 433-453.

336

Konolige, K. (1986), *A Deduction Model of Belief*, San Francisco, CA: Morgan Kaufmann.

Kraus, S. and Lehmann, D. (1988), 'Knowledge, belief and time', *Theoretical Computer Science* 58: 155-174.

Lenzen, W. (1978), 'Recent work in epistemic logic', *Acta Philosophica Fennica* 30: 1-219.

Levesque, H. J. (1984), 'A logic of implicit and explicit belief', in *AAAI'84 Proceedings of the Fourth National Conference on Artificial Intelligence*, Cambridge, Mass.: The MIT Press, pp. 198-202.

Meyer, J. J. Ch. (2001), 'Epistemic logic', Chapter 9 of *Blackwell Guide to Philosophical Logic*, Lou Goble ed., Oxford: Wiley-Blackwell, pp. 183-202.

Meyer, J. J. Ch. (2003), 'Modal epistemic and doxastic logic', Chapter 1 of *Handbook of Philosophical Logic*, volume 10, D. Gabbay and F. Guenthner eds., Amsterdam: Kluwer Academic Publishers, pp. 1-38.

Montague, R. (1968), 'Pragmatics', in *Contemporary Philosophy*, R. Kalibansky ed., Florence, Italy: La Nuova Italia Editrice, pp. 101-121.

Montague, R. (1970), 'Universal grammer', *Theoria* 36: 373-398.

Moore, R. C., and Hendrix, G. (1979), *Computational Models of Beliefs and the Semantics of Belief-Sentences*, Technical Note 187, SRI International, Menlo Park.

Moses, Y. (2008), 'Reasoning about knowledge and belief', in *Handbook of Knowledge Representation*, F. van Harmelen, V. Lifschitz and B. Porter eds., Amsterdam: Elsevier, pp. 621-647.

Moses, Y. and Shoham, Y. (1993), 'Belief as defeasible knowledge', *Artificial Intelligence*, 64(2): 299-321. A preliminary version with the same title appeared in *Proceedings of IJCAI-89*, San Francisco, CA: Morgan Kaufmann, 1989, pp. 1168-1173.

Rantala, V. (1982a), 'Impossible worlds semantics and logical omniscience', *Acta Philosophica Fennica* 35: 18-24.

Rantala, V. (1982b), 'Quantified modal logics: Non-normal worlds and propositional attitudes', *Studia Logica* 41: 41-65.

Sim, K. M. (1997), 'Epistemic Logic and Logical Omniscience: A Survey', *International Journal of Intelligent Systems* 12: 57-81.

Stalnaker, R. (1993), 'A note on non-monotonic modal logic', *Artificial Intelligence* 64: 183-196.

Stalnaker, R. (2006), 'On logic of knowledge and belief', *Philosophical Studies* 128 (1)–A special issue with subtitle: *Bridges between Mainstream and Formal Epistemology*, pp. 169-199.

van Benthem, J. (2006), 'Epistemic logic and epistemology: The state of their affairs', *Philosophical Studies* 128 (1)–A special issue with subtitle: *Bridges between Mainstream and Formal Epistemology*, pp. 49-76.

van der Hoek, W. (1993), 'Systems for knowledge and belief', *Journal of Logic and Computation*, 3(2): 173-195.

van Ditmarsch, H., van der Hoek, W., and Kooi, B. (2008), *Dynamic Epistemic Logic*, Amsterdam: Springer.

Vardi, M. (1986), 'On epistemic logic and logical omniscience', in *Proceedings of the First Conference on Theoretical Aspects of Reasoning about Knowledge*, San Francisco, CA: Morgan Kaufmann, pp. 293-306.

von Wright, G. H. (1951), *An Essay in Modal Logic*, London: Routledge.

von Wright, G. H. (1957), *Logical Studies*, London: Routledge.

Voorbraak, F. (1991), 'The logic of objective knowledge and rational belief', in *Logics in AI*, J. van Eijck ed., Berlin: Springer, pp. 499-515.

Voorbraak, F. (1992), 'Generalized Kripke models for epistemic logic', in *TARK'92 Proceedings of the 4th Conference on Theoretical Aspects of Reasoning about Knowledge*, San Francisco, CA: Morgan Kaufmann, pp. 214-228.

Wansing, H. (1990), 'A general possible worlds framework for reasoning about knowledge and belief', *Studia Logica* 49(4): 523-539.

Williamson, T. (2000), *Knowledge and Its Limits*, Oxford: Oxford University Press.

Williamson, T. (2004), 'Replies to commentators', *Philosophical Books*, 45(4): 313-323.

Williamson, T. (2009), 'Some computational constraints in epistemic logic', in *Logic Epistemology and the Unity of Science* (Cognitive Science Series), D. Gabbay, S. Rahman, J. M. Torres and J. P. van Bendegem eds., Dordrecht: Springer, pp. 437-456.